Springer Monographs in Mathematics

Friedrich Ischebeck
Ravi A. Rao

Ideals and Reality

Projective Modules
and Number of Generators of Ideals

 Springer

Friedrich Ischebeck

Institut für Mathematik
Universität Münster
Einsteinstr. 62
48149 Münster, Germany
e-mail: ischebe@math.uni-muenster.de

Ravi A. Rao

School of Mathematics
Tata Institute of Fundamental Research
Dr Homi Bhabha Road
400005 Mumbai, India
e-mail: ravi@math.tifr.res.in

Library of Congress Control Number: 2004114476

Mathematics Subject Classification (2000): 11C99, 13A99, 13C10, 13C40, 13D15, 55R25

ISSN 1439-7382
ISBN 3-540-23032-7 Springer Berlin Heidelberg New York

Springer is a part of Springer Science+Business Media
springeronline.com
© Springer-Verlag Berlin Heidelberg 2005
Printed in Germany

Typeset by the authors using a Springer LaTeX macro package
Production: LE-TeX Jelonek, Schmidt & Vöckler GbR, Leipzig
Cover design: Erich Kirchner, Heidelberg

Printed on acid-free paper   41/3142YL - 5 4 3 2 1 0

Dedicated to our close friend and former colleague
Hartmut Lindel
who left us much too early

Preface

Besides giving an introduction to Commutative Algebra – the theory of commutative rings – this book is devoted to the study of projective modules and the minimal number of generators of modules and ideals.

The notion of a module over a ring R is a generalization of that of a vector space over a field k. The axioms are identical. But whereas every vector space possesses a basis, a module need not always have one. Modules possessing a basis are called free. So a finitely generated free R-module is of the form R^n for some $n \in \mathbb{N}$, equipped with the usual operations. A module is called projective, iff it is a direct summand of a free one. Especially a finitely generated R-module P is projective iff there is an R-module Q with $P \oplus Q \cong R^n$ for some n. Remarkably enough there do exist nonfree projective modules. Even there are nonfree P such that $P \oplus R^m \cong R^n$ for some m and n. Modules P having the latter property are called stably free. On the other hand there are many rings, all of whose projective modules are free, e.g. local rings and principal ideal domains. (A commutative ring is called local iff it has exactly one maximal ideal.) For two decades it was a challenging problem whether every projective module over the polynomial ring $k[X_1, \ldots, X_n]$ with a field k was free. It was known from the beginning that such a module had to be stably free. The statement that it should be actually free was called Serre's Conjecture. This was proved independently by D. Quillen and A. Suslin in 1976. We give several proofs of it.

Later we show how vector bundles over a compact Hausdorff space X (more generally vector bundles of special type over any topological space) can be interpreted as projective modules over the ring of (real, complex or quaternion) continuous functions on X. This gives the concept of projective modules an intuitive meaning. For instance it is no surprise that nontrivial vector bundles exist – at least once one has seen the Möbius band.

In the second half of the book we study the question what one can say about the minimal number of generators of certain ideals. This often – but not always – is connected with the theory of projective modules. We begin with dimension theory on commutative so named Noetherian rings, i.e. such whose ideals are finitely generated. (This property of rings was first identified and studied by E. Noether, a student of D. Hilbert.) Its fundamental theorem states the equality of two numbers: Let R be a local Noetherian ring with maximal ideal $\mathfrak{m}$. Then the minimal number n of elements $a_1, \ldots, a_n$ such that $\mathfrak{m}$ is a minimal prime over-ideal of $(a_1, \ldots, a_n)$ equals the number of steps of a maximal chain of prime ideals in R, i.e.

$$\mathfrak{p}_0 \subsetneq \mathfrak{p}_1 \subsetneq \cdots \subsetneq \mathfrak{p}_n = \mathfrak{m}$$

with maximal n. This number n is called the (Krull) dimension of R. If $\mathfrak{m}$ itself can be generated by $n = \dim(R)$ elements then R is called a regular local ring. A not necessarily local ring whose ideals are finitely generated is called regular, if all its localizations are so. (To every commutative ring R and every prime ideal $\mathfrak{p}$ of it one can associate in a canonical way a local ring, the localization of R in $\mathfrak{p}$.) An example is $k[X_1, \ldots, X_n]$ with a field k. All of its maximal ideals can be generated by n elements.

Other examples are the domains all of whose ideals are projective modules. Namely these are the regular Noetherian domains of dimension ≤ 1, i.e. whose nonzero prime ideals are maximal. These rings are called Dedekind rings. We give a complete classification of the finitely generated modules, especially of the projective ones over those rings. Also we prove the theorem of the finiteness of the class number, which says that over certain 'classical' Dedekind rings there are only finitely many projective modules upto 'stable isomorphy', i.e. upto 'adding free modules'.

The Forster-Swan Theorem gives upper bounds to the number of generators of modules if one knows these numbers for the localizations. As a consequence one sees that an ideal I of $R = k[X_1, \ldots, X_n]$ can be generated by $n + 1$ generators if R/I is regular.

Improving an old theorem of Kronecker, in 1972 Eisenbud and Evans and independently Storch have shown that every prime ideal in $k[X_1, \ldots, X_n]$ is a minimal prime overideal of an ideal generated by n elements. (In the book we will give a version of this theorem which is not restricted to prime ideals.) Geometrically this means that every set defined by polynomial equations can already be defined by n equations.

The last chapter is dedicated to the question: Under what hypotheses can one describe an algebraic curve in the affine n-space by $n - 1$ equations? i.e. let $\mathfrak{p}$ be a prime ideal of $R = k[X_1, \ldots, X_n]$ with $\dim(R/\mathfrak{p}) = 1$. When is $\mathfrak{p}$ a minimal prime overideal of an ideal, generated by $n - 1$ elements? (Also here the good formulation does not restrict to prime ideals.) We show that the answer is positive in the following two cases:

1. $R/\mathfrak{p}$ is regular, i.e. a Dedekind ring. (Mohan Kumar)
2. k is of positive characteristic. (Cowsik and Nori)

The general answer is still not known. That is the reality today.

The 'book' started as a M. Phil. project of the junior author with Selby Jose, in the course of which they found the expository article [106] of Valla. The initiative and determination of the senior author, and the urging of colleagues, led to *Selby's thesis* becoming into a book which could be used as a graduate course or for an intensive workshop.

The whole book tells the story of a philosophy of J-P. Serre and his vision of relating that philosophy to problems in Affine geometry. A thorough development of this subject till the end of 1980 is done in this book. The intermix of Classical Algebraic K-theory and Complete Intersection problems is emphasized in this text for the first time. The results of Eisenbud-Evans, Swan's connection between vector bundles and projective modules, Lindel's proof of Serre's conjecture appear for the first time in a student text form.

The book is almost self-contained, and serves as an introduction to basic Commutative algebra and its applications in problems of affine algebraic geometry. In a first reading, the student could skip Chapter 5; but a better understanding of the subject and its interconnection with other parts of Mathematics can only be had by eventually perusing this chapter. The reader could also skip the Eisenbud-Evans theorem in Chapter 9, §9.4; though he should know that this is the best achieved via *general position arguments*, the recurrent theme.

The material in this text has been crystallized from various books, research and expository articles. Earlier works on this subject are the notes of Geyer, Ohm, Badescu; survey articles by Valla, Lyubeznik, Murthy, Bass, Suslin; and the books of Lam, Szpiro, Kunz, Mandal respectively. Following the tradition in the books written by J-P. Serre, we have not let the exercises interrupt the flow of reading. We have placed them chapter-wise at the very end. The exercises are challenging; the student who wades through them will be very proficient to work in this active research area. This is because most of the exercises are culled from the research papers of experts.

Interesting developments, not found here, include- Effective methods of Sturmfels, etc.; Bose on Serre's conjecture via Groebner bases, and its relation to problems in Electrical Engineering; numerous applications of the Local-Global techniques due to Bass, Suslin, Asanuma, etc; generalizations of Serre's conjecture: the Bass conjectures, the Bass-Quillen conjecture by Lindel, Popescu, and the Anderson conjecture over monoidal rings due to Gubeladze; surprising developments in the orthogonal groups due to Parimala, Sridharan, Ojanguren; analogous results of Suslin and his students in the classical groups; G. Lyubeznik's work in higher dimension which uses the Cowsik-Nori Theorem as the inductive start; Murthy's results on complete intersection questions; and Mohan Kumar's on Eisenbud-Evans conjectures over affine algebras; later developments, inspired by Nori, by Bhatwadekar, Mandal, Raja Sridharan on complete interestion and Euler classes; parallel development on complete intersections over real algebras by Ojanguren, Ischebeck, etc.; Herzog on monomial curves; results of Warfield, Stafford, etc. on development of Serre's conjecture and the Eisenbud-Evans theorem in non-commutative noetherian rings; examples of Macaulay, Moh; projective varieties, and the corresponding problems there; connections with local cohomology started by Faltings, and developed in Lyubeznik's thesis; scheme theoretic and ideal theoretic questions studied by Hartshorne, Abhyankar amongst others; local algebra - 'symbolic primes' results of Cowsik, Huneke, etc.; motivations from algebraic topology; principal G-bundles; ...

We would like to thank many colleagues for their valuable suggestions which encouraged us, and have helped to improve the material presented. We thank our wives for their support. The senior author acknowledges the help from Deutsche Forschungsgemeinschaft (DFG) and T.I.F.R. which made it possible for him to interact with the junior author. The book was finally completed in the idyllic setting at the I.C.T.P. this summer, whom we thank for their hospitality. Finally, we look forward to hearing from you, to improve the material and its presentation.

September 13, 2004

Friedrich Ischebeck
Ravi A. Rao

Contents

Basic Commutative Algebra 1

Here and in Chapter 6 we give an introduction to Commutative Algebra. This is the theory of commutative rings and their modules, i.e. abelian groups on which a ring operates as a field on a vector space.

A ring in our sense is associative, but not always commutative. Upto one or two exercises, where it is expressively mentioned, a ring always is supposed to have a 1. (This does not excluude the zero ring, where $1 = 0$.) Every ring homomorphism $A \to B$ has to map the 1 of A to the 1 of B. Consistently with this, we require a subring A of a ring B to have the same 1.

1.1 The Spectrum

Throughout this section all rings are supposed to be *commutative*. We introduce the Zariski topology on the set of all **prime ideals** of the ring.

Remember the notion of a prime ideal, which is fundamental for Commutative Algebra. A prime ideal $\mathfrak{p}$ of a ring A is an ideal such that $A/\mathfrak{p}$ is a domain, i.e. has no zero-divisors and is not the zero ring. (So A itself is not a prime ideal of A.)

Lemma 1.1.1 *Let $\mathfrak{p}$ be an ideal of a ring A, different from A.*

a) *The following properties are equivalent:*

(1) *$\mathfrak{p}$ is a prime ideal;*

(2) *for all $a, b \in A$ we have $ab \in \mathfrak{p} \Rightarrow a \in \mathfrak{p}$ or $b \in \mathfrak{p}$;*

(3) *for all ideals $\mathfrak{a}, \mathfrak{b}$ of A one has $\mathfrak{a}\mathfrak{b} \subset \mathfrak{p} \Rightarrow \mathfrak{a} \subset \mathfrak{p}$ or $\mathfrak{b} \subset \mathfrak{p}$,*

b) *If $\mathfrak{p}$ is a prime ideal, then $\mathfrak{a} \cap \mathfrak{b} \subset \mathfrak{p} \Rightarrow \mathfrak{a} \subset \mathfrak{p}$ or $\mathfrak{b} \subset \mathfrak{p}$, for all ideals $\mathfrak{a}$, $\mathfrak{b}$ of A.*

Proof. a) (3) $\Rightarrow$ (2) $\Longleftrightarrow$ (1) is clear. To prove (2) $\Rightarrow$ (3), assume $\mathfrak{a} \not\subset \mathfrak{p}$ and $\mathfrak{b} \not\subset \mathfrak{p}$. Then there are $a \in \mathfrak{a} \setminus \mathfrak{p}$ and $b \in \mathfrak{b} \setminus \mathfrak{p}$. By (2) we have $ab \notin \mathfrak{p}$, whence $\mathfrak{ab} \not\subset \mathfrak{p}$.

b) follows from a) and the fact that $\mathfrak{ab} \subset \mathfrak{a} \cap \mathfrak{b}$. $\square$

(Recall that by $\mathfrak{ab}$ we mean the ideal generated by the set $\{ab \mid a \in \mathfrak{a},\, b \in \mathfrak{b}\}$, i.e. the set of all finite sums $\sum_i a_i b_i$ with $a_i \in \mathfrak{a}$, $b_i \in \mathfrak{b}$.)

The converse of b) is not true: Every ideal of $\mathbb{Z}$ which is generated by a power of a prime number with exponent > 1 is a counterexample.

Remarks 1.1.2 a) Every **maximal ideal** of a ring A, i.e. a maximal one under all ideals different from A, is a prime ideal, since its residue class ring has only the 2 obvious ideals and hence is a field! It is an easy application of Zorn's Lemma, that there exist maximal ideals in every ring $A \neq (0)$. More generally every ideal $\neq A$ is contained in a maximal one.

To apply Zorn's Lemma, it is enough to show that the union I of a **chain** of proper ideals (i.e. a set of ideals $\neq A$ which is totally ordered by inclusion) is also a proper ideal. To see that it is proper, note that $1 \notin I$. Make sure, you have understood the following: The union of a chain of ideals is an ideal, whereas the union of a general set of ideals need not be an ideal!

b) By a **minimal prime ideal** we mean one which is minimal under *all* prime ideals. So in a domain there is exactly one minimal prime ideal, namely (0).

Using Zorn's Lemma (in an 'order reversing' kind) we also easily see that every prime ideal $\mathfrak{q}$ contains a minimal one. Namely, let $\{\mathfrak{p}_i \mid i \in I\}$ be a chain of prime ideals, contained in $\mathfrak{q}$. Then $\bigcap_i \mathfrak{p}_i$ is a prime ideal, since the complements of the $\mathfrak{p}_i$ form a chain of multiplicatively closed sets. So the set of all prime sub-ideals of $\mathfrak{q}$ is inductively ordered by '$\leq$'='$\supset$'.

As a corollary we get: If $\mathfrak{q}$ is a prime over-ideal of an ideal I, i.e. $\mathfrak{q} \supset I$, then there is a minimal prime over-ideal of I, contained in $\mathfrak{q}$.

1.1.3 Let X be the set of all prime ideals of a ring A. We have $X \neq \emptyset$, if $A \neq 0$. We define a topology on the set X called the **Zariski topology**. X together with this topology is denoted by $\mathrm{Spec}(A)$ and called the **spectrum** of A. The Zariski topology of $\mathrm{Spec}(A)$ is defined as follows:

If $E \subset A$, let

$$V(E) := \{\mathfrak{p} \in X \mid \mathfrak{p} \supset E\}.$$

(We write $V(f_1,\dots,f_r) := V(\{f_1,\dots,f_r\})$.) The subsets $V(E)$ are defined to be the **closed subsets** of $\mathrm{Spec}(A)$.

Note that

(a) $V(\emptyset) = V(0) = X, \quad V(A) = V(1) = \emptyset$.

(b) $\bigcap_\alpha V(E_\alpha) = V(\bigcup_\alpha E_\alpha)$.

(c) $\bigcup_{i=1}^n V(E_i) = \bigcup_{i=1}^n V(\mathfrak{a}_i) = V(\bigcap_{i=1}^n \mathfrak{a}_i) = V(\prod_{i=1}^n \mathfrak{a}_i)$, where $\mathfrak{a}_i = (E_i)$ denotes the ideal, generated by E_i.

Hence the collection $\{V(E) \mid E \subset A\}$ – as the collection of closed set – gives a topology (the Zariski toplogy) on X.

Example 1.1.4 Let $A = \mathbb{Z}$. Then $X = \{p\mathbb{Z} \mid p = 0 \text{ or a prime number}\}$. $V((0)) = X$, $V(m\mathbb{Z}) = \{p\mathbb{Z} \mid p|m\}$, for $m \in \mathbb{Z}$. So a subset of $\mathrm{Spec}(\mathbb{Z})$ is closed, if and only if it is either the whole set or a finite set of maximal ideals.

In every ring A a prime ideal is a closed point of $\mathrm{Spec}(A)$ if and only if it is maximal. But also the set of maximal ideals of a ring, equipped with the induced topology, in most cases is not Hausdorff, for instance in the case $A = \mathbb{Z}$

Remarks 1.1.5 Let us study some simple properties of $\mathrm{Spec}(A)$.

a) For $f \in A$, let $D(f) := \{\mathfrak{p} \in X \mid f \notin \mathfrak{p}\} = \mathrm{Spec}(A) \setminus V(f)$. Then $D(f)$ is an open subset of $\mathrm{Spec}(A)$, called the **principal open subset associated with** f.

The collection $\{D(f) \mid f \in A\}$ is a base of $\mathrm{Spec}(A)$ because

$$X \setminus V(E) = X \setminus \bigcap_{f \in E} V(f) = \bigcup_{f \in E} D(f).$$

b) With the help of these basic open sets it is easy to see that $X = \mathrm{Spec}(A)$ is quasi-compact: For, if $X = \bigcup_{\alpha \in I} D(f_\alpha)$, then $\sum_{\alpha \in I} Af_\alpha = A$. Clearly there are finitely many $\alpha_1,\dots,\alpha_n \in I$ and $\lambda_1,\dots,\lambda_n \in A$ such that $1 = \sum_{i=1}^n \lambda_i f_{\alpha_i}$. And it is easily verified that $X = \bigcup_{i=1}^n D(f_{\alpha_i})$.

Now let $(U_i)_{i \in I}$ be an open covering of X. Since every U_i is a union of principal open subsets, and finitely many of the latter cover X, also finitely many of the U_i will cover X.

c) Given a subset E of A we have associated with it a closed set $V(E) \subset \mathrm{Spec}(A)$. Conversely, given a subset $Y \subset \mathrm{Spec}(A)$ we can associate with it the ideal $I(Y) = \bigcap_{\mathfrak{p} \in Y} \mathfrak{p}$. Both the correspondences

$$E \to V(E), \qquad\qquad Y \to I(Y)$$

are inclusion reversing. One has $V(I(Y)) = \overline{Y}$ the closure of Y:

For, if $K = V(J) \supset Y$ for some ideal J, then $I(Y) \subset I(K)$ and $V(I(K)) \supset V(I(Y))$. But $V(I(K)) = V(I(V(J))) = V(J)$. Hence $K \supset V(I(Y))$, i.e. $V(I(Y))$ is the smallest closed set containing Y and so is its closure.

Later we will see that $I(V(E)) = \sqrt{\mathfrak{a}} = \{a \in A \,|\, a^n \in \mathfrak{a} \text{ for some } n\}$ where $\mathfrak{a}$ is the ideal, generated by E.

Definition 1.1.6 *The **maximal spectrum** of a ring A is the set of maximal ideals of A equipped with the induced topology of $\mathrm{Spec}(A)$. (Recall that the maximal ideals are prime.) The maximal spectrum of A is denoted by $\mathrm{Spmax}(A)$.*

1.1.7 The next lemma (and not only that) will be better understood, if we interpret elements of a (commutative) ring A as 'functions' on its spectrum, namely we regard $f \in A$ as the 'function'

$$\mathfrak{p} \mapsto (f \bmod \mathfrak{p}).$$

Especially $f(\mathfrak{p}) = 0$ means $f \in \mathfrak{p}$.

This is motivated by the following observations:

Let k be a field and $A := k[X_1, \ldots, X_n]$ the polynomial ring in n indeterminates (=variables) over k. Then there is a canonical injection

$$k^n \longrightarrow \mathrm{Spec}(A), \qquad (a_1, \ldots, a_n) \mapsto (X_1 - a_1)A + \ldots + (X_n - a_n)A.$$

Namely the ideal $(X_1 - a_1)A + \cdots + (X_n - a_n)A$ is the kernel of the (surjective) k-algebra homomorphism

$$A \longrightarrow k \text{ defined by } X_i \mapsto a_i,$$

hence maximal. And if, say, $a_i \neq b_i$, then $X_i - b_i$ clearly is not in the above kernel. So different n-tuples are mapped to different maximal ideals.

If $a = (a_1, \ldots, a_n)$, write $\mathfrak{m}_a := (X_1 - a_1)A + \ldots + (X_n - a_n)A$. We will see later that for algebraically closed k every maximal ideal of A is of the form $\mathfrak{m}_a$ for some $a \in k^n$. (Hilbert's Nullstellensatz). But already here (for a general field k) we observe the following:

For any $f \in A$ and $a \in k^n$ we have:

$$f(a) = 0 \iff f \in \mathfrak{m}_a.$$

Moreover $f(a)$ maps to $(f \bmod \mathfrak{m}_a)$ via the only possible k-algebra isomorphism $k \to A/\mathfrak{m}_a$.

Compare the following lemma to Urysohn's Lemma, which says that two disjoint closed subsets of a normal topological space can be 'separated' by a continuous real function on the whole space in the way that this takes the value 0 on one and the value 1 on the other subset. By the following lemma one can extend the zero function on a closed set $V(I)$ of the spectrum of a ring to the whole space in such a way that it takes non-zero values on finitely many (not necessarily closed) given points outside $V(I)$. (More generally a function on $V(I)$ which is extendable to $\mathrm{Spec}(A)$ can also be extended in the way that it takes non-zero values on finitely many given points outside $V(I)$.) The exact analogue of Urysohn's Lemma on two disjoint closed set is a direct consequence of the Chinese Remainder Theorem, which we will prove later in Section 1.5.

Lemma 1.1.8 (Prime Avoidance Lemma) *Let I be an ideal and f an element of a ring A.*

a) *Let $\mathfrak{p}_1, \mathfrak{p}_2, \ldots, \mathfrak{p}_r \in \mathrm{Spec}(A)$, such that $f + I \subset \bigcup_{i=1}^{r} \mathfrak{p}_i$, then $(f, I) \subset \mathfrak{p}_{i_0}$ for some i_0.*

b) *Let $\mathfrak{p}_1, \ldots, \mathfrak{p}_r$ be ideals of A, such that at most two of them are not prime. If $I \subset \bigcup_{i=1}^{r} \mathfrak{p}_i$, then $I \subset \mathfrak{p}_{i_0}$ for some i_0.*

Proof. a) Choose a counterexample with minimal r. We must have $r > 1$ and $\mathfrak{p}_i \not\subset \mathfrak{p}_j$ for $i \neq j$. We claim that $f \in \bigcap \mathfrak{p}_i$.

Suppose $f \notin \mathfrak{p}_i$ for some i. Then $f + \mathfrak{p}_i I$ is disjoint from $\mathfrak{p}_i$, so $f + \mathfrak{p}_i I \subset \bigcup_{j \neq i} \mathfrak{p}_j$. By the minimality of r, we infer that $(f, \mathfrak{p}_i I) \subset \mathfrak{p}_j$ for some $j \neq i$. Since $\mathfrak{p}_i \not\subset \mathfrak{p}_j$ and $\mathfrak{p}_j$ is a prime ideal, this gives $(f, I) \subset \mathfrak{p}_j$, a contradiction. So $f \in \bigcap_i \mathfrak{p}_i$. So also $I \subset \bigcup_{i=1}^{r} \mathfrak{p}_i$. Hence the rest of a) follows, once we have shown b).

b) Assume that $I, \mathfrak{p}_1, \ldots, \mathfrak{p}_r$ form a counterexample with minimal r and that $\mathfrak{p}_3, \ldots, \mathfrak{p}_r$ are prime ideals. Then we have $I \cap \mathfrak{p}_i \not\subset \bigcup_{j \neq i} \mathfrak{p}_j$, for every $i = 1, \ldots, r$. Choose $f_i \in (I \cap \mathfrak{p}_i) \setminus \bigcap_{j \neq i} \mathfrak{p}_j$. Then

$$f_1 \cdot \ldots \cdot f_{r-1} + f_r \in I \setminus \bigcup_{i=1}^{r} \mathfrak{p}_i,$$

since either $\mathfrak{p}_r$ is a prime ideal or $r = 2$. This contradiction proves b). $\square$

Remarks 1.1.9 a) In this book we will need part a) of the lemma as well as part b). There is an example of ideals $\mathfrak{a}_1$, $\mathfrak{a}_2$, I and an element f in a ring, such that $f + I \subset \mathfrak{a}_1 \cup \mathfrak{a}_2$, but $f + I \not\subset \mathfrak{a}_i$ for $i = 1, 2$. (See Exercise 1.)

b) The simple looking Prime Avoidance Lemma will be useful in the context of several important theorems proved later. We will come across at least five important applications - the Serre's Splitting Theorem, the Bass Cancellation Theorem, the Forster-Swan Theorem, the Eisenbud-Evans Theorem, and H.

Lindel's ring theoretic proof of the Quillen-Suslin Theorem i.e. Serre's Conjecture.

c) The Prime Avoidance Lemma is a typical argument known as a general position argument. By refining the method used in proving this lemma we will later derive the Forster-Swan bounds (on the number of generators of a module) and also the Eisenbud-Evans Theorems, from which all the previously mentioned theorems can be independently deduced.

d) Let V be a vector space (of finite or infinite dimension) over an infinite field, $U, U_1, \ldots, U_r$ be subspaces and $x_1, \ldots, x_r$ be elements of V with $U \subset \bigcup_{i=1}^{r}(x_i + U_i)$. Then $U \subset x_i + U_i = U_i$ for some i. (You will find hints, how to prove this, in Exercise 2.) So if a ring A contains an infinite subfield, then Lemma 1.1.8 holds for arbitrary ideals $\mathfrak{p}_1, \ldots, \mathfrak{p}_r$, not necessarily prime ones.

e) Lemma 1.1.8 frequently is applied in the following situation:
Let $\mathfrak{a}_1, \ldots, \mathfrak{a}_r$ be ideals and $\mathfrak{p}_1, \ldots, \mathfrak{p}_s$ prime ideals of a ring such that $\mathfrak{a}_i \not\subset \mathfrak{p}_j$ for all i, j. Then there is an $f \in (\mathfrak{a}_1 \cap \ldots \cap \mathfrak{a}_r) \setminus (\mathfrak{p}_1 \cup \ldots \cup \mathfrak{p}_s)$.

Indeed set $I := \mathfrak{a}_1 \cap \ldots \cap \mathfrak{a}_r$ and use Lemma 1.1.1 b).

1.2 Modules

Modules are 'vector spaces over rings', i.e. the concept of a module generalizes that of a vector space, replacing the underlying field by a general ring.

The reader should not be deterred by the lot of – easy – concepts we will introduce here. It makes sense to get only a rough impression by the first reading and to repeat the details later, when they really are used. Also, since our deeper results concern modules only over *commutative rings*, the reader may well restrict his attention to this case, and so avoid several complications, appearing in the non-commutative case.

Definition 1.2.1 *Let R be a (not necessarily commutative) ring. A left R-module M is a (here always additively written) abelian group M together with a map $R \times M \to M$, $(a, x) \mapsto ax$ (the scalar multiplication), satisfying the following conditions.*

(1) $a(x + y) = ax + ay$; $a \in R$, $x, y \in M$

(2) $(a + b)x = ax + bx$; $a, b \in R$, $y \in M$

(3) $(ab)x = a(bx)$; $a, b \in R$, $x \in M$

(4) $1x = x$; $x \in M$

Remark 1.2.2 A **right** R-module is defined in nearly the same way. The only difference is that (c) is replaced by

(3') $(ab)x = b(ax)$.

In this case everyone writes the 'scalars' on the right side, i.e. xa instead of ax with $x \in M$, $a \in R$, so that (3') gets the form of an associativity law:

(3') $x(ab) = (xa)b$.

(But the essential difference between right and left modules is *not the place* where the scalars are written!)

Over commutative rings there is no difference between left and right modules.

Examples 1.2.3 a) A vector space V over a field K is a K-module.

b) An abelian group G is a $\mathbb{Z}$-module. (For $n \in \mathbb{Z}$ and $x \in G$ one defines $nx := \pm(x + \cdots + x)$ where the sum has $|n|$ summands and the sign is that of n. Here we have used additive notation in G.)

c) Let V be a K-vector space together with an endomorphism α. Then (V, α) becomes a module over the polynomial ring $K[X]$ by $f \cdot v := f(\alpha)(v)$. More explicitly let $f = \sum_i a_i X^i$, then $f \cdot v = \sum_i a_i \cdot \alpha^i(v)$, where $\alpha^i := \alpha \circ \cdots \circ \alpha$, i-times.

d) The set $R^n = R \times \cdots \times R$ of n-tuples of elements of a ring R is an R-module in the same manner as K^n is a vector space over a field K. Modules of this type, i.e. isomorphic to R^n will be called **finitely generated free modules**. (The meaning of 'isomorphic' should be clear, and is anyhow defined below.)

Especially R can be regarded as a left or as a right R-module. Its submodules (see below) are its left, resp. right ideals.

The zero module which consists only of the zero element 0 is the special case $n = 0$ of the R^n. We will denote it by 0, not worrying that this symbol also denotes the zero element of every ring and module.

e) If R and S are rings and $\varphi : R \to S$ is a ring homomorphism then S can be considered as an R-module as follows. Define the scalar multiplication by $rs = \varphi(r)s$ for $r \in R, s \in S$. In particular we do so if R is a subring of S.

f) In the situation of e) let M be an S-module. Then M becomes an R-module by $rx := \varphi(r)x$.

g) Let I be a two-sided ideal of R and M an R-module. Assume $IM = 0$, i.e. $cx = 0$ for all $c \in I$, $x \in M$. Then M can also be regarded as an R/I-module by $\bar{a}x := ax$, since $ax = a'x$ if $a \equiv a'(\mathrm{mod}\ I)$. We often will do this tacitly! (Here and also elsewhere $\bar{a}$ denotes the residue class $a + I$.)

h) Let M be an R-module and $R[x]$ the polynomial ring over R. By $M[x]$ we mean the analogously defined 'polynomial module'. Its elements are formal sums

$$\sum_{i=0}^{\infty} m_i x^i, \quad m_i \in M, \quad m_i = 0 \text{ for almost every } i.$$

(By 'almost every' resp. 'almost all' in this book we always mean 'all but finitely many'.) The addition in $M[x]$ is defined componentwise:

$$\sum_i m_i x^i + \sum_i n_i x^i := \sum_i (m_i + n_i) x^i.$$

The scalar multiplication is defined analogously to that in $R[x]$:

$$\left(\sum_i a_i x^i\right)\left(\sum_i m_i x^i\right) := \sum_k \left(\sum_{i+j=k} a_i m_j\right) x^k.$$

Remarks 1.2.4 The most useful and peculiar property of vector spaces is the existence of bases and the uniqueness of their cardinality. So we have a very easy classification of vector spaces. But this is *not at all* so for general modules.

On the other hand there are many definitions, theorems and proofs for modules which are verbatim the same as for the special case of vector spaces. Here is a list.

a) The definition of a **submodule** is clear. The intersection of any family of submodules of a module is again one. The **sum of** finitely many submodules $U_1, \ldots, U_n$ of a module M is

$$\sum_{i=1}^{n} U_i = U_1 + \ldots + U_n := \{\sum_{i=1}^{n} x_i \mid x_i \in U_i\}.$$

Of course one can also define the sum of an infinite family of submodules:

$$\sum_{i \in I} U_i := \{\sum_{i \in I} x_i \mid x_i \in U_i, \ x_i = 0 \text{ for almost all } i\}.$$

Any sum of submodules clearly is a submodule (whether the sum is finite or infinite).

b) If E is any subset of an A-module M, the submodule **generated** by E, is the intersection of all submodules which contain E. It is also the set of all linear combinations of elements of E, *viz.*

$$\{\sum_{e \in E} a_e e \mid a_e \in A, \ a_e = 0 \text{ for almost all } e\}$$

E is called a **generating set** (or a **set of generators**) of M, if the submodule generated by E is M itself.

c) A module M is said to be **finitely generated** if it has a finite set of generators $\{m_1, \ldots, m_n\}$; i.e.

$$M = \{\sum_{i=1}^{n} a_i m_i \,|\, a_i \in A\}.$$

M is called **monogene** or **cyclic**, if it has a generating set consisting of one element. If this element is x, we call x a **generator**. This is equivalent to $M = Ax := \{ax \,|\, a \in A\}$.

d) A module M is called **simple**, if $M \neq \{0\}$ and M and $\{0\}$ are the only submodules of M. Clearly every simple module is monogene, every nonzero element of M being a generator. Note that the $\mathbb{Z}$-modules $\mathbb{Z}$ and $\mathbb{Z}/4$ both are monogene but not simple.

e) A subset E of a module M is called **linearly independent**, if

$$\sum_{e \in E} a_e e = 0 \text{ implies } a_e = 0 \text{ for all } e \in E.$$

(Here we assume tacitly $a_e = 0$ for almost all $e \in E$.) Otherwise we call it **linearly dependent**. The linear independence of E is equivalent to:

$$\sum_{e \in E} a_e e = \sum_{e \in E} b_e e \Longrightarrow a_e = b_e \text{ for all } e \in E.$$

Definitions 1.2.5 a) *A* **basis** *of a module is a linearly independent generating set of it. If E is a basis of M, then every $x \in M$ can uniquely be written as*

$$x = \sum_{e \in E} a_e e, \quad a_e \in A, \quad a_e = 0 \text{ for almost all } e.$$

b) *An R-module F is called* **free**, *if it admits a basis.*

Remarks 1.2.6 a) If R is neither the 0-ring nor a skew-field ($=$ division ring), then there will always exist nonfree R-modules.

b) The module R^n is indeed a free R-module. A basis is formed by the elements

$$e_1 := (1, 0, \ldots, 0), \quad e_2 := (0, 1, 0, \ldots, 0), \ldots, \quad e_n := (0, \ldots, 0, 1)$$

as in the case of vector spaces.

Conversely every free R-module with a finite basis is isomorphic to some R^n.

c) If a non-zero ideal of a *commutative* ring R is free as an R-module, then necessarily it is a principal ideal, generated by a non-zero-divisor. Namely every pair (a, b) in R is linearly dependent, since $ba - ab = 0$.

Definition 1.2.7 *If U is a submodule of an R-module M, then the factor group M/U is also an R-module, the so called* **factor module M modulo U**. *Namely if $x \in M$ and $\bar{z}$ denotes the residue class of any $z \in M$, then $a\bar{x} := \overline{ax}$. This is clearly well defined.*

Remarks 1.2.8 a) Especially R/I is a left R-module if I is a left ideal of the ring R.

Now assume, R is commutative and I different from R and (0). Then R/I is not free, since no subset of R/I can be linearly independent because $x \in R/I$, $a \in I \setminus (0)$ implies $ax = 0$. The case of non-commutative R is a trifle more difficult.

b) Let M, U be as in the definition. The submodules of M/U are exactly those of the form V/U, where V is a module 'between' U and M, i.e. a submodule of M, containing U.

c) Modules over a ring R often reflect the properties of the ring. For instance, if R is a principal (ideal) domain (a domain where every ideal is of type Ra for some $a \in R$), then also the finitely generated modules over R have a nice structure. For more details see the Section 4.1 on elementary divisors.

Homomorphisms of modules, i.e. linear maps are defined as for vector spaces:

Definitions 1.2.9 a) *Let M and N be R-modules. A map $f : M \to N$ is called an* **R-module homomorphism** *or* **R-linear** *if*

$$f(x+y) = f(x) + f(y), \quad f(ax) = af(x)$$

where $x, y \in M$ and $a \in R$

b) *The* **kernel** $\ker(f)$ *and the* **image** $\operatorname{im}(f)$ *of a module homomorphism $f : M \to N$ are defined to be the kernel $(f^{-1}(0))$, resp. the image $(f(M))$ of f as a group homomorphism. (Clearly they are submodules of M, resp. N.) The* **cokernel** $\operatorname{coker}(f)$ *of f is defined to be $N/\operatorname{im}(f)$.*

c) *An injective (i.e. one-to-one) linear map is also called a* **monomorphism**, *a surjective (i.e. onto) linear map an* **epimorphism**.

A bijective linear map is called an **isomorphism**. *This is justified by the fact that the inverse map of a bijective linear map is linear as well. (Prove this! Continuous maps behave differently. The inverse map of a bijective continuous map of topological spaces need not be continuous!)*

Modules M, M' are called **isomorphic** *if there exists an isomorphism $f : M \to M'$.*

d) *An* **endomorphism** *of a module M is a homomorphism $M \to M$. An* **automorphism** *of M is a bijective endomorphism of M.*

Examples 1.2.10 a) If R is a commutative ring, then for any $a \in R$, the map $h_a : M \to M$, given by scalar multiplication by a, i.e. $h_a(x) = ax$ is an R-module homomorphism. This map is often pompously called the **homothesy** of a; but really it is a simple thing!

Note that for a non-commutative R the homothesy of an element is not necessarily R-linear; for e.g. let $M = R$ and a not commute with all elements of R. But, clearly if a belongs to the center of R, i.e. a commutes with all elements of R, then the homothesy of a on every R-module is R-linear.

b) On the other hand, regard R as a *left* R-module and let $a \in R$ (not necessarily in the center of R). Then the map $R \to R$, $x \mapsto xa$ is R-linear. Further all R-linear maps of the left module R into itself are of this kind.

c) More generally, if we consider R^m, R^n as *left* R-modules and write their elements as *rows*, then every R-linear map $R^m \to R^n$ is given by $v \mapsto v\alpha$, where $\alpha \in R^{m \times n}$ is a suitable matrix with m rows and n columns. Conversely every such matrix gives us an R-linear map $R^m \to R^n$.

(If you consider R^m, R^n as *right* R-modules and write their elements as *columns*, then every R-linear map $R^m \to R^n$ is given by $v \mapsto \alpha v$, where α is a matrix with n rows and m columns.)

d) Another generalization of b) is the following: Let M be a left R-module and consider R as a left R-module. The R-linear maps $R \to M$ are in bijective correspondence with the elements of M. Namely to $x \in M$ associate the map $R \to M$, $a \mapsto ax$, and to $f : R \to M$ associate $f(1) \in M$.

The usual Homomorphism Theorem and Isomorphism Theorems hold. We state them here without proofs, since these are the same as for (abelian) groups.

Proposition 1.2.11 (Homomorphism Theorem) *Let $f : M \to N$ be a module homomorphism and $U \subset \ker(f)$ be a submodule. Then f factorizes through the canonical map $\kappa : M \to M/U$, i.e. f is the compositions of linear maps*

$$M \xrightarrow{\kappa} M/U \xrightarrow{f'} N,$$

*where f' is (well-)defined by $f'(x \bmod U) := f(x)$. This is a typical case where we say that f' is **induced** by f.*

Especially, f induces an isomorphism $M/\ker(f) \to \mathrm{im}(f)$.   □

Proposition 1.2.12 (First Isomorphism Theorem) *Let $U \subset V$ be submodules of a module M. Then canonical map $M \to M/V$ induces a homomorphism $M/U \to M/V$. And this induces an isomorphism*

$$(M/U)/(V/U) \xrightarrow{\sim} M/V.$$

□

Proposition 1.2.13 (Second, i.e. Noether Isomorphism Theorem) *Let U, V be submodules of a module M. Then the inclusion map $V \hookrightarrow U + V$ induces an isomorphism*

$$V/(U \cap V) \xrightarrow{\sim} (U + V)/U.$$

□

Remarks 1.2.14 a) Let M, N be R-modules. The set of all R-module homomorphisms from M to N is denoted by $\operatorname{Hom}_R(M, N)$. . It is an abelian group, addition defined by $(f + g)(x) := f(x) + g(x)$.

If R is commutative, $\operatorname{Hom}_R(M, N)$ has the structure of an R-module: $(af)(x) := a(f(x))$. (If R is not commutative, the map $af : M \to N$ defined by $(af)(x) := a(f(x))$ is not always R-linear. Also the trial to define $(af)(x) := f(ax)$ does not succeed.)

Especially $\operatorname{Hom}_R(R^m, R^n)$ can be identified with the additive group of all $m \times n$-matrices, addition defined componentwise.

We write $M^* := \operatorname{Hom}_R(M, R)$ and call it the **dual** of M. If M is a left module, then M^* is a right module and vice versa. Namely for $a \in R$, $f \in M^*$, $x \in M$ define $(fa)(x) := f(x)a$.

For commutative R there is an isomorphism $\operatorname{Hom}_R(R^n, R) \xrightarrow{\sim} R^n$. Namely if one writes the elements of R^n as columns, then as usual the elements of $\operatorname{Hom}_R(R^n, R$ are written as one row matrices. This isomorphism is a special case of the map constructed in b).

This isomorphism is *not canonical* in the following sense: Let F be a finitely generated free R-module. Then by different bases of F different isomorphisms $R^n \xrightarrow{\sim} F$ are defined. And these induce *different* isomorphisms $F^* \xrightarrow{\sim} F$.

b) Note that if R is commutative and the R-module M finitely generated, say by $m_1, \dots, m_n$, then every homomorphism $f \in M^*$ is determined by the values $f(m_i)$. In this way M^* can be regarded as an R-submodule of R^n via the injective R-homomorphism:

$$M^* \to R^n, \ f \mapsto (f(m_i))_{1 \le i \le n}$$

c) Let $M^{**} := (M^*)^*$, the **bidual**. The **evaluation map** $M \times M^* \to R$, defined by $(x, f) \mapsto f(x)$ induces an R-linear map $\alpha : M \to M^{**}$. If M is a free finitely generated R-module, α is an isomorphism. Note that here and also in b) the finite generation is essential.

d) Let $f : M \to M'$ be an R-module homomorphism and N another R-module. Then f induces group (resp. module) homomorphisms

$$f^* : \operatorname{Hom}_R(M', N) \to \operatorname{Hom}_R(M, N) \text{ and } f_* : \operatorname{Hom}_R(N, M) \to \operatorname{Hom}_R(N, M')$$

by $\alpha \mapsto \alpha \circ f$, and $\alpha \mapsto f \circ \alpha$. Note the 'arrow reversing' in the first case.

See also in the section on the Tensor Product for more properties of Hom.

e) The endomorphism of an R-module M form a ring, the multiplication giving by composing homomorphisms: $fg := f \circ g$. This ring is denoted by $\mathrm{End}_R(M)$. The ring $\mathrm{End}_R(R^n)$ can be identified with the ring $\mathrm{M}_n(R)$ of $n \times n$-matrices over R.

One must be cautious with the order of factors: Usually $f \circ g$ is defined by $(f \circ g)(x) := f(g(x))$ (i.e. 'first g, then f'); now if R^n is regarded as a left module of n-tuples, written in rows, and the homomorphisms are given by multiplying a row on the right side by a matrix, then $\mathrm{End}_n(M)$ is isomorphic to the opposite ring – i.e. where the multiplication is reversed – of $\mathrm{M}_n(R)$.

1.2.15 Now we go a step into the field of **Homological Algebra**. At first sight its results do not look of great importance, and its proofs are simple, or even automatic. (Therefore S. Lang in the first edition of his 'Algebra' suggested to 'take any book on Homological Algebra and prove all its theorems without looking at the proofs given in that book.') But this first view is deceiving. Homological Algebra is a strong tool in many mathematical disciplines, as Algebraic Topology, Algebraic Geometry, Class Field Theory, Group Theory, ...

Definitions 1.2.16 a) *A finite or infinite sequence of linear maps (i.e. a* **complex***)*

$$\cdots \longrightarrow M_{n+1} \xrightarrow{f_{n+1}} M_n \xrightarrow{f_n} M_{n-1} \longrightarrow \cdots$$

is called an **exact sequence** *if* $\mathrm{im}(f_{n+1}) = \ker(f_n)$ *for every n where this statement makes sense.*

b) *An exact sequence of the form* $0 \to M \xrightarrow{f} N \xrightarrow{g} L \to 0$ *is called a* **short exact sequence** . *Here exactness means: f is injective, g is surjective and* $\mathrm{im}(f) = \ker(g)$, *i.e. M can be identified with a submodule of N via f, and L via g with the factor module of N modulo that submodule.*

Examples 1.2.17 a) Let U be a submodule of a module M then the sequence

$$0 \longrightarrow U \xrightarrow{i} M \xrightarrow{p} M/U \longrightarrow 0$$

is exact, if i, p are the inclusion map, resp. the canonical projection.

b) Let $m \in \mathbb{Z} \setminus (0)$ and $h_m : \mathbb{Z} \to \mathbb{Z}$ be the multiplication by m, i.e. the homothesy of m, further $\pi : \mathbb{Z} \to \mathbb{Z}/m\mathbb{Z}$ the canonical projection. Then the sequence

$$0 \longrightarrow \mathbb{Z} \xrightarrow{h} \mathbb{Z} \xrightarrow{\pi} \mathbb{Z}/m\mathbb{Z} \longrightarrow 0$$

is exact.

c) Let $M_1 \xrightarrow{f} M_2 \xrightarrow{g} M_3$ be an exact sequence of modules.

If $M_1 = 0$ or more generally $f = 0$, then g is injective.

If $M_3 = 0$ or more generally $g = 0$, then f is surjective.

Especially, the sequence

$$0 \longrightarrow M \xrightarrow{f} N \longrightarrow 0$$

is exact if and only if f is an isomorphism.

An example of a homological result is the so called Snake Lemma.

Lemma 1.2.18 *Let*

$$
\begin{array}{ccccccc}
M' & \xrightarrow{f'} & M & \longrightarrow & M'' & \longrightarrow & 0 \\
\downarrow{\alpha'} & & \downarrow{\alpha} & & \downarrow{\alpha''} & & \\
0 & \longrightarrow & N' & \longrightarrow & N & \xrightarrow{g} & N''
\end{array}
$$

be a commutative diagram with exact rows. Then there is a homomorphism $\delta : \ker(\alpha'') \to \operatorname{coker}(\alpha')$ such that the sequence

$$\ker(\alpha') \to \ker(\alpha) \to \ker(\alpha'') \xrightarrow{\delta} \operatorname{coker}(\alpha') \to \operatorname{coker}(\alpha) \to \operatorname{coker}(\alpha''),$$

– where the maps between the kernels and cokernels are the induced ones– is exact.

Moreover, if f' is injective, so is the induced map $\ker(\alpha') \to \ker(\alpha)$. And if g is surjective, so is $\operatorname{coker}(\alpha) \to \operatorname{coker}(\alpha'')$.

(We leave the explanation of the lemma's name to the imagination of the reader.)

Proof. We only define the *connecting homomorphism* δ and leave the rest to the reader. So let $x \in \ker(\alpha'') \subset M''$. Choose a preimage $y \in M$. Then $\alpha(y) \in \ker(g)$ has a unique preimage $y' \in N'$. Let $\delta(x)$ be the class of y' in $\operatorname{coker}(\alpha')$. It is well defined. □

Remark 1.2.19 Let M be a finitely generated R-module, say generated by $x_1, \ldots, x_n$. Define $f_0 : R^n \to M$ by $f_0(a_1, \ldots, a_n) := \sum_{i=1}^{n} a_i x_i$. Then f_0 is surjective. Hence we get an exact sequence

$$R^n \xrightarrow{f_0} M \longrightarrow 0.$$

Now assume that also ker f_0 is finitely generated, say by $y_1, \ldots, y_m$ and define $f_1 : R^m \to R^n$ by $f_1(a_1, \ldots, a_m) := \sum_{i=1}^{m} a_i y_i$. Then clearly the sequence

$$R^m \xrightarrow{f_1} R^n \xrightarrow{f_0} M \longrightarrow 0$$

is exact. One can continue this construction, provided the kernel of the last constructed map is finitely generated.

If one allows to use infinitely generated free modules, then for any module M over every ring R one can construct an infinite exact sequence

$$\cdots \longrightarrow F_2 \longrightarrow F_1 \longrightarrow F_0 \longrightarrow M \longrightarrow 0.$$

with free R-modules F_i, a **free resolution of** M.

It may happen that this process will "stop" with the zero module after some stage, in which case we say that the module M has a **finite free resolution**

$$0 \longrightarrow F_n \longrightarrow \cdots \longrightarrow F_1 \longrightarrow F_0 \longrightarrow M \longrightarrow 0.$$

We will see in Chapter 3 that this is the case for all finitely generated modules over a polynomial ring $k[x_1, \ldots, x_n]$ with a field k. It is also true for modules over this ring which are not finitely generated. But we will not prove it.

1.3 Localization

In this section every ring is supposed to be *commutative*.

We duplicate the construction of the rational numbers from the integers. The rational number field $\mathbb{Q}$ is the field got from the ring of integers $\mathbb{Z}$ 'by inverting all non-zero integers.'

1.3.1 Multiplicatively Closed Subsets

Definition 1.3.1 *Let R be a ring. A subset S of R is said to be **multiplicatively closed** or **multiplicative** if*

(1) $1 \in S$.

(2) $s_1, s_2 \in S$ *implies* $s_1 s_2 \in S$.

Examples 1.3.2 a) If $s \in R$, take $S = \{s^n \mid n \in \mathbb{N}\}$.

b) If $\mathfrak{p}$ is a prime ideal of R, take $S = R \setminus \mathfrak{p}$.

c) If $\mathfrak{p}_\alpha$ are prime ideals of R, take $S = R \setminus \bigcup_\alpha \mathfrak{p}_\alpha$.

d) Let S be the set of all non-zero-divisors of R.

e) If I is an ideal of R, take $S = 1 + I$.

f) Let $(S_i)_{i \in X}$ be a family of multiplicative subsets of R, the the intersection of the S_i as well as their product $\prod_{i \in X} S_i$, i.e. the set, generated multiplicatively by the union $\bigcup_{i \in X} S_i$ are multiplicative.

Lemma 1.3.3 (Krull) *Let S be a multiplicatively closed subset and I an ideal of a ring R with $S \cap I = \emptyset$. Then in the set of ideals $\mathfrak{a}$ with $I \subset \mathfrak{a} \subset R \setminus S$ exist maximal elements and these are prime ideals.*

Proof. The existence of maximal elements in the above set of ideals follows directly by Zorn's Lemma. Let $\mathfrak{p}$ be one of these. If $x \notin \mathfrak{p}$, $y \notin \mathfrak{p}$ then there are $s_1 \in (x, \mathfrak{p}) \cap S$, $s_2 \in (y, \mathfrak{p}) \cap S$ by the maximality of $\mathfrak{p}$. Let $s_1 = \lambda x + p_1$, $s_2 = \mu y + p_2$, for $\lambda, \mu \in R$, $p_1, p_2 \in \mathfrak{p}$. If $xy \in \mathfrak{p}$ then $s_1 s_2 \in \mathfrak{p}$. But $\mathfrak{p} \cap S = \emptyset$ and so $xy \notin \mathfrak{p}$. $\qquad\square$

Definitions 1.3.4 a) *The* **radical** *of an ideal I is the set*

$$\sqrt{I} := \{a \in R \mid a^n \in I \text{ for some } n \in \mathbb{N}\}.$$

It is easy to verify – using the Binomial Theorem – that $\sqrt{I}$ is an ideal of R. (This is not generally so for non-commutative rings.)

b) *An ideal is called a* **radical ideal** *if it is its own radical. (Of course, $\sqrt{\sqrt{I}} = \sqrt{I}$ and so the radical of any ideal is a radical ideal.)*

c) *$\sqrt{(0)}$ is also called the* **nilradical** *of R and denoted by $\mathrm{Nil}(R)$.*

d) *An element a of a ring R is called* **nilpotent** *if it belongs to $\mathrm{Nil}(R)$, i.e. $a^n = 0$ for a suitable $n \in \mathbb{N}$.*

e) *A ring R is called* **reduced** *if $\mathrm{Nil}(R) = (0)$.*

Corollary 1.3.5 *Let I be an ideal of a ring R. Then $\sqrt{I} = \bigcap_{\mathfrak{p} \in V(I)} \mathfrak{p}$.*

Proof. Clearly $\sqrt{I}$ is a subset of the right hand side ideal. In view of the canonical bijection between prime ideals containing I and prime ideals of R/I we may as well work in R/I. Hence we can assume that $I = (0)$. Let a belong to the right hand side and let $S = \{a^n \mid n \geq 0\}$. If a were not nilpotent, i.e. $(0) \cap S = \emptyset$, by Lemma 1.3.3 there would be a prime ideal $\mathfrak{p} \subset R \setminus S$. But since a belongs to every prime ideal, we would get $a \in \mathfrak{p}$, a contradiction to $a \in S$. Therefore, $a^m = 0$ for some m. $\qquad\square$

Using Remark 1.1.2 b) we conclude that $\sqrt{I}$ is also the intersection of the minimal prime over-ideals of I.

Corollary 1.3.6 *The correspondence $\mathfrak{a} \mapsto V(\mathfrak{a})$ gives a bijective correspondence between radical ideals of R and closed subsets of* $\mathrm{Spec}(R)$.

More explicitly, the map I from the set of closed sets of $\mathrm{Spec}(R)$ to the set of radical ideals which assigns to the closed set X the intersection $I(X) := \bigcap_{\mathfrak{p} \in X} \mathfrak{p}$ provides the inverse map to the above correspondence as is evident from:

Corollary 1.3.7 *If $\mathfrak{a}$ is an ideal of R then $I(V(\mathfrak{a})) = \sqrt{\mathfrak{a}}$.*

1.3.2 Rings and Modules of Fractions

We now construct the **ring of fractions** (or **localization**) of R w.r.t. S and the **module of fractions** (or **localization**) of an R-module M w.r.t. a multiplicative set S:

Define an equivalence relation on $M \times S$ by

$$(m, s) \sim (m', s') \iff \text{ there is an } s'' \in S \text{ with } s''s'm = s''sm'.$$

(If S consists of non-zero-divisors, this is equivalent to $s'm = sm'$. But the latter does not define an equivalence relation in general. It may lack transitivity. We will not restrict ourselves to the case where there are no zero-divisors in S. The important multiplicative complements of prime ideals often contain zero-divisors.)

Define the analogous equivalence relation on $R \times S$.

Let $S^{-1}M$ or M_S, resp. $S^{-1}R$ or R_S denote the set of equivalence classes $M \times S/\sim$, resp. $R \times S/\sim$ under this equivalence relation.

The equivalence class of (x, s), $x \in M$, resp. R, $s \in S$ will be denoted by $\dfrac{x}{s}$ or fairly often – for reasons of typographical convenience – by x/s. We can make $S^{-1}M$, resp. $S^{-1}R$ an additive group by defining addition

$$\frac{m}{s} + \frac{m'}{s'} := \frac{s'm + sm'}{ss'}$$

for $m, m' \in M$ or in R, $s, s' \in S$. This is easily checked to be well defined.

We define a (commutative) ring structure on $S^{-1}R$ by defining multiplication as follows:

$$\frac{r}{s} \cdot \frac{r'}{s'} := \frac{rr'}{ss'}$$

for $r, r' \in R$, $s, s' \in S$. The ring axioms can easily be checked.

We make $S^{-1}M$ an $S^{-1}R$-module analogously:

$$\frac{r}{s} \cdot \frac{x}{s'} := \frac{rx}{ss'}$$

for $r \in R$, $x \in M$, $s, s' \in S$. Note that, if M is a finitely generated R-module then $S^{-1}M$ is a finitely generated $S^{-1}R$-module.

1.3.8 Notations: a) For $f \in R$ we write $R_f := S^{-1}R$ and $M_f = S^{-1}M$ where $S = \{1, f, f^2, \dots\}$. Also the notation $R[1/f]$ is in use.

b) If $S = R \setminus \mathfrak{p}$, where $\mathfrak{p}$ is a prime ideal of R, then $S^{-1}M$ is denoted by $M_{\mathfrak{p}}$, and $S^{-1}R$ by $R_{\mathfrak{p}}$. (We do not worry that this notation is not quite consistent with that of a) and c).)

c) We often write R_S and M_S instead of $S^{-1}R$, resp. $S^{-1}M$. Especially, if $S = 1 + I$ we use the notations M_{1+I} and R_{1+I}.

d) If S is the set of all non-zero-divisors in R, $S^{-1}R$ is called the **total quotient ring** of R or its **full ring of fractions**. It is denoted by $Q(R)$. If R is a domain, then $Q(R)$ is the quotient field of R.

1.3.9 Remarks and Definitions. a) Let M be an R-module and $S \subset R$ be multiplicative. There are canonical maps $i_{R,S} : R \to S^{-1}R$ and $i_{M,S} : M \to S^{-1}M$, defined by $x \mapsto \frac{x}{1}$ for $x \in R$ or in M. In general these maps are neither injective nor surjective. We often write x_S instead of $\frac{x}{1}$, especially when there are several multiplicative sets involved.

The map $i_{R,S}$ is a ring-homomorphism. Therefore $S^{-1}M$ is also an R-module. (See Example 1.2.3 f).) And $i_{M,S}$ is an R-module homomorphism. The kernel of $i_{M,S}$ consists of the $x \in M$ with $sx = 0$ for some $s \in S$, analogously if $M = R$. So if $0 \in S$ (or more generally, if $sM = \{0\}$ fore some $s \in S$), then $S^{-1}M \cong 0$.

b) Let $f : M \to N$ be an R-module homomorphism. We define the (clearly well defined) $S^{-1}R$-module homomorphism

$$S^{-1}f : S^{-1}M \longrightarrow S^{-1}N \text{ by } \frac{m}{s} \mapsto \frac{f(m)}{s}.$$

This is the unique R-module homomorphism making the following diagram **commutative**:

$$
\begin{array}{ccc}
M & \xrightarrow{\ f\ } & N \\
{\scriptstyle i_{M,S}}\downarrow & & \downarrow{\scriptstyle i_{N,S}} \\
S^{-1}M & \xrightarrow{\ S^{-1}f\ } & S^{-1}N
\end{array}
$$

c) The property to be **commutative** for the above diagram means $S^{-1}f \circ i_{M,S} = i_{N,S} \circ f$. We think the reader will understand by this example the meaning of commutativity of any diagram!

Remarks 1.3.10 a) The assignments $M \mapsto S^{-1}M$ and $f \mapsto S^{-1}f$ form a so-called functor, i.e. $S^{-1}\mathrm{id}_M = \mathrm{id}_{S^{-1}M}$ and $S^{-1}(g \circ f) = (S^{-1}g) \circ (S^{-1}f)$, whenever $g \circ f$ is defined.

b) This functor is R-linear; which means that the map $f \mapsto S^{-1}f$

$$\mathrm{Hom}_R(M, N) \longrightarrow \mathrm{Hom}_{S^{-1}R}(S^{-1}M, S^{-1}N)$$

is R-linear. (Note that the 'Hom-group' on the right is an $S^{-1}R$-module, hence an R-module via $i_{R,S}$.)

c) Every R-linear map $S^{-1}M \to S^{-1}N$ is also $S^{-1}R$-linear, therefore $\mathrm{Hom}_{S^{-1}R}(S^{-1}M, S^{-1}N) = \mathrm{Hom}_R(S^{-1}M, S^{-1}N)$, both as sets and $S^{-1}R$-modules.

d) There is a canonical $S^{-1}R$-linear map $S^{-1}\mathrm{Hom}_R(M, N) \to \mathrm{Hom}_{S^{-1}R}(S^{-1}M, S^{-1}N)$, which in important cases, but not always, is an isomorphism. We will discuss this later, when we introduce the notion of a finitely presented module.

Lemma 1.3.11 (Exactness of Localization) *If the sequence of R-module homomorphisms $M' \xrightarrow{f} M \xrightarrow{g} M''$ is exact, then the sequence*

$$S^{-1}M' \xrightarrow{S^{-1}f} S^{-1}M \xrightarrow{S^{-1}g} S^{-1}M''$$

is exact as well.

Proof. $S^{-1}g \circ S^{-1}f = S^{-1}(g \circ f) = 0$ and so $\mathrm{im}(S^{-1}f) \subset \ker(S^{-1}g)$. If $x = m/s \in \ker(S^{-1}g)$ then $s''g(m) = 0$ for some $s'' \in S$, i.e. $g(s''m) = 0$ and so $s''m \in \ker(g) = \mathrm{im}(f)$. Let $s''m = f(m')$. Then $S^{-1}f(m'/s'') = m$. □

Corollary 1.3.12 *Let S be a multiplicatively closed subset of R. Let*

$$0 \to M' \xrightarrow{f} M \xrightarrow{g} M'' \to 0$$

be an exact sequence of R-modules. Then

$$0 \to S^{-1}M' \xrightarrow{S^{-1}f} S^{-1}M \xrightarrow{S^{-1}g} S^{-1}M'' \to 0$$

is exact.

In particular if N is sub-module of M then $S^{-1}N$ can be regarded as a sub-module of $S^{-1}M$ and $S^{-1}(M/N) \cong S^{-1}M/S^{-1}N$ canonically. Localization and taking quotients commute.

In particular, if I is an ideal of R then $S^{-1}(R/I) \cong S^{-1}R/S^{-1}I$ as R-modules and also as rings canonically. Further $S^{-1}(R/I) \cong \overline{S}^{-1}(R/I)$, if $\overline{S}$ denotes the image of S in R/I. □

Corollary 1.3.13 *Let S be a multiplicatively closed subset of R. Let M, N be R-modules and $f : M \to N$ be a R-linear map. Then $\ker(S^{-1}f) = S^{-1}\ker(f)$, and $\operatorname{im}(S^{-1}f) = S^{-1}\operatorname{im}(f)$.*

Proof. Apply Lemma 1.3.11 to the exact sequences

$$0 \to \ker(f) \to M \xrightarrow{f} N \quad \text{and} \quad \ker(f) \to M \xrightarrow{f} \operatorname{im}(f) \to 0.$$

$\square$

Lemma 1.3.14 *Let U_i for $i \in I$ be submodules of an R-module M.*

a) $S^{-1} \sum_{i \in I} U_i = \sum_{i \in I} U_i$.

b) *If I is finite, then* $S^{-1} \bigcap_{i \in I} U_i = \bigcap_{i \in I} S^{-1} U_i$.

If $I = S$ is the set of positive integers and $U_i = i\mathbb{Z}$, $M = \mathbb{Z}$, then $\bigcap_{i \in I} S^{-1} U_i = \mathbb{Q}$ and $S^{-1} \bigcap_{i \in I} U_i = 0$.

Proof. a) Both sides of the claimed equality are submodules of $S^{-1}M$. Note that the elements of $\sum_{i \in I} S^{-1} U_i$ are essentially finite sums $\sum_{i \in I} u_i$ with u_i in $S^{-1} U_i$, i.e. all but finitely many summands are 0. So there is a common denominator. Hence they belong to $S^{-1} \sum_{i \in I} U_i$. The converse inclusion is trivial.

b) We may assume $I = \{1, 2\}$. Clearly $S^{-1}(U_1 \cap U_2) \subset S^{-1} U_i$ for $i = 1, 2$, whence $S^{-1}(U_1 \cap U_2) \subset (S^{-1}U_1) \cap (S^{-1}U_2)$. Conversely, let $u_1/s_1 = u_2/s_2$ in $(S^{-1}U_1) \cap (S^{-1}U_2)$ with u_i in U_i, $s_i \in S$. Then there is an $s \in S$ with $ss_2 u_1 = ss_1 u_2 =: v$. Since $v \in U_1 \cap U_2$, we have $u_1/s_1 = v/(ss_1 s_2) \in S^{-1}(U_1 \cap U_2)$. $\square$

1.3.3 Localization Technique

There are many properties of rings, modules and homomorphisms which can be checked 'locally'. This means, some property holds, say, for M if and only if it holds for $M_\mathfrak{p}$ for all prime (or only all maximal) ideals $\mathfrak{p}$.

Lemma 1.3.15 *Let M be an R-module. Then $M = 0$ if and only if $M_\mathfrak{p} = 0$ for all maximal ideals $\mathfrak{p}$ of R.*

Proof. One direction is trivial. So assume $M \neq 0$ and take an $x \in M \setminus \{0\}$. The ideal $\operatorname{Ann}(x) := \{a \in R \mid ax = 0\}$, called the **annhilator** of x, is different from R, since $1x \neq 0$. So by Lemma 1.3.3 there is a maximal ideal $\mathfrak{p} \supset \operatorname{Ann}(x)$. It is enough to disprove $M_\mathfrak{p} = 0$. If it were so, in $M_\mathfrak{p}$ we would have the equality $x/1 = 0/1$, i.e. there would be an $s \in R \setminus \mathfrak{p}$ with $sx = 0$, contradicting $\operatorname{Ann}(x) \subset \mathfrak{p}$. $\square$

Another illustration: Given an R-linear map $f : M \to N$ of R-modules M, N one can test whether it is injective (resp. surjective, resp. 0) by checking this locally.

Corollary 1.3.16 *An R-linear map $f : M \to N$ is injective (resp. surjective, resp. the zero map) if and only if $f_{\mathfrak{p}} : M_{\mathfrak{p}} \to N_{\mathfrak{p}}$ is injective (resp. surjective, resp. zero) for all maximal ideals $\mathfrak{p}$ of R.*

Proof. Let $K = \ker(f)$. One has the exact sequence

$$0 \to K \xrightarrow{i} M \xrightarrow{f} N.$$

From Corollary 1.3.13 we get $K_{\mathfrak{p}} \cong \ker(f_{\mathfrak{p}})$. Therfore we have the equivalences f injective $\iff$ $K = 0$ $\iff$ $K_{\mathfrak{p}} = 0$ for all $\mathfrak{p}$ $\iff$ $f_{\mathfrak{p}}$ injective for all $\mathfrak{p}$. The other assertions follow in the same way. $\qquad\square$

Corollary 1.3.17 *Let U, V be submodules of an R-module M such that $U_{\mathfrak{p}} = V_{\mathfrak{p}}$ for every maximal ideal $\mathfrak{p}$ of R. Then $U = V$.*

Proof. $U = V$ is equivalent to $(U+V)/U = (U+V)/V = 0$. Use the Lemmas 1.3.14 and 1.3.15. $\qquad\square$

Proposition 1.3.18 *Let M be a finitely generated R-module and let $\mathfrak{p} \in \mathrm{Spec}(R)$ such that $M_{\mathfrak{p}} = 0$. Then there is an $s \in R \setminus \mathfrak{p}$ with $M_s = 0$.*

Proof. Let $x_1, \dots, x_r$ generate M. Since $M_{\mathfrak{p}} = 0$, there are $s_1, \dots, s_n \in R \setminus \mathfrak{p}$ with $s_i x_1$ for $i = 1, \dots, n$. The element $s := s_1 \cdots s_n$ fulfills the claim. $\qquad\square$

1.3.4 Prime Ideals of a Localized Ring

We next discuss the structure of prime ideals in a localized ring. One has the natural ring homomorphism $i = i_{R,S} : R \to S^{-1}R$ given by $i(r) = r/1$. Remember that i need not be injective and that

$$\ker(i) = \{r \in R \mid sr = 0 \text{ for some } s \in S\}.$$

(The study of $\ker(i)$ plays an important role in Commutative Algebra - see [1] Theorem 10.17, Corollary 10.21).

Lemma 1.3.19 *Let $\varphi : R \to S$ be a ring homomorphism. If $\mathfrak{p}$ is a prime ideal of S then $\varphi^{-1}(\mathfrak{p})$ is a prime ideal of R.*

Proof. $S/\mathfrak{p}$ is a domain, and φ induces an injective homomorphism $R/\varphi^{-1}(\mathfrak{p}) \to S/\mathfrak{p}$, whence $R/\varphi^{-1}(\mathfrak{p})$ is a domain, too. $\qquad\square$

Corollary 1.3.20 *Let $\varphi : R \to S$ be a ring homomorphism. Then φ induces a continuous map $\varphi^* : \mathrm{Spec}(S) \to \mathrm{Spec}(R)$ given by $\varphi^*(\mathfrak{p}) := \varphi^{-1}(\mathfrak{p})$ $(= \mathfrak{p} \cap R$ if $R \subset S$).*

Proof. It is easy to check that $\varphi^{*-1}(D(f)) = D(\varphi(f))$, for any $f \in R$. □

Theorem 1.3.21 *There is a bijective correspondence between the prime ideals of $S^{-1}R$ and the prime ideals of R which do not intersect S, given by $i^* : \mathfrak{p} \mapsto i^{-1}(\mathfrak{p}) = \mathfrak{p} \cap R$.*

Proof. By Lemma 1.3.19 the preimage $i^{-1}(\mathfrak{p})$ is a prime ideal of R. Since $\mathfrak{p}$ is a proper ideal of $S^{-1}R$, $i^{-1}(\mathfrak{p}) \cap S$ must be empty.

If $\mathfrak{p}'$ is a prime ideal of R, then $S^{-1}\mathfrak{p}'$ is a prime ideal of $S^{-1}R$ as $S^{-1}R/S^{-1}\mathfrak{p}' \xrightarrow{\sim} S^{-1}(R/\mathfrak{p}')$ is a domain, provided $S \cap \mathfrak{p}' = \emptyset$. One defines the inverse of the above map i^* viz. $i_* : \mathfrak{p}' \mapsto S^{-1}\mathfrak{p}'$. To show i^* and i_* are inverses of each other we need to show $i^{-1}(S^{-1}\mathfrak{p}') = \mathfrak{p}'$, and $S^{-1}i^{-1}(\mathfrak{p}) = \mathfrak{p}$, for primes $\mathfrak{p}'$ of R, $\mathfrak{p}$ of $S^{-1}R$ with $\mathfrak{p}' \cap S = \emptyset$.

"$i^{-1}S^{-1}\mathfrak{p}' = \mathfrak{p}'$": $x = a/s \in S^{-1}R$, $a \in \mathfrak{p}'$, $s \in S$ implies $s'(sx-a) = 0 \in \mathfrak{p}'$ for some $s' \in S$, and this implies $sx - a \in \mathfrak{p}'$ as $s' \notin \mathfrak{p}'$, whence $x \in \mathfrak{p}'$ (as $x \in R$ and $s \notin \mathfrak{p}'$).

"$S^{-1}i^{-1}(\mathfrak{p}) = \mathfrak{p}$": $x = \frac{a}{s} \in \mathfrak{p}$, $a \in R$, $s \in S$ implies $s'(sx - a) = 0 \in \mathfrak{p}$ for some $s' \in S$, which implies $sx - a \in \mathfrak{p}$ as $s' \notin \mathfrak{p}$. Clearly $sx \in i^{-1}(\mathfrak{p})$ and so $a \in i^{-1}(\mathfrak{p})$. Hence $x = \frac{a}{s} \in S^{-1}i^{-1}(\mathfrak{p})$. □

Corollary 1.3.22 *Let R be a ring and $f \in R$. Let further $i : R \to R_f$ be the natural map. Then the map $i^* : \mathrm{Spec}(R_f) \to \mathrm{Spec}(R)$ induces a homeomorphism $\mathrm{Spec}(R_f) \cong D(f)$.*

Proof. By Theorem 1.3.21 we see $i^*(\mathrm{Spec}(R_f)) = D(f)$. The map $j^* : \mathfrak{p} \mapsto \mathfrak{p}_f$ from $D(f)$ to $\mathrm{Spec}(R_f)$ is a continuous map as $j^{*-1}(D(fg)) = D(fg)$ for all $g \in R$. Then j^* is the inverse of $i^* : \mathrm{Spec}(R_f) \to D(f)$ by Theorem 1.3.21 □

Corollary 1.3.23 *Let R be a ring and let $\mathfrak{p}, \mathfrak{p}_1, \dots, \mathfrak{p}_n$ be prime ideals of R. Let $S = R \setminus \bigcup_{i=1}^{n} \mathfrak{p}_i$. Then $R_\mathfrak{p}$ has a unique maximal ideal and $S^{-1}R$ has only finitely many maximal ideals, namely the $S^{-1}\mathfrak{p}_i$, $1 \leq i \leq n$.* □

Corollary 1.3.24 *Let R be reduced. Then the set of zero divisors of R equals the union of minimal prime ideals.*

Proof. Let $\mathfrak{p}$ be a minimal prime ideal and $a \in \mathfrak{p}$. Then $\mathfrak{p}R_\mathfrak{p}$ is the only prime ideal of $R_\mathfrak{p}$, and so $a/1$ in nilpotent in this ring. This means, there is an $s \in R \setminus \mathfrak{p}$ and an $n \geq 1$ with $sa^n = 0$, but $sa^{n-1} \neq 0$. So a is a zero divisor.

Conversely let $a \notin \mathfrak{p}$ for every minimal prime ideal $\mathfrak{p}$ and $ab = 0$. But then $b \in \mathfrak{p}$ for every minimal prime ideal. Since R is reduced, this implies $b = 0$. □

Definition 1.3.25 *A ring R which has a unique maximal ideal $\mathfrak{m}$ is called a **local ring** and is sometimes denoted by $(R, \mathfrak{m})$. A ring which has only finitely many maximal ideals is called a **semilocal ring**.*

1.4 Integral Ring Extensions

In this section all rings are supposed to be *commutative*. The ring of fractions is an **over-ring** of a ring which is intimately related to it. We introduce some over-rings, or ring extensions, which enjoy nice properties related to the ring. It is found that their study can often give back interesting information of the parent ring.

1.4.1 Integral Elements

Definition 1.4.1 *Let $f : R \to A$ be a ring homomorphism (by which we consider A as an R-algebra). (In most cases R will be a subring of A and f the inclusion.)*

a) *A is called **finite** over R (or f is called finite), if A is finitely generated as an R-module.*

b) *A is called **of finite type** over R (or f is called of finite type), if A is finitely generated as an R-algebra, i.e. if and only if f factorizes as*

$$R \xrightarrow{\kappa} R[X_1, \dots, X_n] \xrightarrow{g} A,$$

where κ is the canonical imbedding into a polynomial algebra in finitely many indeterminates and g is surjective.

Remark 1.4.2 If A is finite over R, it is clearly also of finite type, a generating set of A as an R-module being also a generating set of A as an R-algebra. On the other hand any polynomial R-algebra in finitely many (and at least one) indeterminates is of finite type, but not finite over R if $R \not\cong 0$.

Definitions 1.4.3 a) *A polynomial $f \in R[X]$ is called **monic**, if it is not zero and its leading coefficient is 1, i.e. if f is of the form*

$$f = X^n + a_{n-1}X^{n-1} + \cdots + a_1 X + a_0$$

b) *The **annihilator** of an $x \in M$ is the set*

$$\mathrm{Ann}(x) := \{a \in R \mid ax = 0\}.$$

Clearly $\text{Ann}(M)$ and $\text{Ann}(x)$ are ideals of R. (In the case where R is not necessarily commutative, $\text{Ann}(M)$ is a two sided ideal, whereas the annihilator of an element $x \in M$, i.e. $\text{Ann}(x)$ is a left ideal.)

Proposition 1.4.4 *Let $R \subset A$ be a ring extension and $\alpha \in A$. The following statements are equivalent:*

(i) *α is a zero of a monic polynomial $f \in R[X]$.*

(ii) *$R[\alpha]$, i.e. the subring of A, generated by $R \cup \{\alpha\}$, (which clearly consists of all polynomials in α with coefficients in R) is finite over R.*

(iii) *There is a ring B between $R[\alpha]$ and A, which is finite over R.*

(iv) *There exists an $R[\alpha]$-module M, which has a trivial annihilator and is finitely generated as an R-module.*

Proof. (i)$\Rightarrow$ (ii). Any element of $R[\alpha]$ is of the form $g(\alpha)$ where $g \in R[X]$ is a polynomial. But since f is monic, in $R[X]$ we can write $g = qf + g'$ with $\deg(g') < n$. And then $g(\alpha) = g'(\alpha)$, which shows that $R[\alpha]$ is generated by $1, \alpha, \dots, \alpha^{n-1}$ as an R-module.

(ii) $\Rightarrow$ (iii). Set $B = R[\alpha]$.

(iii) $\Rightarrow$ (iv). Set $M = B$.

(iv) $\Rightarrow$ (i). Let $x_1, \dots, x_n$ be a generating set for the R-module M. Let $a_{ij} \in R$ be so that $\alpha x_i = \sum_j a_{ij} x_j$. Then for the matrix $\beta := (a_{ij} - \delta_{ij}\alpha)$ we get

$$\beta \begin{pmatrix} x_1 \\ \vdots \\ x_n \end{pmatrix} = \begin{pmatrix} 0 \\ \vdots \\ 0 \end{pmatrix}$$

in the module M^n. (Recall the meaning of Kronecker's Symbol $\delta_{ij} = 0$ for $i \neq j$, $\delta_{ii} = 1$.) By Cramer's Rule $\text{adj}(\beta)\beta = \det(\beta)I_n$. (By I_n we denote the $n \times n$ unit matrix.) Therefore $\det(\beta)x_i = 0$, whence $\det(\beta)M = 0$ since the x_i generate M. So $\det(\beta)$ itself is zero, M having a trivial annihilator. By the Leibniz Formula this gives an equation of the form

$$\alpha^n + a_{n-1}\alpha^{n-1} + \dots + a_0$$

the a_ν being combinations of certain a_{ij}, so lying in R. $\qquad\square$

Definition 1.4.5 *In the above situation α is called **integral** over R, if it fulfills the equivalent conditions (i) to (iv). The ring A is called **integral** over R, if every $\alpha \in A$ is integral over R.*

Proposition 1.4.6 a) *Every finite ring extension is integral.*

b) *Let $A \subset B$ and $B \subset C$ be finite ring extensions. Then so is $A \subset C$.*

c) *Let $R \subset A$ be any ring extension. The set B of elements of A which are integral over R is a subring of A (clearly containing R).*

Proof. a) This is clear by condition (iii).

b) Let $x_1, \ldots, x_m$ and $y_1, \ldots, y_n$ be finite generating sets of the A-module B, resp. the B-module C. Then the $x_i y_j$ with $1 \leq i \leq m$, $1 \leq j \leq n$ generate C as an A-module. The reader will easily see this, if he remembers the proof of it in the case where A, B, C are fields.

c) Let $\alpha, \beta \in A$ be integral over R. We have to show that $\alpha \pm \beta$, $\alpha\beta$ are integral too. They are so by a), since they are elements of $R[\alpha, \beta]$, which is finite over R by b). (Note that β is integral over $R[\alpha]$, if it is so over R.) $\quad\square$

Examples 1.4.7 a) The ring $\mathbb{Z} + \mathbb{Z}i$ of Gauss numbers is finite, hence integral over $\mathbb{Z}$.

b) More generally, let I be an ideal of the polynomial algebra $R[X]$ over a ring R. If I contains a monic polynomial, then the ring extension $R/(I \cap R) \subset R[X]/I$ is finite, hence integral. Namely as an R-module it is generated by $1, x, \ldots, x^{n-1}$ where x denotes the residue class of X and n is the degree of a monic polynomial in I.

1.4.8 Note that a ring extension is finite if and only if it is integral and of finite type.

There are non-finite integral extensions, for e.g. $\mathbb{Z} \subset A$, where A is the subring of $\mathbb{C}$ consisting of all complex numbers which are integral over $\mathbb{Z}$. (If it were finite, A would lie in a finite dimensional subspace of the $\mathbb{Q}$-vector space $\mathbb{C}$. But there are linearly independent subsets of arbitrarily big finite cardinality. Namely if α is a root of the polynomial $X^n - 2$, which is irreducible by Eisenstein's Criterion, then $\alpha, \alpha^2 \ldots, \alpha^n$ are linearly independent over $\mathbb{Q}$.)

A ring extension $A \subset B$ is integral, if and only if B is a union of finite subextensions.

Definitions 1.4.9 a) *Let $A \subset B$ be a ring extension. The ring of all elements of B which are integral over A is called the **integral closure of A in B**.*

b) *Let A be a domain. The integral closure of A in $Q(A)$ is simply called the **integral closure of A**.*

c) *A domain which coincides with its integral closure is said to be **integrally closed**.*

Examples 1.4.10 a) Every factorial domain A is integrally closed. Especially the polynomial ring $k[x_1, \ldots, x_n]$ is so, if k is a field or a principal domain.

Namely let $a/b \in Q(A)$ be integral over A with a, b mutually prime. We have an equation of the form

$$\left(\frac{a}{b}\right)^n + a_{n-1}\left(\frac{a}{b}\right)^{n-1} + \cdots + a_1\frac{a}{b} + a_0 = 0.$$

Multiplying by b^n we get

$$a^n = -ba_{n-1} - \cdots - b^n a_0.$$

So every prime factor of b is one of a^n, hence of a, since A is factorial. But b is prime to a; therefore b has no prime factor, i.e. $a/b \in A$.

b) The ring $\mathbb{Z} + \mathbb{Z}i$ of Gauss numbers is integral over $\mathbb{Z}$ and integrally closed, since it is a Euclidean domain, as one knows. Hence it is the integral closure of $\mathbb{Z}$ in $\mathbb{Q}(i)$.

c) In Algebraic Number Theory the integral closures of $\mathbb{Z}$ in general finite field extensions of $\mathbb{Q}$ are considered. These are not always principal domains. They are so called Dedekind rings . We refer our readers to Chapter 8 on this class of rings.

To check the integrality of elements the following proposition is highly useful.

Proposition 1.4.11 *Let A be an integrally closed domain and $Q(A) \subset L$ a field extension. An element $\alpha \in L$ is integral over A if and only if it is algebraic and its minimal polynomial f over $Q(A)$ has coefficients in A.*

By definition the minimal polynomial of α over $Q(A)$ is the – uniquely determined – monic polynomial $f \in Q(A)[X]$ of minimal degree with $f(\alpha) = 0$.

Proof. If f fulfills the above condition, α is integral over A by Proposition 1.4.4 (i).

Conversely, if α is integral over A, so clearly are its conjugates in the algebraic closure of L. And the (non-leading) coefficients of f are (the elementary symmetric) polynomials in the conjugates of α, so are integral over A and belonging to $Q(A)$. Since A is integrally closed they even belong to A.   □

The following two propositions will be of great use for us.

Proposition 1.4.12 *Let $R \subset A$ be an integral ring extension and $r \in R$ be a unit in A. Then r is already a unit in R.*

Proof. Let $\alpha \in A$ be an inverse of r. We have to show $\alpha \in R$. But α fulfills an equation of the form

$$\alpha^n = -a_0 - a_1\alpha - \cdots - a_{n-1}\alpha^{n-1}.$$

Multiplying this equation by r^{n-1} we derive $\alpha \in R$.   □

Proposition 1.4.13 *Let $A \subset B$ be an integral extension of domains. Then A is a field if and only if B is one.*

Proof. Assume, B is a field. Then A is clearly one by Proposition 1.4.12.

Now assume, A is a field. It is enough to show that $A[\beta]$ is a field for every $\beta \in B$. But $B' = A[\beta]$ is a finite dimensional A-vector space and a domain. Multiplying by a non-zero element b of B' is an injective A-linear endomorphism of B', hence a bijective one. So there is a $b' \in B'$ with $bb' = 1$.   □

1.4.2 Integral Extensions and Prime Ideals

Proposition 1.4.14 *Let $A \subset B$ be an integral extension of rings and S a multiplicative subset of A, then $S^{-1}B$ is integral over $S^{-1}A$.*

Proof. Let $x/s \in S^{-1}B$ where $x \in B$ and $s \in S$. But as B is integral over A, x is a zero of a monic polynomial, say,

$$x^n + a_1x^{n-1} + \cdots + a_n = 0$$

with $a_i \in A$. But then

$$(x/s)^n + (a_1/s)(x/s)^{n-1} + \cdots + (a_n/s^n) = 0.$$

This shows that x/s is integral over $S^{-1}A$.   □

Proposition 1.4.15 *Let $A \subset B$ be an integral extension of rings, I an ideal of B and $J = A \cap I$. Then B/I is integral over A/J.*   □

Proposition 1.4.16 *Let $A \subset B$ be an integral extension. Let $\mathfrak{p}$ be a prime ideal of B and $\mathfrak{p}' = \mathfrak{p} \cap A$. Then $\mathfrak{p}$ is maximal if and only if $\mathfrak{p}'$ is maximal.*

Proof. By Proposition 1.4.15, $B/\mathfrak{p}$ is integral over $A/\mathfrak{p}'$ and both of these rings are integral domains. So by Proposition 1.4.13, $A/\mathfrak{p}$ is a field if and only if $B/\mathfrak{p}'$ is a field, i.e. $\mathfrak{p}$ is a maximal ideal if and only if $\mathfrak{p}'$ is one.   □

Corollary 1.4.17 *Let $A \subset B$ be an integral extension. Let $\mathfrak{p} \subsetneq \mathfrak{q}$ be different prime ideals of B. Then $\mathfrak{p} \cap A \neq \mathfrak{q} \cap A$.*

If two different prime ideals of B do not contain each other they may have the same intersection with A.

Proof. Let $S = A \setminus \mathfrak{q} \cap A$. By Proposition 1.4.14, $S^{-1}A \subset S^{-1}B$ is an integral extension. Clearly, $S^{-1}\mathfrak{p} \subsetneq S^{-1}\mathfrak{q}$, and $S^{-1}\mathfrak{p} \cap A = S^{-1}(\mathfrak{p} \cap A)$, $S^{-1}\mathfrak{q} \cap A = S^{-1}(\mathfrak{q} \cap A)$. Therefore, it suffices to assume that A is a local ring with maximal ideal $\mathfrak{q} \cap A$. If $\mathfrak{p} \cap A = \mathfrak{q} \cap A$, then $\mathfrak{p}$ would be maximal by Proposition 1.4.16; hence $\mathfrak{p} = \mathfrak{q}$, a contradiction.   □

Proposition 1.4.18 (Lying Over Theorem) *Let $A \subset B$ be an integral extension and let $\mathfrak{p}$ be a prime ideal of A. Then there exists a prime ideal $\mathfrak{p}'$ such that $\mathfrak{p}' \cap A = \mathfrak{p}$.*

Proof. By Propositon 1.4.14, $B_{\mathfrak{p}} := (A \setminus \mathfrak{p})^{-1}B$ is integral over $A_{\mathfrak{p}}$. Now consider the diagram,

$$\begin{array}{ccc} A & \xrightarrow{\ i\ } & B \\ \downarrow{\alpha} & & \downarrow{\beta} \\ A_{\mathfrak{p}} & \xrightarrow{\ j\ } & B_{\mathfrak{p}} \end{array}$$

where j is induced by the inclusion map i and α and β are the canonical homomorphisms. Let $\mathfrak{m}$ be a maximal ideal of $B_{\mathfrak{p}}$, then $\mathfrak{n} = \mathfrak{m} \cap A_{\mathfrak{p}}$ is a maximal ideal by Proposition 1.4.16, hence is the unique maximal ideal of the local ring $A_{\mathfrak{p}}$, i.e. $\mathfrak{m} \cap A_{\mathfrak{p}} = \mathfrak{p}A_{\mathfrak{p}}$. If $\mathfrak{p}' = \beta^{-1}(\mathfrak{m})$, then $\mathfrak{p}'$ is a prime ideal and we have $\mathfrak{p}' \cap A = \alpha^{-1}(\mathfrak{n}) = \mathfrak{p}$.   □

Theorem 1.4.19 (Going-Up Theorem) *Let $A \subset B$ be an integral ring extension,*

$$\mathfrak{p}_1 \subsetneq \cdots \subsetneq \mathfrak{p}_n$$

be a chain of prime ideals of A and $\mathfrak{p}'_1 \in \mathrm{Spec}(B)$ lie over $\mathfrak{p}_1$, i.e. $\mathfrak{p}'_1 \cap A = \mathfrak{p}_1$ Then there is a chain

$$\mathfrak{p}'_1 \subsetneq \mathfrak{p}'_2 \subsetneq \cdots \subsetneq \mathfrak{p}'_n$$

of prime ideals in B, with $\mathfrak{p}'_i$ lying over $\mathfrak{p}_i$.

Proof. Using induction, we need only construct $\mathfrak{p}'_2$. The induced ring extension $A/\mathfrak{p}_1 \hookrightarrow B/\mathfrak{p}'_1$ is integral and hence there is prime ideal of $B/\mathfrak{p}'_1$ over $\mathfrak{p}_2/\mathfrak{p}_1$. This is of the form $\mathfrak{p}'_2/\mathfrak{p}'_1$ with a prime ideal $\mathfrak{p}'_2 \in \mathrm{Spec}(B)$. That $\mathfrak{p}'_2$ is the prime ideal, searched for.   □

Corollary 1.4.20 *Let $A \subset B$ be as above. Then:*

a) $\dim\ A = \dim\ B$.

b) *Let $\mathfrak{p} \in \mathrm{Spec}(A)$.*

1. *For every prime ideal $\mathfrak{p}'$ of B over $\mathfrak{p}$ one has* $\dim(B/\mathfrak{p}') = \dim(A/\mathfrak{p})$ *and* $\operatorname{ht} \mathfrak{p}' \leq \operatorname{ht} \mathfrak{p}$.
2. *If* $\operatorname{ht} \mathfrak{p} < \infty$, *there exists a prime ideal $\mathfrak{p}'$ of B over $\mathfrak{p}$ with* $\operatorname{ht} \mathfrak{p} = \operatorname{ht} \mathfrak{p}'$.

Proof. a) By Corollary 1.4.17 a chain of n different prime ideals in B gives such a chain in A. Hence $\dim A \geq \dim B$. The converse inequality follows from the Going-Up Theorem.

b) 1. is a consequence of Corollary 1.4.17 too, whereas 2. will follow from Theorem 1.4.19. $\qquad\square$

Theorem 1.4.21 (Going-Down Theorem) *Let $A \subset B$ be an integral extension of domains, A integrally closed,*

$$\mathfrak{p}_1 \subsetneq \cdots \subsetneq \mathfrak{p}_n$$

a chain of prime ideals of A and $\mathfrak{p}'_n \in \operatorname{Spec}(B)$ lying over $\mathfrak{p}_n$. Then there is a chain

$$\mathfrak{p}'_1 \subsetneq \cdots \subsetneq \mathfrak{p}'_{n-1} \subsetneq \mathfrak{p}'_n$$

of prime ideals in B, with $\mathfrak{p}'_i$ lying over $\mathfrak{p}_i$.

Proof. Going down step by step, we need only construct $\mathfrak{p}'_{n-1}$. Set $S_0 := A \setminus \mathfrak{p}_{n-1}$, $S_1 := B \setminus \mathfrak{p}'_n$ and $S := S_0 \cdot S_1$. These are multiplicatively closed.

CLAIM: $\mathfrak{p}_{n-1} B \cap S = \emptyset$.

From this the theorem follows. Namely by Krull's Lemma 1.3.3 there is a prime ideal $\mathfrak{p}'_{n-1} \in \operatorname{Spec}(B)$ with $\mathfrak{p}_{n-1} B \subset \mathfrak{p}'_{n-1} \subset B \setminus S$. By $\mathfrak{p}'_{n-1} \cap S_1 = \emptyset$ we see $\mathfrak{p}'_{n-1} \subset \mathfrak{p}'_n$, and by $\mathfrak{p}'_{n-1} \cap S_0 = \emptyset$ we see $\mathfrak{p}'_{n-1} \cap A = \mathfrak{p}_{n-1}$.

PROOF OF THE CLAIM: Assume there were an $\alpha \in \mathfrak{p}_{n-1} B \cap S$ and let

$$f := X^n + a_1 X^{n-1} + \cdots + a_{n-1} X + a_n$$

be the minimal polynomial of α over $Q(A)$. (Since α is integral over A, it is algebraic over $Q(A)$.)

We will show that $a_i \in \mathfrak{p}_{n-1}$. Namely these are polynomials in the conjugates of α. And if B' arises from B by adjoining the conjugates of α, then the latter belong to $\mathfrak{p}_{n-1} B'$, and so $a_i \in A \cap \mathfrak{p}_{n-1} B'$. Since B' is integral over A we have $A \cap \mathfrak{p}_{n-1} B' = \mathfrak{p}_{n-1}$; for $\mathfrak{p}_{n-1} B'$ is contained in any prime ideal of B' over $\mathfrak{p}_{n-1}$.

Since $\alpha \in S = S_0 \cdot S_1$, we can write $\alpha = rs$ with $r \in S_0 \subset A$, $s \in S_1 \subset B$. The minimal polynomial of $s = \alpha/r$ then is

$$X^n + b_1 X^{n-1} + \cdots + b_n \tag{1.1}$$

with $b_i = a_i/r^i$ Since s is integral over A we have $b_i \in A$. Then $r^i b_i = a_i \in \mathfrak{p}_{n-1}$ and $r \in S_0 = A \setminus \mathfrak{p}_{n-1}$ imply $b_i \in \mathfrak{p}_{n-1}$. So Equation (1.1) says $s^n \in \mathfrak{p}_{n-1} B$, which contradicts $S_1 \cap \mathfrak{p}_{n-1} B = \emptyset$ as S_1 is multiplicative. $\qquad\square$

Corollary 1.4.22 *Under the assumptions of the Going-Down Theorem for every prime ideal $\mathfrak{p}$ of B we have* ht $\mathfrak{p} =$ ht$(\mathfrak{p} \cap A)$.   □

1.5 Direct Sums and Products

To study modules over a ring, the notion of addition and product of modules over a ring R is necessary and useful.

Definition 1.5.1 a) *Let $M_1, \ldots, M_n$ be finitely many A-modules. Their* **direct sum** *(or* **direct product***), written in two manners*

$$M_1 \oplus \ldots \oplus M_n = \bigoplus_{i=1}^{n} M_i$$

is defined to be their cartesian product with the obvious operations making it an A-module:

$$(x_1, \ldots, x_n) + (y_1, \ldots, y_n) := (x_1 + y_1, \ldots, x_n + y_n),$$

$$a(x_1, \ldots, x_n) := (ax_1, \ldots, ax_n).$$

b) *For an A-module M and $n \in \mathbb{N}$ we write*

$$M^n := \underbrace{M \oplus \ldots \oplus M}_{n-\text{times}} = \bigoplus_{i=1}^{n} M_i$$

with $M_i = M$ for all i.

The latter is in accordance with the notation R^n above. Nevertheless there is a notational ambiguity. If I is a (left, right or two-sided) ideal of A, by I^n usually we denote the additive subgroup of A generated by all products $b_1 \cdot \ldots \cdot b_n$ with $b_i \in I$ (which will be a left, right, resp. two-sided ideal again).

1.5.2 Given a direct sum $M = \bigoplus_i M_i$ for every i we have an 'inclusion map'

$j_i : M_i \to M, \quad x \mapsto (0, \ldots, 0, x, 0, \ldots, 0)$, the x placed in the i-th position,

and a '**projection**'

$$pr_i : M \to M_i, \quad (x_1, \ldots, x_n) \mapsto x_i.$$

Clearly $pr_i \circ j_i = \text{id}_{M_i}$.

Write $M_i' := j_i(M_i)$. It is a submodule of M, isomorphic to M_i. We have $\sum_i M_i' = M$ and $M_i' \cap \sum_{j \neq i} M_j' = 0$ for every i. The converse statement (in some sense) is:

Proposition 1.5.3 *Let* $M'_1, \ldots, M'_n$ *be submodules of a module* M. *The following statements are equivalent.*

(1) a) $\sum_{i=1}^{n} M'_i = M$, *and* b) $M'_i \cap \sum_{j \neq i} M'_j = 0$ *for every* i.

(2) *The map*

$$\bigoplus_{i=1}^{n} M'_i \to M, \quad (x_1, \ldots, x_n) \mapsto x_1 + \ldots + x_n$$

is an isomorphism.

Indeed (1) a) is equivalent to the surjectivity of the map in (2), whereas b) is equivalent to its injectivity. □

In this situation we say that M is the direct sum of the submodules $M_1, \ldots, M_n$ and write

$$M = \bigoplus_{i=1}^{n} M_i = M_1 \oplus \ldots \oplus M_n.$$

There is an easy but useful criterion for a module to be a direct summand of another module.

Lemma 1.5.4 *Let*

$$N \xrightarrow{j} M \xrightarrow{p} N$$

be module homomorphisms with $p \circ j = \mathrm{id}_N$. *Then* j *is injective,* p *is surjective and* $M = j(N) \oplus \ker(p)$.

Proof. Since $p \circ j$ is injective, j is so. Analogously p is surjective. Let $x \in M$, then $x = j \circ p(x) + (x - j \circ p(x))$. But $j \circ p(x) \in j(N)$ and $x - j \circ p(x) \in \ker(p)$, since $p(x - j \circ p(x)) = p(x) - p \circ j \circ p(x) = p(x) - p(x) = 0$. So $M = j(N) + \ker(p)$. At last let $x = j(y) \in j(N) \cap \ker(p)$, then $0 = p(x) = p \circ j(y) = y$. So $x = j(0) = 0$, whence $j(N) \cap \ker(p) = 0$. □

Corollary 1.5.5 *Let*

$$0 \longrightarrow M' \xrightarrow{i} M \xrightarrow{p} M'' \longrightarrow 0$$

be an exact sequence of modules. The following statements are equivalent:

(1) *There is a homomorphism* $q : M \to M'$ *with* $q \circ i = \mathrm{id}_{M'}$;

(2) *There is a homomorphism* $j : M'' \to M$ *with* $p \circ j = \mathrm{id}_{M''}$;

(3) *There is a commutative diagram*

$$
\begin{array}{ccccc}
M' & \xrightarrow{\ i\ } & M & \xrightarrow{\ p\ } & M'' \\
\downarrow{\scriptstyle\mathrm{Id}} & & \downarrow{\scriptstyle\simeq} & & \downarrow{\scriptstyle\mathrm{Id}} \\
M' & \xrightarrow{\ j_1\ } & M' \oplus M'' & \xrightarrow{\ \mathrm{pr}_2\ } & M''
\end{array}
$$

where the middle vertical arrow denotes an isomorphism. □

Definition 1.5.6 *The exact sequence of the Corollary 1.5.4 is called* **split exact** *if the equivalent properties* (i) *to* (iii) *hold. Also in this situation we say that the sequence* **splits.** *Further we say that an epimorphism p (resp. a monomorphism i)* **splits** *if it fits in a split exact sequence.*

Now let $(M_i)_{i \in I}$ be a (not necessarily finite) family of modules. We generalize the above definition in two different ways.

Definitions 1.5.7 a) *The* **direct sum**

$$\bigoplus_{i \in I} M_i$$

is defined to be the set of those families $(x_i)_{i \in I}$ *where* $x_i = 0$ *for nearly all, i.e. for all but finitely many* $i \in I$.

b) *The* **direct product**

$$\prod_{i \in I} M_i$$

is defined to be the set of **all** *families* $(x_i)_{i \in I}$.

In both cases the module structure is defined as above.

c) *Let* I *be some set and* M *a module. Then*

$$M^{(I)} := \bigoplus_{i \in I} M_i \quad and$$

$$M^I := \prod_{i \in I} M_i,$$

where $M_i = M$ *for all* $i \in I$.

Note that an A-module M clearly is free, if and only if there is a set I, such that $M \cong A^{(I)}$.

Remarks 1.5.8 a) Proposition 1.5.3 generalizes without difficulties to infinitely many direct summands. So if $(M_i')_{i \in I}$ is a family of submodules of a module M which fulfills the equivalent statements of Proposition 1.5.3, we say that M is the direct sum of its submodules M_i', $i \in I$ and write

$$M = \bigoplus_{i \in I} M_i'.$$

b) *A direct summand of a free module, i.e. a module P such that $P \oplus Q$ is free for some module Q, will be called* **projective**. *The study of projective modules – beginning in Chapter 2 – is one of the themes of our book. There do exist nonfree projective modules, but our central theorem will be that projective modules over polynomial rings $k[X_1, \ldots , X_n]$ over a field k are actually free. This was called "Serre's Conjecture" before it was proved, and is sometimes called so today.*

Definition 1.5.9 *Let $(A_i)_{i \in I}$ be a finite or infinite family of rings. Their direct product*

$$\prod_{i \in I} A_i$$

is defined to be, as a set their Cartesian product, with component-wise addition and multiplication – analogously to the addition in Definition 1.5.1. In the case of a finite $I = \{1, \ldots , n\}$ we also write

$$\prod_{i=1}^{n} A_i = A_1 \times \ldots \times A_n.$$

Remarks 1.5.10 a) Let I be infinite and $A_i \not\simeq 0$ for all i. Then the set of families

$$\{(a_i)_{i \in I} \mid a_i = 0 \text{ for almost all } i\}$$

lacks the 1-element and so is no ring in our sense.

b) Some people call – for finite I – the direct product of rings a direct sum and use the notations $\bigoplus$ or $\oplus$. But one of the authors heavily dislikes this, since the direct product of rings is indeed what usually is called the direct product in the category of rings, whereas the direct sum in the category of rings would be the tensor product (over $\mathbb{Z}$).

Definition 1.5.11 *Two-sided ideals I, J of a ring A are called* **comaximal**, *if $I + J = A$ or equivalently if there are $a \in I$, $b \in J$ with $a + b = 1$.*

Remarks 1.5.12 a) For commutative A this means that I, J are not contained in a common prime ideal, in other words that $V(I) \cap V(J) = \emptyset$.

b) Let I, J_i be comaximal for $i = 1, 2$, then I, $J_1 J_2$ are comaximal. Namely let $a_i \in I$, $b_i \in J_i$ such that $a_1 + b_1 = a_2 + b_2 = 1$. Then $1 = (a_1 + b_1)(a_2 + b_2) = (a_1 a_2 + a_1 b_2 + b_1 a_2) + b_1 b_2 \in I + J_1 J_2$.

Consequently, if I, J are comaximal, so are I^n, J^m.

Lemma 1.5.13 (Chinese Remainder Theorem) *Let R be a ring, $I_1, \ldots, I_n$ be two-sided ideals of R and $\eta : R \to \prod_{i=1}^{n} R/I_i$ be the natural homomorphism defined by $\eta(r) = (r + I_1, \ldots, r + I_n)$. Then η is surjective if and only if the $I_1, \ldots, I_n$ are mutually pairwise comaximal. In this case we have*

$$R/(I_1 \cap \ldots \cap I_n) \cong \prod_{i=1}^{n} (R/I_i).$$

If the I_i pair-wise commute, especially if R is commutative, then we have further that $I_1 \cdots I_n = I_1 \cap \cdots \cap I_n$ in the above case.

Proof. Assume $I_i + I_j = R$ for $i \neq j$. Let $1 = u_i + v_i$, $u_i \in I_1$, $v_i \in I_i$ for $i \neq 1$, and let $r_1 = \prod_{i=1}^{n} v_i = \prod_{i=2}^{n} (1 - u_i) = 1 + u$ for some $u \in I_1$. Then $\eta(r_1) = e_1 := (1 + I_1, \ 0 + I_2, \ldots, 0 + I_n)$. Analogously, $e_i \in \text{im}(\eta)$ for all i, and so η is onto.

Conversely, assume η is onto. If $\eta(a_i) = e_i$ then $1 - a_i \in I_i$ and $a_i \in I_j$ if $j \neq i$. Therefore, $1 = (1 - a_i) + a_i \in I_i + I_j$.

Clearly $\ker(\eta) = I_1 \cap \cdots \cap I_n$.

We prove the last assertion by induction on n, the case $n = 1$ being trivial. If $I + J = R$ we compute

$$I \cap J = (I + J)(I \cap J) = I(I \cap J) + J(I \cap J) \subset IJ,$$

whence $I \cap J = IJ$. The induction step is

$$I_1 \cap \cdots \cap I_n = (I_1 \cdots I_{n-1}) \cap I_n = I_1 \cdots I_n.$$

$\square$

Compare this lemma to Urysohn's Lemma, which says that for any pair V, W of disjoint closed subsets of a normal topological space X, there is a continuous real function being 0 on V and 1 on W. By the above lemma (in the commutative case) for closed sets $V(I)$, $V(J)$ with $V(I) \cap V(J) = \emptyset$ we find an $f \in A$ with $f \equiv 1 \bmod I$, $f \equiv 0 \bmod J$.

Definition 1.5.14 *An element e of a (not necessarily commutative ring) is called* **idempotent** *if $e^2 = e$.*

Remark 1.5.15 Let $A = A_1 \times \cdots \times A_n$ be a finite direct product of rings and $e_i := (0, \ldots, 0, 1, 0, \ldots, 0)$, where the i-th component is the 1. Then

$$\text{(i)} \quad e_i^2 = e_i \qquad \text{(ii)} \quad e_i e_j = 0 \text{ for } i \neq j \qquad \text{(iii)} \quad \sum_{i=1}^{n} e_i = 1. \qquad (*)$$

Further the e_i belong to the center of A.

There is a converse.

Proposition 1.5.16 *Let $e_1, \ldots, e_n$ in the center of a ring R be finitely many elements fulfilling $(*)$. Then the ideals Re_i are also rings with 1-element e_i and the map*

$$\eta : R \longrightarrow Re_1 \times \cdots \times Re_n, \qquad a \mapsto (ae_1, \ldots, ae_n)$$

is an isomorphism of rings.

Proof. Since e_i is idempotent, $(ae_i)e_i = ae_i$. So e_i is the 1-element of Re_i and Re_i is a ring. It is not hard to show that the map $\eta_i : R \to Re_i$, $a \mapsto ae_i$ is a surjective ring homomorphism.

Using (ii) and $\sum_i ae_i = a$, which holds by (iii), one shows $\ker(\eta_i) = \sum_{j \neq i} Re_j$. Therefore the different $\ker(\eta_i)$ are mutually comaximal. On the other hand $\ker(\eta_1) \cap \cdots \cap \ker(\eta_n) = (0)$, since $a \in \ker(\eta_i)$ for all i implies $ae_i = 0$ for all i, hence $a = \sum_i ae_i = 0$. So by Lemma 1.5.13 the proposition follows. $\qquad \square$

1.5.17 An analogous results holds for modules. Let $M = M_1 \oplus \cdots \oplus M_n$ be a finite direct sum. Let

$$e_i : M \to M, \qquad (x_1, \ldots, x_n) \mapsto (0, \ldots, 0, x_i, 0, \ldots, 0),$$

the x_i standing in the i-th place.

Clearly the $e_1, \ldots, e_n$ are endomorphisms satisfying $(*)$, where (iii) is to be read $\sum_i e_i = \text{id}_M$. (The e_i need not belong to the center of the endomorphism ring of M.) Conversely,

Proposition 1.5.18 *Let M be a module and $e_1, \ldots, e_n$ be endomorphisms of M satisfying $(*)$. Then*

$$M = \bigoplus_{i=1}^{n} e_i(M).$$

Proof. From (iii) we get $\sum_i e_i(M) = M$. Let $x \in e_i(M) \cap (\sum_{j \neq i} e_i(M))$. We have to show $x = 0$. From $x = e_i(y)$ we get $e_i(x) = e_i^2(y) = e_i(y) = x$. From $x = \sum_{j \neq i} e_j(y_j)$ we get $e_i(x) = \sum_{j \neq i} e_i e_j(y_j) = 0$. Hence $x = 0$. $\qquad\square$

The theory of modules over a finite direct product of rings can be completely reduced to the theories of those over the individual factors. This is the content of the following theorem.

Theorem 1.5.19 *Let $A_1, \dots, A_n$ be rings and $A = A_1 \times \cdots \times A_n$. Let further $e_1, \dots, e_n \in A$ be the idempotents defined by this splitting as above. Then any A-module M admits the splitting*

$$M = M_1 \oplus \cdots \oplus M_n, \quad \text{where } M_i := e_i M.$$

M_i is an A_i-module in an obvious way. Conversely if M_i is an A_i-module for $i = 1, \dots, n$ then the group $M_1 \oplus \cdots \oplus M_n$ is an A-module in a canonical and obvious way.

For any A-submodule U of M we have $U = \bigoplus_i e_i U$. Conversely, if U_i is an A_i-submodule of M_i, then $U := U_1 + \cdots + U_n = U_1 \oplus \cdots \oplus U_n$ is an A-submodule of M. Further $M/U \cong M_1/U_1 \oplus \cdots \oplus M_n/U_n$.

If $f : M \to N$ is an A-module homomorphism and $N_i := e_i N$, then $f(M_i) \subset N_i$, so that

$$\mathrm{Hom}_A(M, N) = \mathrm{Hom}_{A_1}(M_1, N_1) \oplus \cdots \oplus \mathrm{Hom}_{A_n}(M_n, N_n).$$

Proof. Note that the e_i belong to the center of A, so that multiplication by them are endomorphisms. Apply Proposition 1.5.18. Let $a \in A_i$ and $x \in M_i \subset M$. Then the only sensible definition of ax is

$$ax := (0, \dots, 0, a, 0, \dots, 0)x, \quad a \text{ in the } i^{th} \text{ place.}$$

Since $e_i U \subset U$ by the definition of a submodule we get for U the analogous splitting as for M.

We leave the rest to the reader. $\qquad\square$

Note 1.5.20 Consider any non-zero module E. Then the diagonal submodule $D := \{(x, x) \mid x \in E\}$ of $E \oplus E$ is not of the form $e_1 D \oplus e_2 D$, where e_i are the endomorphisms of $E \oplus E$, defined by $e_1(x_1, x_2) = (x_1, 0)$, $e_2(x_1, x_2) = (0, x_2)$. In general the submodules of direct sums of modules cannot easily be derived from submodules of the individual summands. This is different in the case of ideals in a direct product of rings.

Corollary 1.5.21 *Let $A = A_1 \times \cdots \times A_n$ as in Theorem 1.5.19 and I a left (resp. right, resp. two-sided) ideal of A, then $I = I_1 \times \cdots \times I_n$, where I_j is a suitable left (resp. right, resp. two-sided) ideal of A_j for $j = 1, \dots, n$. Conversely such a product is an ideal of A. In this situation there is a canonical isomorphism $A/I \cong (A_1/I_1) \times \cdots \times (A_n/I_n)$.* $\qquad\square$

Proposition 1.5.22 *let $A = A_1 \times \cdots \times A_n$ be a commutative ring. Then every prime ideal of A is of the form*

$$\mathfrak{p} = A_1 \times \cdots \times A_{i-1} \times \mathfrak{p}_i \times A_{i+1} \times \cdots \times A_n,$$

for some $i \in \{1, \dots, n\}$ and some $\mathfrak{p}_i \in \mathrm{Spec}(A_i)$. Conversely every ideal of this form is prime. So $\mathrm{Spec}(A)$ may be identified with the disjoint union of the $\mathrm{Spec}(A_i)$. Moreover it is the 'coproduct' of the $\mathrm{Spec}(A_i)$ in the category of topological spaces; i.e. a subset $U \subset \mathrm{Spec}(A) = \bigcup_{i=1}^{n} \mathrm{Spec}(A_i)$ is closed in $\mathrm{Spec}(A)$, if and only if $U \cap \mathrm{Spec}(A_i)$ is closed in $\mathrm{Spec}(A_i)$ for every i.

Proof. Note that every product of rings, where more than one factor is nontrivial, has zero divisors $\neq 0$. (In $B \times C$ we have $(1,0)(0,1) = (0,0)$.)

Let $I = I_1 \times \cdots \times I_n$ be an ideal of A. Then $A/I \cong (A_1/I_1) \times \cdots \times (A_n/I_n)$ is a domain, if and only if $A_j = I_j$ for all but one j, say i and A_i/I_i is a domain, i.e. I_i is a prime ideal of A_i.

We prove the last assertion. A subset T of $\mathrm{Spec}(A)$ is closed, if and only if it is of the form $T = V(I)$ for some ideal $I = I_1 \times \cdots \times I_n$ of A. But clearly $T \cap \mathrm{Spec}(A_j) = V(I_j)$ in $\mathrm{Spec}(A_j)$. □

So every decomposition of a commutative ring $A = B \times C$ gives rise to a partition $\mathrm{Spec}(A) = \mathrm{Spec}(B) \cup \mathrm{Spec}(C)$ where $\mathrm{Spec}(B)$ and $\mathrm{Spec}(C)$ are closed and open in $\mathrm{Spec}(A)$ and mutually disjoint. There is a converse.

Proposition 1.5.23 *Let A be a commutative ring and $\mathrm{Spec}(A)$ be the disjoint union of two closed (hence open) subsets U_1 and U_2. Then A can be decomposed: $A = A_1 \times A_2$ with $\mathrm{Spec}(A_i) = U_i$.*

Proof. Since the U_j are closed there are ideals I_1, I_2 of A with $U_j = V(I_j)$. Since $U_1 \cap U_2 = \emptyset$, we have $I_1 + I_2 = A$. So there are $a_j \in I_j$ with $a_1 + a_2 = 1$. Now, since $U_1 \cup U_2 = \mathrm{Spec}(A)$ we have $I_1 \cap I_2 \subset \mathrm{Nil}(A) = \sqrt{(0)}$. Therefore from $a_1 a_2 \in I_1 \cap I_2$ we get $(a_1 a_2)^n = 0$ for suitable n. There is no prime ideal containing both a_1 and a_2, because $a_1 + a_2 = 1$. Hence there is also no prime ideal containing a_1^n and a_2^n. So the ideals (a_1^n) and (a_2^n) are comaximal, whence $V(a_1^n) \cap V(a_2^n) = \emptyset$. Since $(a_j^n) \subset I_j$, whence $V(a_j^n) \supset U_j$, we get $V(a_j^n) = U_j$. Now we apply the Chinese Remainder Theorem, to see

$$(A/(a_1^n)) \times (A/(a_2^n)) \cong A/(a_1^n) \cap (a_2^n) = A/(a_1^n a_2^n) = A.$$

□

Corollary 1.5.24 *Let A be a commutative ring. The following statements are equivalent:*

(i) $\mathrm{Spec}(A)$ *is connected.*

(ii) *If $A \cong A_1 \times A_2$, one of the rings A_1, A_2 is a zero ring.*

(iii) *A possesses no idempotent elements but 0 and 1.* □

Corollary 1.5.25 *Let A have at least one of the following properties:*

(i) *A is local,*

(ii) *A has only one minimal prime ideal.*

Then A is not directly decomposable, i.e. there are no non-zero rings A_1, A_2 with $A \cong A_1 \times A_2$. $\square$

1.6 The Tensor Product

The category of modules over a ring has a much more natural concept of product, in a categorical sense, called the tensor product.

1.6.1 Definition

Let R be a ring (not necessarily commutative), M a right and N a left R-module. A map $f : M \times N \to P$ where P is an abelian group is called a **balanced map** if

$$f(m_1 + m_2, n) = f(m_1, n) + f(m_2, n)$$
$$f(m, n_1 + n_2) = f(m, n_1) + f(m, n_2)$$
$$f(ma, n) = f(m, an)$$

where $a \in R$, $m, m_1, m_2 \in M$ and $n, n_1, n_2 \in N$.

Given a right R-module M and a left R-module N, there exists a unique (upto a unique isomorphism) abelian group $M \otimes_R N$, (sometimes also written $M \otimes N$), called the **tensor product** of M and N over R, and a balanced map $\psi : M \times N \to M \otimes_R N$ with the '**universal property**' that for any balanced map $f : M \times N \to P$ with any abelian group P, there exists a unique group homomorphism $g : M \otimes N \to P$ such that $g \circ \psi = f$, i.e. the following diagram commutes:

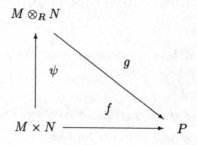

We show now that a tensor product of a right R-module M and a left R-module N over R exists. Let F be the free abelian group (i.e. free $\mathbb{Z}$-module) with the set $M \times N$ as basis (i.e. $F \cong \mathbb{Z}^{(M \times N)}$) and let H be the subgroup of F generated by all elements of the types

$(x + y, z) - (x, z) - (y, z)$, $(x, z + t) - (x, z) - (x, t)$, $(xa, y) - (x, ay)$,

where $x, y \in M$, $z, t \in N$ and $a \in R$. Let $M \otimes_R N := F/H$ and $\varphi : M \times N \to M \otimes_R N$ be the composite $M \times N \hookrightarrow F \to M \otimes_R N$, where $F \to M \otimes_R N$ is the canonical epimorphism. Then, it is easily seen that $M \otimes_R N$ is a tensor product of M and N over R.

The uniqueness is a formal consequence of the universal property: Let $\psi : M \times N \to M \otimes_R N$ and $\psi' : M \times N \to M \otimes'_R N$ be two tensor products with the accompanying balanced maps. By the universal properties of ψ and ψ' there are unique group homomorphisms $\alpha : M \otimes_R N \to M \otimes'_R N$ and $\beta : M \otimes'_R N \to M \otimes_R N$ with $\alpha \circ \psi = \psi'$ and $\beta \circ \psi' = \psi$. Then $(\beta \circ \alpha) \circ \psi = \mathrm{id}_{M \otimes N} \circ \psi$. By the uniqueness claim in the universal property we get $\beta \circ \alpha = \mathrm{id}_{M \otimes N}$. Analogously we have $\alpha \circ \beta = \mathrm{id}_{M \otimes' N}$. So α, β are isomorphisms and unique. The tensor product with the accompanying balanced map is unique up to a *unique* isomorphism.

For $(x, y) \in M \times N$, we denote its image under φ in $M \otimes_R N$ by $x \otimes y$. (These are called the **pure tensors**). Note that any element of $M \otimes_R N$ is of the form $\sum_i x_i \otimes y_i$ (which are called **mixed tensors**, with $x_i \in M$, $y_i \in N$). But this description is not at all unique!

This non-uniqueness is the main source of difficulties, when using the tensor product. Therefore we give here recipes and rules how to work with it.

Remark 1.6.1 How to show

$$\sum_{i=1}^{n} x_i \otimes y_i = \sum_{i=1}^{m} x'_i \otimes y'_i ? \qquad (*)$$

Answer: Use successively identities of the following types

$$(x + x') \otimes y = x \otimes y + x' \otimes y, \quad x \otimes (y + y') = x \otimes y + x \otimes y'$$

$$xa \otimes y = x \otimes ay \ \text{ with } a \in R,$$

to convert one side in $(*)$ into the other.

Conversely, the existence of such a series of transformations is necessary for the equation $(*)$. But since it is not easy to show that such transformations do not exist, here is another recipe:

How to show

$$\sum_{i=1}^{n} x_i \otimes y_i \neq \sum_{i=1}^{m} x_i' \otimes y_i' \ ?$$

Answer: Find an abelian group G and a balanced map $\varphi : M \times N \to G$, such that

$$\sum_{i=1}^{n} \varphi(x_i, y_i) \neq \sum_{i=1}^{m} \varphi(x_i', y_i').$$

Examples 1.6.2 a) Always $x \otimes 0 = 0 \otimes y = 0$. Namely $x \otimes 0 + x \otimes 0 = x \otimes (0 + 0) = x \otimes 0$. Consequently $M \otimes 0 \cong 0 \otimes N \cong 0$.
(Note that the 0 above has six different meanings: the 0-element of M, N, $M \otimes N$, the trivial left, right module or group.)

Another consequence is $(-x) \otimes y = -(x \otimes y) = x \otimes (-y)$.

b) $\mathbb{Q}/\mathbb{Z} \otimes_{\mathbb{Z}} \mathbb{Q}/\mathbb{Z} \cong 0$.

Namely let $x, y \in \mathbb{Q}/\mathbb{Z}$. There is an $n \in \mathbb{Z} \setminus \{0\}$ with $nx = 0$ and a $y' \in \mathbb{Q}/\mathbb{Z}$ with $ny' = y$. Then $x \otimes y = x \otimes ny' = nx \otimes y' = 0 \otimes y' = 0$.

c) The balanced map $R \times M \to M$ given by $(a, x) \mapsto ax$ induces a canonical isomorphism $R \otimes_R M \xrightarrow{\sim} M$.

d) More generally, let I be a right ideal of R. Then the balanced map

$$(R/I) \times M \to M/IM, \quad (\bar{a}, x) \mapsto \overline{ax}$$

induces a group isomorphism
$(R/I) \otimes_R M \xrightarrow{\sim} M/IM$. (Here the overbar denotes the residue class, and IM is the subgroup of M generated by all ax with $a \in I$ and $x \in M$. If I is a two-sided ideal, then IM and hence M/IM are R-modules.)

e) Let R be commutative and $S \subset R$ be a multiplicative subset. Consider the map

$$S^{-1}R \times M \longrightarrow S^{-1}M, \quad \left(\frac{a}{s}, x\right) \mapsto \frac{ax}{s}.$$

It is well defined and balanced, and hence induces a homomorphism
$f : (S^{-1}R) \otimes_R M \to S^{-1}M$.

CLAIM: f is an isomorphism.

An inverse map is $g : S^{-1}M \to (S^{-1}R) \otimes_R M$, defined by $m/s \mapsto (1/s) \otimes m$. This is well defined. Namely if $m/s = m'/s'$, there is a $t \in S$ with $ts'm = tsm'$. So

$$(1/s) \otimes m = (1/sts') \otimes ts'm = (1/s'ts) \otimes tsm' = (1/s') \otimes m'.$$

Clearly $f \circ g = \mathrm{id}_{S^{-1}M}$. Further $g \circ f((a/s) \otimes m) = (1/s) \otimes am = (a/s) \otimes m$. Since $(S^{-1}R) \otimes_R M$ is generated by elements of the form $(a/s) \otimes m$, we see that g is inverse to f.

Remarks 1.6.3 a) Let $f : R \to S$ be a ring homomorphism and M a left R-module. Then S is a right R-module by $sr := sf(r)$. Hence we can form the tensor product $S \otimes_R M$, and this is an S-module by $a(b \otimes m) := (ab) \otimes m$ for $a, b \in S$.

Note that for a general element $\sum s_i \otimes m_i \in S \otimes_R M$ this means $a(\sum s_i \otimes m_i) = \sum (as_i) \otimes m_i$. Using the above remark, one sees easily, that this is well-defined.

If I is a twosided ideal of R, the isomorphism $(R/I) \otimes_R M \overset{\sim}{\to} M/IM$ of example d) above is R/I-linear. The same holds for the isomorphism f in example e).

b) Now let R be commutative and M, N any R-modules – right or left having no different meaning here. Then $M \otimes_R N$ is an R-module by $r(m \otimes n) := (rm) \otimes n = m \otimes (rn)$.

(Note, that this does not work for noncommutative R. Namely the attempt to define $r(m \otimes n) := m \otimes (rn)$ does not make much sense, since for $r, s \in R$ this would imply $m \otimes rsn = r(m \otimes sn) = r(ms \otimes n) = ms \otimes rn = m \otimes srn$. But in general the equation $m \otimes rsn = n \otimes srn$ does not hold in $M \otimes_R N$.)

For commutative R the canonical map $\psi : M \times N \to M \otimes_R N$ is not only balanced, but even bilinear, i.e. $\psi(rm, n) = \psi(m, rn) = r\psi(m, n)$. And we may define the tensor product by a universal property with respect to bilinear instead of balanced maps. Also we may prove its existence – over a commutative ring R – to be the factor *module F/U*, where F is a free R-module with basis $M \times N$ and U is generated by all elements of the form $(m + m', n) - (m, n) - (m', n)$, $(m, n + n') - (m, n) - (m, n')$, $(rm, n) - r(m, n)$, $(m, rn) - r(m, n)$.

Finally we clearly have a canonical isomorphism $M \otimes N \cong N \otimes M$, when R is commutative.

c) The module structures on $S \otimes_R N$, resp. $M \otimes_R N$ in a) and b) are special cases of the following: Let M be an S-R-bimodule, i.e. an abelian group, which has as well a left S- as a right R-module structure, with $(sm)r = s(mr)$. Then $M \otimes_R N$ is an S-module by $s(m \otimes n) := (sm) \otimes n$.

d) The balanced map $(M \oplus N) \times P \to (M \otimes_R P) \oplus (N \otimes_R P)$ given by $((x, y), z) \mapsto (x \otimes z, y \otimes z)$ induces a canonical R-isomorphism $(M \oplus N) \otimes P \overset{\sim}{\to} (M \otimes P) \oplus (N \otimes P)$.

This holds also for infinite direct sums: we have a canonical isomorphism

$$\left(\bigoplus_{i \in I} M_i \right) \otimes_R N \cong \bigoplus_{i \in I} (M_i \otimes_R N).$$

(The analogue for infinite direct products does not generally hold.)

Especially we have: Let F be a free R-module with basis $\{e_i\}_{i \in I}$ and let $R \to S$ be a ring homomorphism. Then $S \otimes_R F$ is a free S-module with $\{1 \otimes e_i\}_{i \in I}$ as a basis.

e) Let $R[x]$ be the polynomial R-algebra. Then we have a canonical $R[x]$-linear isomorphism $R[x] \otimes_R M \cong M[x]$ for every R-module M.

1.6.2 Functoriality

Definition 1.6.4 *Let R be a ring, M, M' right, N, N' left R-modules, further $f : M \to M'$ and $g : N \to N'$ be R-linear maps. These induce a diagram*

$$
\begin{array}{ccc}
M \times N & \xrightarrow{\psi} & M \otimes_R N \\
\downarrow{\scriptstyle f \times g} & & \\
M' \times N' & \xrightarrow{\psi'} & M' \otimes_R N'
\end{array}
$$

Since $\psi' \circ (f \times g)$ clearly is balanced, by definition of the tensor product there is a unique group homomorphism $f \otimes g : M \otimes_R N \to M' \otimes_R N$ which makes the above diagram commutative – i.e. $(f \otimes g) \circ \psi = \psi' \circ (f \times g)$.

We say, $f \otimes g$ is induced by f and g.

Sometimes we write $M \otimes g := \mathrm{id}_M \otimes g$.

Remarks 1.6.5 a) We have $\mathrm{id}_M \otimes \mathrm{id}_N = \mathrm{id}_{M \otimes N}$ and $\mathrm{id}_M \otimes (g' \circ g) = (\mathrm{id}_M \otimes g') \circ (\mathrm{id}_M \otimes g)$, whenever $g' \circ g$ is defined. This means, for a given right R-module M, the assignments $N \mapsto M \otimes_R N$, $g \mapsto \mathrm{id}_M \otimes g$ form a so called **functor** from the **category** of left R-modules and R-linear maps to that of abelian groups and group homomorphisms. One may write this functor as $M \otimes_R ?$.

Further the map

$$
\mathrm{Hom}_R(N, N') \to \mathrm{Hom}_{\mathbb{Z}}(M \otimes N, M \otimes N'), \ g \mapsto \mathrm{id}_M \otimes g
$$

is a group homomorphism. This is expressed by: the tensor product is an **'additive functor'**.

If R is commutative the above map is even R-linear: The tensor product $M \otimes_R ?$ is an 'R-**linear functor**' for commutative R.

Analogously these statements hold if one interchanges the roles of M and N.

b) For $f : M \to M'$, $g : N \to N'$ as above, the following square is commutative:

$$
\begin{array}{ccc}
M \otimes N & \xrightarrow{f \otimes \mathrm{id}_N} & M' \otimes N \\
\downarrow{\scriptstyle \mathrm{id}_M \otimes g} & & \downarrow{\scriptstyle \mathrm{id}_{M'} \otimes g} \\
M \otimes N' & \xrightarrow{f \otimes \mathrm{id}_{N'}} & M' \otimes N'
\end{array}
$$

This is expressed by the phrase, that the tensor product is a **'bifunctor'**.

1.6.3 To what Extent do Hom and $\otimes$ Preserve Exactness

The assumption that
$$M' \longrightarrow M \longrightarrow M''$$
is an exact sequence of modules, does not generally imply the exactness of either of the following induced sequences:
$$\operatorname{Hom}(N, M') \to \operatorname{Hom}(N, M) \to \operatorname{Hom}(N, M''),$$

$$\operatorname{Hom}(M'', N) \to \operatorname{Hom}(M, N) \to \operatorname{Hom}(M', N), \qquad N \otimes M' \to N \otimes M \to N \otimes M''.$$

For example let the ring be $\mathbb{Z}$, $N := \mathbb{Z}/2\mathbb{Z}$ and consider the exact sequence $\mathbb{Z} \to \mathbb{Z}/2\mathbb{Z} \to 0$ in the first, resp. $0 \to \mathbb{Z} \xrightarrow{\cdot 2} \mathbb{Z}$ in the second and third case. (By $\cdot 2$ we mean the homothesy of 2.)

Here we will show that partial results do hold.

Proposition 1.6.6 a) *Let*

$$M' \longrightarrow M \to M'' \longrightarrow 0 \quad resp. \quad 0 \longrightarrow N' \longrightarrow N \longrightarrow N''$$

be exact sequences of R-modules. Then the induced sequences

$$0 \longrightarrow \operatorname{Hom}(M'', N) \longrightarrow \operatorname{Hom}(M, N) \longrightarrow \operatorname{Hom}(M', N)$$

resp.

$$0 \longrightarrow \operatorname{Hom}(M, N') \longrightarrow \operatorname{Hom}(M, N) \longrightarrow \operatorname{Hom}(M, N'')$$

are exact as well.
This is the so called **left exactness** *of the* Hom-*functor.*

b) *Moreover, if for a (not necessarily exact) sequence*

$$(*) \qquad M' \xrightarrow{f} M \xrightarrow{g} M'' \longrightarrow 0$$

the induced sequence

$$0 \longrightarrow \operatorname{Hom}(M'', N) \xrightarrow{g^*} \operatorname{Hom}(M, N) \xrightarrow{f^*} \operatorname{Hom}(M', N)$$

is exact for every module N, then $(*)$ is exact.

Proof. a) If one understands the meaning of the hypotheses and the assertions, the proof is automatic.

b) First set $N := M''/\operatorname{im}(g)$. Then the canonical map $\kappa : M'' \to N$ belongs to the kernel of g^*. So the injectivity of g^* implies the surjectivity of g.

Secondly set $N := M''$. Then $f^* \circ g^* = 0$ implies $g \circ f = f^* \circ g^* (\operatorname{id}_{M''}) = 0$.

At last set $N := M/\operatorname{im}(f)$ and let $\kappa : M \to N$ be the canonical map. Then $f^*(\kappa) = \kappa \circ f = 0$, i.e. $\kappa \in \ker(f^*)$. So by hypothesis there is a linear map $\lambda : M'' \to N$ with $\kappa = \lambda \circ g$. Now if $x \in \ker(g)$, then $\kappa(x) = 0$, i.e. $x \in \operatorname{im}(f)$. $\square$

There is a useful connection between the tensor product and the Hom-functor. Namely there is a canonical isomorphism

$$\text{Hom}_{\mathbb{Z}}(M \otimes_R N,\ G) \xrightarrow{\sim} \text{Hom}_R(N,\ \text{Hom}_{\mathbb{Z}}(M, G)).$$

(Here G is an abelian group, M is a right, N a left R-module.)

The principal idea how to construct this, is that a balanced map $M \times N \to G$ means: every $x \in N$ gives – in an R-linear way – a $\mathbb{Z}$-linear map $M \to G$. We leave the details to the reader.

Note that $\text{Hom}_{\mathbb{Z}}(M, G)$ becomes a left R-module, if M is a right one.

The isomorphism f is '**natural**', which means that it is compatible with maps $M \to M'$, $N \to N'$ and $G \to G'$: The induced squares are commutative; for e.g. the induced square

$$
\begin{array}{ccc}
\text{Hom}_{\mathbb{Z}}(M' \otimes_R N,\ G) & \xrightarrow{\sim} & \text{Hom}_R(N,\ \text{Hom}_{\mathbb{Z}}(M', G)) \\
\downarrow & & \downarrow \\
\text{Hom}_{\mathbb{Z}}(M \otimes_R N,\ G) & \xrightarrow{\sim} & \text{Hom}_R(N,\ \text{Hom}_{\mathbb{Z}}(M, G))
\end{array}
$$

is commutative.

Let $f : N \to N'$ be an epimorphism, then one easily shows that $M \otimes f$ is so too. One needs a bit more effort to prove the following.

Proposition 1.6.7 *Let*

$$N' \longrightarrow N \longrightarrow N'' \longrightarrow 0$$

be exact. Then the induced sequence

$$M \otimes N' \longrightarrow M \otimes N \longrightarrow M \otimes N'' \longrightarrow O$$

is exact too. This is called the **right exactness** *of the tensor product.*

Proof. Assume that
$$M' \longrightarrow M \longrightarrow M'' \longrightarrow 0$$
is exact. By the left exactness of the Hom-functor we get that

$$0 \to \text{Hom}_R(N, \text{Hom}_{\mathbb{Z}}(M'', G)) \to \text{Hom}_R(N, \text{Hom}_{\mathbb{Z}}(M, G)) \to \text{Hom}_R(N, \text{Hom}_{\mathbb{Z}}(M', G))$$

is exact, and hence

$$0 \to \text{Hom}_{\mathbb{Z}}(M'' \otimes_R N,\ G) \to \text{Hom}_{\mathbb{Z}}(M \otimes_R N,\ G) \to \text{Hom}_{\mathbb{Z}}(M' \otimes_R N,\ G)$$

is exact for every abelian group G. So the sequence

$$M' \otimes_R N \longrightarrow M \otimes_R N \longrightarrow M'' \otimes_R N \longrightarrow 0$$

is exact. $\qquad\qquad\square$

Let us give a first application.

Definition 1.6.8 *An R-module M is called* **finitely presented**, *if and only if there are $m, n \in \mathbb{N}$ and an exact sequence of module homomorphisms*

$$R^m \xrightarrow{\alpha} R^n \longrightarrow M \longrightarrow 0 \qquad (*)$$

The right exactness of the tensor product enables us to describe $M \otimes_R N$ for a finitely presented right R-module M as a cokernel of a group homomorphism which can easily be described. Namely from an exact sequence as $(*)$ we get the exact sequence of abelian groups

$$N^m \xrightarrow{\alpha'} N^n \longrightarrow M \otimes_R N \longrightarrow 0,$$

where α' is given by the same matrix as α. (Note that the matrix operates from the left, since M is a right module and so R^m, R^n are regarded as right modules.)

If one allows infinite matrices, one can do the same for general modules.

1.6.4 Flat Algebras

Definition 1.6.9 *A module E is called* **flat**, *if and only if tensoring with it is exact. This means:*

$$M' \longrightarrow M \longrightarrow M'' \ exact \implies M' \otimes E \longrightarrow M \otimes E \longrightarrow M'' \otimes E \ exact,$$

the analogue for right modules.

An R-algebra A (for commutative R) is called flat, if and only if it is flat as an R-module.

Examples 1.6.10 $S^{-1}R$ and $R[X]$ *are flat R-algebras. The first follows from Example 1.6.2 e) combined with Lemma 1.3.11, the second by Example 1.6.2 c) combined with Remark 1.6.3 e).*

Proposition 1.6.11 *Let R be a commutative ring, M a finitely presented, N any R-module and A a flat R-algebra. Then the canonical map*

$$\mathrm{Hom}_R(M, N) \otimes_R A \longrightarrow \mathrm{Hom}_A(M \otimes_R A, N \otimes_R A)$$

is an isomorphism of groups.

Proof. By hypothesis there is an exact sequence

$$R^m \longrightarrow R^n \longrightarrow M \longrightarrow 0.$$

By the right exactness of the tensor product this gives an exact sequence

$$A^m \longrightarrow A^n \longrightarrow M \otimes_R A \longrightarrow 0.$$

Since A is flat over R the left exactness of Hom gives a commutative diagram with exact sequences

$$0 \to \quad \operatorname{Hom}_R(M,N) \otimes_R A \quad \to \quad \operatorname{Hom}_R(R^n,N) \otimes_R A \to \operatorname{Hom}_R(R^m,N) \otimes_R A$$

$$\downarrow \qquad\qquad\qquad \downarrow \qquad\qquad\qquad \downarrow$$

$$0 \to \operatorname{Hom}_A(M \otimes_R A, N \otimes_R A) \to \operatorname{Hom}_A(A^n, N \otimes_R A) \to \operatorname{Hom}_A(A^m, N \otimes_R A)$$

Since the second and the third vertical arrow are isomorphisms, so is the first one. $\qquad\square$

1.6.5 Exterior Powers

Let R be a commutative ring, M an R-module and $n \geq 1$ an integer. Let $\bigotimes^n M$ denote the tensor product of M with itself, n times. An R-homomorphism f of $\bigotimes^n M$ into an R-module N is said to be **alternating**, if for $x_1, \dots, x_n \in M$, we have $f(x_1 \otimes x_2 \otimes \cdots \otimes x_n) = 0$, whenever $x_i = x_j$ for some i, j, $i \neq j$, $1 \leq i, j \leq n$.

An n-**fold exterior power** of M is a pair (E_n, ψ), where E_n is an R-module and $\psi : \bigotimes^n M \to E_n$ is an alternating R-homomorphism, such that for any alternating R-homomorphism f of $\bigotimes^n M$ into an R-module N, there exists a unique R-homomorphism $f' : E_n \to N$ which makes the following diagram commute:

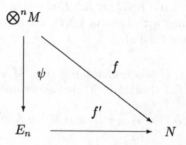

It is easy – and analogously to the uniqueness of the tensor product – to see that the n-fold exterior power of M is unique, if it exists. We now show that

n-fold exterior power of M exists. Let P be the submodule of $\bigotimes^n M$ generated by all elements of the for $x_1 \otimes x_2 \otimes \cdots \otimes x_n$ where $x_1, x_2, \ldots, x_n \in M$ and $x_i = x_j$ for some i, j with $i \neq j$. Let $\bigwedge^n M = (\bigotimes^n M)/P$ and $\psi : \bigotimes^n M \to \bigwedge^n M$ be the canonical homomorphism. Then it is easy to see that the pair $(\bigwedge^n M, \psi)$ (or $\bigwedge^n M$) is the n-fold exterior power of M. For $x_1, \ldots, x_n \in M$, the image of $x_1 \otimes \cdots \otimes x_n$ in $\bigwedge^n M$ under ψ is denoted by $x_1 \wedge \cdots \wedge x_n$.

We set $\bigwedge^0 M = R$ so that $\bigwedge^n M$ is defined for all $n \in \mathbb{N}$. Note that $\bigwedge^1 M = M$.

1.6.12 Some properties of exterior powers: a) Let $R \to S$ be a ring homomorphism and let M be an R-module. Composing the S-isomorphism

$$\overset{n}{\bigotimes}(S \otimes_R M) = (S \otimes_R M) \otimes_S (S \otimes_R M) \otimes_S \cdots \otimes_S (S \otimes_R M) \cong S \otimes_R (\overset{n}{\bigotimes} M)$$

with the S-homomorphism $1_S \otimes \psi : S \otimes_R (\bigotimes^n M) \to S \otimes_R (\bigwedge^n M)$, we have an alternating S-homomorphism $\bigotimes^n (S \otimes_R M) \to S \otimes_R (\bigwedge^n M)$, which induces an S-isomorphism

$$\overset{n}{\bigwedge}(S \otimes_R M) \overset{\sim}{\to} S \otimes_R (\overset{n}{\bigwedge} M).$$

b) Let M, N be R-modules. Then for $0 \leq i \leq n$, the map

$$\underbrace{M \times \cdots \times M}_{m-\text{times}} \times \underbrace{N \times \cdots \times N}_{(n-m)\text{times}} \to \overset{n}{\bigwedge}(M \oplus N)$$

given, for $x_1, \ldots, x_m \in M$ and $y_1, \ldots, y_{n-m} \in N$, by

$$((x_1, \ldots, x_m), (y_1, \ldots, y_{n-m})) \mapsto (x_1, 0) \wedge \cdots \wedge (x_m, 0) \wedge (0, y_1) \wedge \cdots \wedge (0, y_{n-m})$$

induces an R-homomorphism $(\bigwedge^m M) \otimes (\bigwedge^{n-m} N) \to \bigwedge^n (M \oplus N)$ and we have an isomorphism

$$\overset{n}{\bigwedge}(M \oplus N) \cong \sum_{m=0}^{n} (\overset{m}{\bigwedge} M \otimes \overset{n-m}{\bigwedge} N).$$

c) If F is a free R-module with a basis of n elements, then $\bigwedge^n F \cong R$ and $\bigwedge^i F = 0$ for $i > n$.

2

Introduction to Projective Modules

Projective modules are one of the main themes in our book.

2.1 Generalities on Projective Modules

Proposition 2.1.1 *Let A be a ring and P an A-module. The following properties are equivalent:*

(1) *for every diagram of A-modules*

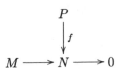

    *with exact line, there is a homomorphism $g : P \to M$ – a so-called **lift** of f – making the diagram commutative;*

(2) *to every epimorphism $f : M \to P$ exists a **cross section**, i.e. a homomorphism $s : P \to M$, such that $f \circ s = id_P$ – this means that P is isomorphic to a direct summand of M;*

(3) *P is a direct summand of some free A-module.*

Proof. (1) $\Rightarrow$ (2). Apply (1) to the situation $N = P$ and $f = id_P$.

(2) $\Rightarrow$ (3). This follows, since there is a surjective homomorphism $F \to P$ with a free module F.

(3) $\Rightarrow$ (1). First assume, P is free and B a basis of P. The restriction $f|B$ can be lifted to a map $\gamma : B \to M$, since $M \to N$ is surjective. The linear extension $g : P \to M$ of γ fulfills the assertion.

Now let $P \oplus Q$ be free. Then f extends to a linear map $f' : P \oplus Q \to N$, $(p, q) \mapsto f(p)$. Let $g' : P \oplus Q \to M$ be a lift of f'. Then $g := g'|P$ is a lift of f.   □

Remark 2.1.2 If the reader has accustomed himself to "Hom" as a functor and its left exactness, it will be clear to him that the properties of the proposition are also equivalent to

(4) *the functor* $\operatorname{Hom}_A(P, -)$ *is exact;*

(5) *the functor* $\operatorname{Hom}_A(P, -)$ *maps epimorphisms to epimorphisms.*

Definition 2.1.3 *A module with the equivalent properties of the proposition is called* **projective**.

Proposition 2.1.4 *For a projective A-module P the following properties are equivalent:*

(1) *P is finitely generated;*

(2) *P is a direct summand of a finitely generated free module;*

(3) *For some $n \in \mathbb{N}$ there is an exact sequence*

$$A^n \longrightarrow A^n \longrightarrow P \longrightarrow 0.$$

(So P is even finitely presented.)

Proof. (1) $\Rightarrow$ (2). Since P is finitely generated, there is a surjective homomorphism $f : A^n \to P$ for some $n \in \mathbb{N}$. By Proposition 2.1.1 (2) there is a section s of f, which identifies P with a direct summand of A^n.

(2) $\Rightarrow$ (3). If $A^n = P \oplus Q$, the projection $A^n \to Q$ and the canonical embedding $Q \to A^n$ compose to a map whose image Q is the kernel of the projection $A^n \to P$.

(3) $\Rightarrow$ (1) is trivial.   □

Remarks 2.1.5 a) A direct sum of (finitely or infinitely many) projective modules is projective. (This need not be so for direct products.)

b) If A is commutative and P and Q are projective A-modules, then so is $P \otimes_A Q$. Namely remember the "distributivity" of the tensor product w.r.t. direct sums. By this $F \otimes G$ is a free A-module, if F and G are so. And if $P \oplus P' = F$ and $Q \oplus Q' = G$ we get $(P \otimes Q) \oplus (P \otimes Q') \oplus (P' \otimes Q) \oplus (P' \otimes Q') \cong F \otimes G$, whence $P \otimes Q$ is a direct summand of a free module.

c) If P is projective and A commutative, then for $n \in \mathbb{N}$ the module $\bigwedge^n P$ is also projective. This is clearly true for free P, hence for a direct summand P of a free module by Paragraph 1.6.12.

d) Every projective A–module is flat. This is true first for $P = A$, then for any free P and finally for a general projective P.

e) Let P be a finitely generated projective left A-module, then $P^* = \mathrm{Hom}_A(P, A)$ is a finitely generated projective right A-module.

This can be derived from the following property of direct sums. Let $(M_i)_{i \in I}$ be a family of A-modules. The families $(f_i : M_i \to N)_{i \in I}$ of A-module homomorphisms are in bijective correspondence to the A-module homomorphisms $\bigoplus_{i \in I} M_i \to N$, which for $x_i \in M_i$ is defined by

$$(f_i)_{i \in I} \mapsto f \text{ where } f((x_i)_{i \in I}) := \sum_{i \in I} f_i(x_i).$$

Let $P \oplus Q = A^n$. Then $\mathrm{Hom}(P, A) \oplus \mathrm{Hom}(Q, A) \cong \mathrm{Hom}(P \oplus Q, A) = \mathrm{Hom}(A^n, A) \cong \mathrm{Hom}(A, A)^n \cong A^n$.

f) Every finitely generated projective module is **reflexive**, i.e. the canonical map $P \to P^{**}$ is bijective.

Namely with the above notations clearly $P \oplus Q = A^n$ is reflexive and $(P \oplus Q)^{**} = P^{**} \oplus Q^{**}$. The canonical isomorphism $P \oplus Q \to (P \oplus Q)^{**}$ maps P to P^{**} and Q to Q^{**}. So $P \to P^{**}$ must be an isomorphism.

Corollary 2.1.6 *Let A be a ring. The isomorphism classes of finitely generated projective A-modules form a commutative monoid (semigroup with a neutral element 0), the addition being defined by*

$$[P] + [Q] := [P \oplus Q].$$

([P] denotes the isomorphism class of P.)

Moreover, if A is commutative, they form a commutative semiring, the multiplication defined by

$$[P] \cdot [Q] := [P \otimes_A Q].$$

*(A **semiring** fulfills the axioms of a ring with the exception that its additive structure only is a commutative monoid, not necessarily a commutative group.) This semiring admits a canonical involution, given by $[P] \mapsto [P^*]$.*

Proof. The only question is, if the isomorphism classes of finitely generated projective A-modules form a set. But they all can be realized as quotients of one A-module, namely of $A^{(\mathbb{N})}$. The rest is clear. $\qquad\square$

Definition 2.1.7 *The set (monoid, semiring) of isomorphism classes of finitely generated projective A-modules will be denoted by $\mathbb{P}(A)$.*

Remark 2.1.8 Let $f : A \to B$ be a ring homomorphism. We define

$$\mathbb{P}(f) : \mathbb{P}(A) \longrightarrow \mathbb{P}(B) \quad \text{by} \quad [P] \longmapsto [B \otimes_A P].$$

In this way $\mathbb{P}$ becomes a so called 'functor' from the 'category' of rings to that of monoids, resp. semirings. (See Section 5.1.)

2.2 Projective Modules over Local Rings, Rank

Here all rings are supposed to be *commutative*; except in the last example.

We next show that all finitely generated projective modules over a local ring are free. For instance this is clear when the local ring is a field. The general case is reduced to this case by means of 'Nakayama's Lemma', i.e. Corollary 2.2.4.

Definition 2.2.1 *The* **Jacobson radical** *$Jac(A)$ of a ring A is the intersection of all maximal ideals of A.*

Lemma 2.2.2 *An element $x \in A$ belongs to $\mathrm{Jac}(A)$ if and only if $1 - ax$ is a unit for all $a \in A$.*

Proof. Let $x \in \mathrm{Jac}(A)$. So $ax \in \mathrm{Jac}(A)$, whence $1 - ax$ does not belong to any maximal ideal. Hence $A(1 - ax) = A$. i.e. there is a b with $b(1 - ax) = 1$.

Conversely if $x \notin \mathrm{Jac}(A)$, there is a maximal ideal $\mathfrak{m}$ of A with $x \notin \mathfrak{m}$. Since $A/\mathfrak{m}$ is a field, there is an $a \in A$ with $ax \equiv 1 \pmod{\mathfrak{m}}$, i.e. $1 - ax \in \mathfrak{m}$, whence $A(1 - ax) \neq A$. $\square$

Lemma 2.2.3 *Let M be a finitely generated R-module and I an ideal of R such that $IM = M$. Then there exists an element $i \in I$ such that $(1+i)M = 0$.*

Proof. (See [39].) We prove the result by induction on the minimal number $\mu := \mu(M)$ of generators of the module M. If $\mu(M) = 0$ take $i = 0$. Let $m_1, \dots, m_\mu$ be generators of M. Let $M' = M/Rm_\mu$. By induction there is an $x \in I$ such that $(1 + x)M' = 0$, i.e. $(1 + x)M \subset Rm_\mu$. Therefore, as $M = IM$, $(1+x)IM = I(1+x)M \subset Im_\mu$. Hence $(1+x)m_\mu = ym_\mu$ for some $y \in I$, i.e. $(1 + x - y)m_\mu = 0$. Thus, $(1 + x - y)(1 + x)M = 0$ and clearly $(1 + x - y)(1 + x) = 1 + i$ for some $i \in I$. $\square$

Corollary 2.2.4 (Nakayama's Lemma) *Let M be a finitely generated R-module and N a sub-module of M, further $I \subset \mathrm{Jac}(A)$ an ideal of A.*

a) *If $IM = M$ then $M = 0$.*

b) *If $N + IM = M$ then $N = M$.*

Proof. a) Clear from Lemma 2.2.3.

b) This will follow by noting that $I \cdot (M/N) = (N + IM)/N = M/N$. Now apply (i) to the finitely generated R-module M/N and I to get $M/N = 0$, i.e. $M = N$. $\qquad\square$

Proposition 2.2.5 *Let A be local (see Definition 1.3.25) and P a finitely generated projective A-module. Then P is free.*

Proof. Let $\mathfrak{m}$ be the maximal ideal, $k := A/\mathfrak{m}$ and $x_1, \ldots, x_n$ elements of P, whose residue classes form a basis of the k–vector space $P/\mathfrak{m}P$. Then P is generated by $\{x_1, \ldots, x_n\} \cup \mathfrak{m}P$, hence by Nakayama's Lemma it is already generated by $\{x_1, \ldots, x_n\}$. So one gets an exact sequence:

$$0 \longrightarrow Q \longrightarrow A^n \xrightarrow{f} P \longrightarrow 0,$$

where f is defined by $x_1, \ldots, x_n$ and Q the kernel. This sequence splits, because P is projective. Hence first Q is also finitely generated and secondly the sequence remains exact after tensoring. Especially one gets the exact sequence

$$0 \longrightarrow k \otimes Q \longrightarrow k^n \xrightarrow{k \otimes f} k \otimes P \longrightarrow 0.$$

By construction $k \otimes f$ is an isomorphism, whence $Q/\mathfrak{m}Q = k \otimes Q = 0$. Again by "Nakayama" we get $Q = 0$, which proves the proposition. $\qquad\square$

Remark 2.2.6 The proposition holds also for non finitely generated projective modules. See [35]. It also holds in the non-commutative case, if A has exactly one maximal left ideal, which then is automatically two-sided.

Lemma 2.2.7 *Let M be a finitely presented, N a finitely generated A-module and $f : N \to M$ an epimorphism, then $K := \ker(f)$ is finitely generated*

Proof. By hypothesis there are natural numbers m, n and an exact sequence

$$A^m \longrightarrow A^n \xrightarrow{f} M \longrightarrow 0.$$

Let $g : A^r \to M$ be any linear map. Since A^r is projective and f surjective, there is a homomorphism $g' : A^r \to A^n$ with $f \circ g' = g$. We get an exact sequence

$$0 \longrightarrow \ker(f) \oplus A^r \xrightarrow{\alpha} A^n \oplus A^r \xrightarrow{(f,g)} M \longrightarrow 0,$$

where $\alpha(x, y) := (x - g'(y), y)$ and $(f, g)(x, y) := f(x) + g(y)$. Therefore, for big enough r, we may construct a a commutative diagram with exact lines and surjective β

$$\begin{array}{ccccccc}
A^{m+r} & \longrightarrow & A^{n+r} & \longrightarrow & M & \longrightarrow & 0 \\
\downarrow{\scriptstyle \alpha} & & \downarrow{\scriptstyle \beta} & & \downarrow{\scriptstyle \mathrm{id}} & & \\
0 \longrightarrow K & \longrightarrow & N & \longrightarrow & M & \longrightarrow & 0
\end{array}$$

Since by the Snake Lemma α is surjective too, the lemma is proven. □

Proposition 2.2.8 *Let A be a ring and P be finitely presented A-module. The following properties are equivalent:*

(1) *P is projective;*

(2) *for every $\mathfrak{m} \in \operatorname{Spmax} A$ the $A_{\mathfrak{m}}$-module $P_{\mathfrak{m}}$ is free;*

(3) *there are $f_1, \dots, f_n \in A$, such that $A = \sum A f_i$ and P_{f_i} is free over A_{f_i} for every i.*

Proof. (1) $\Rightarrow$ (2) is clear.

(2) $\Rightarrow$ (3). Let $\mathfrak{m}$ be a maximal ideal of A. The $A_{\mathfrak{m}}$-module $P_{\mathfrak{m}}$ has a finite basis of the form $x_1/1, \dots, x_n/1$. Let $A^n \to P$ be defined by $x_1, \dots, x_n$ and C its cokernel. We have $C_{\mathfrak{m}} = 0$. Since C is finitely generated, there exists already an $f \in A \setminus \mathfrak{m}$ with $C_f = 0$. So we have an epimorphism $A_f^n \to P_f$. For its kernel K we have $K_{\mathfrak{m}} = 0$. Since K is finitely generated by the above lemma, we get $K_g = 0$ for some $g \in A \setminus \mathfrak{m}$. So P_{fg} is free. In this way we get a – maybe infinite – family $(f_i)_{i \in I}$ in A, such that P_{f_i} is free over A_{f_i} and to every maximal ideal $\mathfrak{m}$ there is an i with $f_i \notin \mathfrak{m}$. The latter means

$$\sum A f_i = A,$$

i.e. there are $g_i \in A$, $g_i = 0$ for nearly all i with $\sum_i f_i g_i = 1$. So we can replace I by a finite subset.

(3) $\Rightarrow$ (2). Let $\mathfrak{m} \in \operatorname{Spmax}(A)$. Since $\sum A f_i = A$, there is an i with $f_i \notin \mathfrak{m}$. So $P_{\mathfrak{m}}$ as a localization of P_{f_i} is free over $A_{\mathfrak{m}}$.

(2) $\Rightarrow$ (1). Let $M \to M''$ be an epimorphism. We have to show that an epimorphism $\operatorname{Hom}_A(P, M) \to \operatorname{Hom}_A(P, M'')$ is induced. It is enough to see that $\operatorname{Hom}_A(P, M)_{\mathfrak{m}} \to \operatorname{Hom}(P, M'')_{\mathfrak{m}}$ is surjective for every maximal ideal $\mathfrak{m}$. Since P is finitely presented, $\operatorname{Hom}_A(P, N)_{\mathfrak{m}} \cong \operatorname{Hom}_{A_{\mathfrak{m}}}(P_{\mathfrak{m}}, N_{\mathfrak{m}})$ in a natural way. So (i) follows, since $P_{\mathfrak{m}}$ is projective over $A_{\mathfrak{m}}$. □

Lemma 2.2.9 *Let $A \neq 0$ be a (commutative) ring. Then $A^m \cong A^n$ implies $m = n$ for all $m, n \in \mathbb{N}$.*

Proof. Since A possesses maximal ideals, there exist homomorphisms $A \to k$, where k is a field. So $A^m \cong A^n$ implies $k^m \cong k^n$. (Justify!) Hence $m = n$. □

Remark 2.2.10 If A is the non-commutative (!) endomorphism ring of an infinite dimensional vector space, then $A \cong A^2$ as A-modules, hence $A^m \cong A^n$ for all $m, n > 0$. (Exercise!)

Definitions 2.2.11 a) *Let $A \not\cong 0$ be a ring. The* **rank** *of a finitely generated free A-module A^m is defined by* $\mathrm{rk}_A(A^m) = \mathrm{rk}(A^m) = m$.

b) *Let P be a finitely generated projective A-module. Then we define the map* $\mathrm{rk}(P) : \mathrm{Spec}(A) \to \mathbb{N}$ *by* $\mathrm{rk}(P)(\mathfrak{p}) = \mathrm{rk}_{A_\mathfrak{p}}(P_\mathfrak{p})$. *We write also* $\mathrm{rk}_\mathfrak{p}(P)$ *instead of* $\mathrm{rk}(P)(\mathfrak{p})$. $\mathrm{rk}(P)$ *is called the* **rank (map)**.

c) *By a projective module of* **rank** $\leq n$, *resp. of* **rank** n *we mean a finitely generated projective module whose rank is bounded by n, resp. whose rank equals n everywhere. It is clear what we mean by a projective module of* **constant rank**. *(Especially it has to be finitely generated)*

Proposition 2.2.12 *Let P be a finitely generated projective A-module. The map $\mathrm{rk}(P) : \mathrm{Spec}(A) \to \mathbb{N}$ is locally constant. Especially if $\mathrm{Spec}(A)$ is connected, i.e. if A is not a direct product of nontrivial rings, then $\mathrm{rk}(P)$ is constant.*

Proof. Let $\mathfrak{p} \in \mathrm{Spec}(A)$. By Proposition 2.2.8 there is an $f \in A \setminus \mathfrak{p}$, such that P_f is free over A_f. So $\mathrm{rk}_\mathfrak{p}(P) = \mathrm{rk}_\mathfrak{q}(P)$ for every $\mathfrak{q} \in D(f)$. (Remember $D(f) = \{\mathfrak{p} \in \mathrm{Spec}(A) | f \notin \mathfrak{p}\}$.) $\square$

Note 2.2.13 Let A be a direct product of infinitely many fields k_i, $i \in I$. For any subset J of I the direct sum $P = \bigoplus_{j \in J} k_j$ is a projective A-module. If J is infinite, P is not finitely generated, and the proposition may happen to be not valid.

Proposition 2.2.14 *Let A be a semilocal ring and P a (finitely generated) projective A-module of constant rank. Then P is free.*

Proof. Let J be the intersection of the maximal ideals of A. Then P/JP is free over A/J. The rest is as in the proof of Proposition 2.2.8. $\square$

Proposition 2.2.15 *Let P be a projective A-module of constant rank n which can be generated by n elements $p_1, \dots, p_n$. Then P is free with basis $p_1, \dots, p_n$.*

Proof. The generating system $p_1, \dots, p_n$ defines an epimorphism $f : A^n \to P$. This induces an isomorphism $f_\mathfrak{p} : A_\mathfrak{p}^n \xrightarrow{\sim} P_\mathfrak{p}$ for every prime ideal $\mathfrak{p}$, since $P_\mathfrak{p} \cong A_\mathfrak{p}^n$. For $A_\mathfrak{p}^n \cong A_\mathfrak{p}^n \oplus \ker(f_\mathfrak{p})$. So f must be an isomorphism. $\square$

2.3 Special Residue Class Rings

First we extend some definitions, already used in the commutative case, to not necessarily commutative rings.

Definitions 2.3.1 *Let A be any ring.*

a) *The* **Jacobson radical** $\mathrm{Jac}(A)$ *of A is the intersection of all its maximal left ideals.*

b) *An element $x \in A$ is called* **strongly nilpotent** *if there is an $n \in \mathbb{N}$, such that every product $x_1 \cdots x_m$ in A vanishes, provided that at least n of its factors are equal to x.*

c) *The* **nilradical** $\mathrm{Nil}(A)$ *of A is the set of all its strongly nilpotent elements.*

Remarks 2.3.2 a) The simple left A-modules are up to isomorphism the modules A/m with a maximal left ideal m. Namely the latter clearly are simple. Now let E be simple and $x \in E \setminus \{0\}$. Then the map $R \to E$, $1 \mapsto x$ is surjective, its image being a non-zero submodule of the simple E, and its kernel must be a maximal left ideal.

(Note that over a noncommutative ring A modules A/I and A/J with different left ideals I, J may happen to be isomorphic. For example let $A = \mathrm{M}_n(k)$ be the $n \times n$-matrix ring over a field k with $n > 1$ and I, resp. J consist of the matrices whose first, resp. second column is zero. Then A/I and A/J both are isomorphic to k^n with its natural A-module structure. This left A-module is simple and I and J are maximal left ideals.)

b) $J := \mathrm{Jac}(A)$ is a two-sided ideal of A.

Namely let $a \in A$. We only have to show $Ja \subset \mathrm{m}$ for every maximal left ideal m. This is clear for $a \in \mathrm{m}$. Otherwise let $\mathrm{n} := \{x \in A \mid xa \in \mathrm{m}\}$, which clearly is a left ideal. Once we have shown that n is maximal, $Ja \subset \mathrm{n}a \subset \mathrm{m}$, and we are done.

Consider the left linear map $f : A \to A/\mathrm{m}$, $x \mapsto (xa \bmod \mathrm{m})$. Then $\ker(f) = \mathrm{n}$. Further f is surjective, because $a \notin \mathrm{m}$ and A/m is simple, i.e. has only the two obvious submodules. Since therefore A/n must also be simple, there is no left ideal properly contained between n and A.

c) Every proper left ideal is contained in a maximal one, as one easily sees with the help of Zorn's Lemma. We use this to show:

An element $x \in A$ belongs to $\mathrm{Jac}(A)$ if and only if $1 - ax$ is left invertible for every $a \in A$.

Namely let $x \in \mathrm{Jac}(A)$. So $ax \in \mathrm{Jac}(A)$, whence $1 - ax$ does not belong to any maximal left ideal. Hence $A(1 - ax) = A$, i.e. there is a b with $b(1 - ax) = 1$.

Conversely if $x \notin \text{Jac}(A)$, there is a maximal left ideal $\mathfrak{m}$ of A with $x \notin \mathfrak{m}$. Since the left A-module $A/\mathfrak{m}$ is simple, there is an $a \in A$ with $ax \equiv 1 \pmod{\mathfrak{m}}$, i.e. $1 - ax \in \mathfrak{m}$, whence $A(1 - ax) \neq A$.

d) The Lemma of Nakayama and its proof are word-by-word the same as in the commutative case. See Corollary 2.2.4.

e) If $a'a = aa'' = 1$ then $a' = a'aa'' = a''$. (In this case we write $a^{-1} := a'(= a'')$ and call a invertible.) We use this to show:

Jac(A) is also the intersection of the maximal right ideals.

By symmetry it is enough to show that the latter contains $\text{Jac}(A)$, i.e. that $1 - xa$ is right invertible for all $x \in \text{Jac}(A)$, $a \in A$, i.e. – $\text{Jac}(A)$ being also a right ideal – that $1 - x$ is right invertible for all $x \in \text{Jac}(A)$. Since it is left invertible there is a y with $y(1 - x) = 1$. So $1 - y = -yx \in \text{Jac}(A)$. Therefore there is an y' with $1 = y'(1 - (1 - y)) = y'y$. So y is invertible, hence $1 - x = y^{-1}$ too.

f) In a commutative ring the the concepts of nilpotent and strongly nilpotent elements coincide. If $x, y \in A$ are strongly nilpotent, so are $x + y$, ax, xa for every $a \in A$. Hence Nil(A) is a two-sided ideal of A.

If I is a nilpotent left or right ideal of A, all elements of I are strongly nilpotent. So $I \subset \text{Nil}(A)$.

If all entries of an $n \times n$-matrix α are strongly nilpotent elements of A, so is α in the ring $M_n(A)$.

g) Always we have $\text{Nil}(A) \subset \text{Jac}(A)$.

Namely let $x \in \text{Nil}(A)$ and $a \in A$; then $(ax)^n = 0$ for some n. Therefore $(1 - ax) \sum_{0 \leq i \leq n-1} (ax)^i = 1 - (ax)^n = 1$, and the criterion of b) is fulfilled.

Proposition 2.3.3 *Let I be a two-sided ideal of a ring A contained in $\text{Jac}(A)$, further P, P' be finitely generated projective A-modules such that $P/IP \cong P'/IP'$. Then $P \cong P'$.*

Proof. We have two obvious epimorphisms

$$P \longrightarrow P'/IP' \quad \text{and} \quad P' \longrightarrow P'/IP'.$$

By the projectivity of P we get a commutative diagram

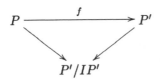

From this we get $f(P) + IP' = P$, hence $f(P) = P'$, by Nakayama's Lemma. By the surjectivity of f and the projectivity of P' we see, $\ker(f)$ is a direct summand of P (hence also finitely generated). Since $\ker(f)$ is mapped to zero in P/IP, we conclude – again by Nakayama's Lemma – $\ker(f) = 0$.   □

This means that $\mathbb{P}(A) \to \mathbb{P}(A/I)$ is injective, if $I \subset \mathrm{Jac}(A)$. It is not always surjective. (The isomorphism class of every so called stably free (A/I)-module is in the image, as we will see later.)

2.3.4 Lifting idempotents. Recall that the decompositions of a module into direct sums of two submodules are in bijective correspondence with its idempotent endomorphisms. Namely to the decomposition $M = M_1 \oplus M_2$ attach the idempotent endomorphism $id_{M_1} \oplus 0_{M_2}$; and to the idempotent endomorphism e attach the decomposition $M = e(M) \oplus \ker(e)$. (Check that one gets a direct decomposition.) *Therefore every (finitely generated) projective module is isomorphic to the image of an idempotent endomorphism of a (finitely generated) free module.*

Now, let $\varphi : A \to B$ be a ring homomorphism. If P is finitely generated projective over A, say $P \cong e(A^n)$ for some idempotent $e \in M_n(A)$, then $B \otimes P \cong \varphi(e)B^n$. So, to show the surjectivity of $\mathbb{P}(\varphi)$, it is sufficient to show, that every idempotent $e \in M_n(B)$ for any n has an idempotent preimage in $M_n(A)$.

Proposition 2.3.5 *Let $I \subset \mathrm{Nil}(A)$ be a two-sided ideal of a ring A and $\varphi : A \to A/I$ the canonical surjection. Then $\mathbb{P}(\varphi)$ is bijective.*

Proof. The surjectivity follows from the last proposition and $\mathrm{Nil}(A) \subset \mathrm{Jac}(A)$.

Now to the surjectivity of $\mathbb{P}(\varphi)$. We show that every idempotent $\bar{e} \in M_n(A/I)$ has an idempotent preimage in $M_n(A)$. Let e be any preimage of $\bar{e}$. Then $e^2 - e \in M_n(I)$. Let J be the two-sided ideal, generated by all entries of $e^2 - e$. (Let c_{ij} be these entries, then $J = \sum_{i,j} Ac_{ij}A$.) Note that $(e \bmod M_n(J))$ already is idempotent. Now J, generated by finitely many strongly nilpotent elements, is nilpotent. Since we may go stepwise, from A/J to A/J^2 to A/J^4 to $\ldots$, we may assume $J^2 = (0)$. Then also $M_n(J)^2 = (0)$.

We search for an $x \in M_n(J)$, such that $e + x$ is idempotent. Since $(e + x)^2 = e^2 + ex + xe + x^2 = e^2 + ex + xe$, we have to solve the equation

$$x - ex - xe = e^2 - e \tag{2.1}$$

by an element $x \in M_n(I)$.

The element $1 - 2e$ is a unit in $M_m(A)$. For, $(1 - 2e)^2 = 1 - 4e + 4e^2 = 1 + 4(e^2 - e)$ is a unit, since $e^2 - e$ is nilpotent. The element $(1 - 2e)^{-1}(e^2 - e)$ belongs to $M_n(I)$, and since it commutes with e, it solves Equation (2.1).   □

2.4 Projective Modules of Rank 1

In this section all rings are supposed to be *commutative*.

We will study projective modules of rank 1. These are isomorphic to ideals in many cases. And we are able to find many nonfree examples of them.

Proposition 2.4.1 *Let A be a ring and M be a finitely presented A-module. The following statements are equivalent:*

(1) *The "evaluation" map $M \otimes M^* \to A$, $(x, \alpha) \mapsto \alpha(x)$ is an isomorphism.*

(2) *There is an A-module N with $M \otimes N \cong A$.*

(3) *$M_{\mathfrak{m}} \cong A_{\mathfrak{m}}$ as $A_{\mathfrak{m}}$-modules for every maximal ideal $\mathfrak{m}$ of A.*

(4) *$M_{\mathfrak{p}} \cong A_{\mathfrak{p}}$ as $A_{\mathfrak{p}}$-modules for every $\mathfrak{p} \in \operatorname{Spec} A$.*

(5) *M is projective of rank 1.*

Proof. (1) $\Rightarrow$ (2) is clear.

(2) $\Rightarrow$ (3). For every maximal ideal $\mathfrak{m}$ we have $M_{\mathfrak{m}} \otimes_{A_{\mathfrak{m}}} N_{\mathfrak{m}} \cong A_{\mathfrak{m}}$. So without restriction of generality we may assume that A is local with maximal ideal $\mathfrak{m}$ and residue class field k. We get $(M/\mathfrak{m}M) \otimes (N/\mathfrak{m}N) \cong k$, hence $M/\mathfrak{m}M \cong k$. So by Nakayama's Lemma M is monogene, i.e. $M \cong A/I$ for some ideal I. This implies $I \cdot (M \otimes N) = 0$. Since $M \otimes N \cong A$, we get $I = 0$, i.e. $M \cong A$.

(3) $\Leftrightarrow$ (4) is clear.

(3) $\Rightarrow$ (5) follows from Proposition 2.2.8.

(5) $\Rightarrow$ (1): As M is finitely presented $(M_{\mathfrak{m}})^* \cong (M^*)_{\mathfrak{m}}$. Therefore $M \otimes_A M^* \cong A$ is equivalent to $M_{\mathfrak{m}} \otimes_{A_{\mathfrak{m}}} M_{\mathfrak{m}}^* \cong A_{\mathfrak{m}}$ for all maximal ideals $\mathfrak{m}$. But the latter is so if $M_{\mathfrak{m}} \cong A_{\mathfrak{m}}$. $\qquad\square$

Corollary 2.4.2 *The isomorphism classes of projective A-modules of rank 1 form an abelian group, the multiplication being defined by $[P] \cdot [Q] := [P \otimes_A Q]$. The neutral element is $[A]$, and the inverse of $[P]$ is $[P]^{-1} := [P^*]$.* $\qquad\square$

Definitions 2.4.3 a) *The group of isomorphism classes of projective A-modules of rank 1 is called the **Picard group** of A and denoted by* Pic A.
b) *Let $Q(A)$ be the total ring of fractions of A, i.e. $Q(A) = S^{-1}A$ where S is the set of the non-zero-divisors of A. A **fractional ideal** of A is an A-submodule I of $Q(A)$ for which there exists an $s \in S$, such that $sA \subset I \subset s^{-1}A$. (Note that $sA \subset I \subset t^{-1}A$ implies $stA \subset I \subset (st)^{-1}A$. Therefore we made the definition with only one s.) Ideals in the usual sense are sometimes called **integral ideals**, in order to distinguish them from fractional ideals.*

c) *Let* I, J *be fractional ideals of* A. *Then we define:*

$$I:J := \{x \in S^{-1}A \mid xJ \subset I\} \quad and \quad I^{-1} := A:I.$$

As for integral ideals, the product IJ *of fractional ideals* I, J *is the additive group generated by all products* ab *with* $a \in I$, $b \in J$.

Note 2.4.4 Every fractional ideal as an A-module is isomorphic to an integral fractional ideal. Namely if $sA \subset I \subset s^{-1}A$, Then $I \cong sI$ and $s^2A \subset sI \subset A$.

Lemma 2.4.5 *If* I, J *are fractional ideals, then so are* $I:J$ *and* IJ.

Proof. There are non-zero-divisors $s, t \in A$, such that $sA \subset I \subset s^{-1}A$ and $tA \subset J \subset t^{-1}A$. From $J \subset t^{-1}A$, i.e. $tJ \subset A$ and $sA \subset I$, we get $stJ \subset I$, hence $st \in I:J$. Let $x \in I:J$, i.e. $xJ \subset I \subset s^{-1}A$. So $xtA \subset xJ \subset I \subset s^{-1}A$, whence $x \in (st)^{-1}A$.

For IJ we get immediately $stA \subset IJ \subset (st)^{-1}A$.   □

Proposition 2.4.6 *Let* I, J *be fractional ideals of a ring. The map*

$$\varphi: I:J \longrightarrow \mathrm{Hom}_A(J, I) \qquad x \mapsto (j \mapsto xj)$$

is an isomorphism of A-*modules.*

Proof. Clearly φ is a homomorphism. Since J contains a non-zero-divisor, φ is injective.
To show surjectivity, let $\alpha: J \to I$ be an A-linear map and $t \in S$ with $tA \subset J \subset t^{-1}A$. Set $x := t^{-1}\alpha(t)$. Now for any $j \in J$ we get (using $tj \in A$):

$$\alpha(j) = t^{-2}(t^2\alpha(j)) = t^{-2}\alpha(t^2 j) = t^{-2}\alpha(tj \cdot t) = t^{-2} \cdot tj \cdot \alpha(t) = t^{-1}\alpha(t) \cdot j = xj.$$

    □

2.4.7 By the definition of I^{-1} we see immediately:

a) $II^{-1} \subset A$.

b) If there is a fractional ideal J with $IJ = A$, then $J \subset I^{-1}$, whence $II^{-1} = A$. Aso we have $J = I^{-1}$ in this case, as one sees by multiplying the equation $IJ = A$ by I^{-1}.

Definition 2.4.8 *An* **invertible ideal** *of a ring* A *is a fractional ideal* I *such that* $II^{-1} = A$.

Proposition 2.4.9 *The following properties of a fractional ideal* I *of a commutative ring* A *are equivalent:*

(1) I *is invertible;*

(2) I *is projective as an A-module;*

(3) I *is projective of rank 1. (Especially I is finitely generated).*

Proof. (1) $\Rightarrow$ (2). Since $1 \in II^{-1}$ by (i), there are $a_1, \dots, a_n \in I$, $b_1, \dots, b_n \in I^{-1}$ with $\sum a_i b_i = 1$. We define homomorphisms:

$$f : A^n \longrightarrow I \quad e_i \mapsto a_i, \quad \text{and}$$

$$g : I \longrightarrow A^n \quad x \mapsto (b_1 x, \dots, b_n x).$$

(Here $e_1, \dots, e_n$ is the canonical basis of A^n. And g maps I to A^n, since $b_i \in I^{-1}$.)
We get $fg(x) = \sum a_i b_i x = x$, i.e. $fg = id_I$. So I is isomorphic to a direct summand of A^n, hence projective and finitely generated.

(2) $\Rightarrow$ (1). I is isomorphic to a direct summand of some $A^{(E)}$. So let

$$f : A^{(E)} \longrightarrow I \quad e_i \mapsto a_i \quad \text{and}$$

$$g : I \longrightarrow A^{(E)} \quad x \mapsto \sum g_i(x) e_i$$

be homomorphisms with $fg = id_I$. To every $i \in E$ by Proposition 2.4.6 there is a $b_i \in I^{-1} \cong \text{Hom}(I, A)$ with $g_i(x) = b_i x$ for $x \in I$. For $x \in I$ we have $g_i(x) = 0$ hence $b_i x = 0$ for nearly every $i \in E$. Since there is a unit of $S^{-1}A$ in I, we have $b_i = 0$ for nearly every i.
Now $x = fg(x) = \sum a_i b_i x$ for $x \in I$. Again, since some of these x are units in $S^{-1}A$, we see $\sum a_i b_i = 1$.

This implies $1 \in II^{-1}$, hence (1).

(3) $\Rightarrow$ (1) is weaker than (2) $\Rightarrow$ (1).

(1) $\Rightarrow$ (3). By the proof of (1) $\Rightarrow$ (2) we know already that I is projective and finitely generated. So it suffices to show $I \otimes I^{-1} \cong A$. This is implied by $I \otimes I^{-1} \cong II^{-1}$. Now $I \subset s^{-1}A \Rightarrow sI \subset A$. Hence I is isomorphic to an integral ideal in A. Since further I^{-1} is invertible, whence projective, whence flat, the proof will be finished by the following observation.

Lemma 2.4.10 *Let J be an ideal of a ring A and E be an A-module.*

a) *By $j \otimes e \mapsto je$ a surjective homomorphism $\varphi : J \otimes E \to JE$ is defined.*

b) *If E is flat (especially if E is projective), φ is bijective.*

(This lemma holds also for a noncommutative ring, if J is a right ideal and E a left module.)

Proof of the lemma. a) is trivial.

b) Consider the commutative diagram

$$
\begin{array}{ccccccccc}
0 & \longrightarrow & J \otimes E & \longrightarrow & A \otimes E & \longrightarrow & A/J \otimes E & \longrightarrow & 0 \\
 & & \downarrow & & \downarrow & & \downarrow & & \\
0 & \longrightarrow & JE & \longrightarrow & E & \longrightarrow & E/JE & \longrightarrow & 0.
\end{array}
$$

The lines are exact, the upper by the flatness of E. Since the middle vertical arrow and the right one are isomorphisms, so is the left one. (Use the Snake Lemma). $\qquad\square$

Definitions 2.4.11 a) *Let* $\mathrm{Inv}(A)$ *denote the group of invertible ideals of A.*

b) *Let T be some multiplicative subset of A consisting of non-zero-divisors. Let* $\mathrm{Inv}(A,T)$ *denote the subgroup of* $\mathrm{Inv}(A)$ *consisting of such invertible ideals I of A, that $tA \subset I \subset t^{-1}A$ for some $t \in T$.*

c) *Let* $\mathrm{Prin}(A)$ *denote the group of invertible principal ideals of A, i.e. of those fractional ideals of A, which are generated as A-modules by a single unit of $Q(A)$.*

d) $\mathrm{Prin}(A,T) := \mathrm{Inv}(A,T) \cap \mathrm{Prin}(A)$.

It is easily seen that we indeed have groups in the above situations.

By Proposition 2.4.9 there is a canonical map

$$\alpha : \mathrm{Inv}(A,T) \longrightarrow \mathrm{Pic}(A) .$$

It is a homomorphism by Lemma 2.4.10. Its kernel is $\mathrm{Prin}(A,T)$.

Proposition 2.4.12 *With the above notations there is a canonical exact sequence*

$$0 \to \mathrm{Prin}(A,T) \to \mathrm{Inv}(A,T) \to \mathrm{Pic}(A) \to \mathrm{Pic}(T^{-1}A) .$$

Proof. Only the exactness at $\mathrm{Pic}A$ is not trivial.

'im $\subset$ ker'. For $I \in \mathrm{Inv}(A,T)$ there is a $t \in T$ with $At \subset I \subset At^{-1}$. This implies

$$T^{-1}A = T^{-1}At \subset T^{-1}I \subset T^{-1}At^{-1} = T^{-1}A, \text{ i.e. } T^{-1}I = T^{-1}A.$$

'ker $\subset$ im'. Let $[P] \in \mathrm{Pic}(A)$ with $[T^{-1}P] = 0$ in $\mathrm{Pic}(T^{-1}A)$. Then there is an isomorphism $g : T^{-1}P \to T^{-1}A$. Let $p \in P$, $t \in T$ be such that $g(p/t) = 1$ in $T^{-1}A$. Non-zero-divisors of A are such of P, the latter being a direct summand of a free module. So we get a monomorphism $P \to T^{-1}P \cong T^{-1}A$. Its image is an A-submodule I. It suffices to show $sA \subset I \subset s^{-1}A$ for some $s \in T$. But $g(p/1) = (t/1)$, whence $t \in I$, with the above t. On the other hand $I \cong P$ is finitely generated, say generated by $a_1/t_1, \ldots, a_n/t_n$. Then $s := t \cdot t_1 \cdots t_n$ will do. $\qquad\square$

Remark 2.4.13 Assume, there is a multiplicative set T, consisting of non zero divisors of A such that $T^{-1}A$ is semilocal. Then for such a T by Proposition 2.4.12 we have

$$\mathrm{Pic}(A) \cong \mathrm{Inv}(A,T)/\mathrm{Prin}(A,T).$$

Namely, if B is semilocal, then $\mathrm{Pic}(B) = 0$ by Proposition 2.2.14. This means that evry projective A-module of rank 1 is isomorphic to some integral invertible ideal. Since $\mathrm{Inv}(A,T) \subset \mathrm{Inv}(A)$ we have then also $\mathrm{Pic}(A)/\mathrm{Inv}(A) = \mathrm{Pic}(A)$.

Let now T be the set of all non-zero-divisors. $T^{-1}A$ then is semilocal, if for instance A is reduced with only finitely many minimal prime ideals, for e.g. a domain. See Corollary 1.3.24.

The same holds for rings whose ideals are finitely generated, which were first studied 'abstractly' by E. Noether, who showed that they also satisfy the property that ascending chains of ideals will terminate. For more details we refer the reader to Chapter 6. We will see there that in such a ring, the so-called **Noetherian ring**, the set of zero-divisors is a union of finitely many prime ideals.

Corollary 2.4.14 *If A is a factorial domain,* $\mathrm{Pic}(A) = 0$.

The converse does not hold, since there are many non-factorial local domains.

Proof. We show, that any invertible ideal $I \subset A$ is principal. Let x be the greatest common divisor of its elements. It is sufficient to show $J := x^{-1}I = A$. But since the greatest common divisor of J is 1, one sees easily that $J^{-1} = A$, whence $J = A$. $\qquad\square$

Examples 2.4.15 We give examples of nonfree projective modules of rank one.

a) Let $A := \mathbb{Z} + \mathbb{Z}\sqrt{-5} \subset \mathbb{C}$. It is clearly a subring of $\mathbb{C}$. The quotient field of A is $\mathbb{Q}(\sqrt{-5}) = \mathbb{Q} + \mathbb{Q}\sqrt{-5}$. Consider the ideal $\mathfrak{a} := (3, 1 + \sqrt{-5}) = (2 - \sqrt{-5}, 1 + \sqrt{-5})$ of A.

CLAIM: $\mathfrak{a}$ is invertible!
Proof. An element $a + b\sqrt{-5} \in \mathbb{Q}(\sqrt{-5})$ belongs to $\mathfrak{a}^{-1}$ if and only if $3(a + b\sqrt{-5}) \in A$ and $(1 + \sqrt{-5})(a + b\sqrt{-5} \in A$. The first condition means $3a, 3b \in \mathbb{Z}$. If we write $a = a'/3$, $b = b'3$ with $a', b' \in \mathbb{Z}$, the second condition means $a' + b' \equiv 0$ (3) and $a' - 5b' \equiv 0$ (3); but that means $a' + b' \equiv 0$ (3). So for e.g. $(-1 + \sqrt{-5})/3$, $(2 + \sqrt{-5})/3 \in \mathfrak{a}^{-1}$. And we compute $(2 - \sqrt{-5})(2 + \sqrt{-5})/3 + (1 + \sqrt{-5})(-1 + \sqrt{-5})/3 = 3 - 2 = 1$.

On the other hand $\mathfrak{a}$ is not principal. Consider the norm

$$N : \mathbb{Q}(\sqrt{-5}) \to \mathbb{Q}, \qquad a + b\sqrt{-5} \mapsto a^2 + 5b^2.$$

We know that it is multiplicative, i.e. $N(\alpha\beta) = N(\alpha)N(\beta)$. By this, one easily sees that $\alpha \in A$ is a unit of A if and only if $N(\alpha) = \pm 1$. Therefore 3 and $1 + \sqrt{-5}$ are irreducible, since the factors 2 and 3 of their norms 9, resp. 6 are not norms themselves. So the only principal ideal of A which contains $\mathfrak{a}$ is A. And $\mathfrak{a} \neq A$, since $\mathfrak{a}^{-1} \neq A$.

b) The following examples are similar to the above one. Let k be a field with $\mathrm{char}(k) \neq 2$ and $f \in k[X]$ be a polynomial of degree > 1 which has only simple roots in an algebraic closure of k. Our ring is $A := k[X,Y]/(Y^2 - f)$ which clearly can also be written as $A = k[X] \oplus k[X]y$, where the multiplication is defined by $y^2 = f$. (The ring A is the "coordinate ring" of a conic ($\deg(f) = 2$), resp. of an elliptic curve ($\deg(f) = 3$ or 4), resp. a hyperelliptic curve ($\deg(f) \geq 5$).)

Now let $a, b \in k$ with $b^2 = f(a)$. (This means the point $(a, b) \in k^2$ belongs to the curve.) For the sake of simplicity we further suppose $b \neq 0$ and $f'(a) \neq 0$. Consider the ideal $\mathfrak{a} = (X - a, y - b)$ of A. We have

$$Y^2 - f = Y^2 - f(X) - (b^2 - f(a)) \in (X - a, Y - b)$$

and so $A/\mathfrak{a} \cong k[X,Y]/(X - a, Y - b) \cong k$. Especially $\mathfrak{a} \neq A$. We will show that $\mathfrak{a}$ is invertible and $\mathfrak{a}^{-1} = (\alpha, \beta)$ with

$$\alpha = \frac{y + b}{2b(X - a)} \quad \text{and} \quad \beta = -\frac{1}{2b}.$$

We have $\alpha \cdot (X - a) \in A$ and also $\alpha \cdot (y - b) = \dfrac{y^2 - b^2}{2b(X - a)} = \dfrac{f(X) - f(a)}{2b(X - a)} \in A$. This gives $\alpha\mathfrak{a} \subset A$. The inclusion $\beta\mathfrak{a} \subset A$ is trivial. Further we have

$$\alpha \cdot (X - a) + \beta \cdot (y - b) = \frac{y + b - (y - b)}{2b} = 1.$$

This proves our statement.

Now we will show that in many cases $\mathfrak{a}$ is not principal. Again we do this using the norm

$$N : A \to k[X], \quad g(X) + h(X)y \mapsto g(X)^2 - h(X)^2 y^2 = g(X)^2 - h(X)^2 f(X).$$

i) First we consider the case where $\deg(f)$ is odd and ≥ 3. An element $g(X) + h(X)y$ is a unit of A, if its norm is in $k^\times$. This implies first that $h(X) = 0$ and then $g(X) \in k^\times$. So $k^\times$ is the unit group of A. Now $N(X - a) = (X - a)^2$ and one sees, by arguing with degrees, that $c \cdot (X - a)$ is not a norm for any $c \in k^\times$. So $X - a$ is irreducible, i.e. $(X - a)$ is maximal amongst all principal ideals $\neq A$.

Now $N(y - b) = b^2 - f(X) = f(a) - f(X)$, which is divisible by $X - a$, but not by $(X - a)^2 = N(X - a)$, since we supposed $f'(a) \neq 0$. Therefore $y - b$ is not contained in $(X - a)$, whence $\mathfrak{a}$ cannot be principal.

ii) Secondly let k be an ordered field, e.g. any subfield of $\mathbb{R}$ and $f(X)$ of even degree (≥ 2) with negative leading coefficient. Then for all non zero polynomials $g(X), h(X) \in k[X]$ the leading coefficients of $g(X)^2$ and $-h(x)^2 f(X)$ are positive. Hence $\deg(N(g(X) + h(X)y) = \mathrm{Max}\{2\deg(g(X)), 2\deg(h(X)) + 2\}$. By the same arguments as in i) one sees that the invertible ideal $\mathfrak{a}$ is not principal.

iii) As a special case of ii) let us consider the field $k = \mathbb{R}$ and the polynomial $f(X) = -X^2 + 1$. Then $A = \mathbb{R}[X, Y]/(X^2 + Y^2 - 1)$ is the coordinate ring of the circle. The projective module $\mathfrak{a}$ may be called Möbius-module, since it is related to the Möbius band. (Note that we have excluded the 4 points $(\pm 1, 0)$, $(0, \pm 1)$ of the circle for simplicity. Any orthogonal transformation different from $\pm\mathrm{id}$ of $\mathbb{R}^2$ will put these points to ones we have considered.)

Let us see, what happens in this case, if we extend the groundfield to $\mathbb{C}$. Then the ideal $I := (X - a, Y - b)$ of $\mathbb{C}[X, Y]$ coincides with the ideal $J := (X^2 + Y^2 - 1, X + Yi - a - bi)$, i.e. $I/(X^2 + Y^2 - 1)$ is principal. Namely – remember $a^2 + b^2 = 1$ –

$$X^2 + Y^2 - 1 = X^2 + Y^2 - (a^2 + b^2) \tag{2.2}$$
$$= (X + a)(X - a) + (Y + a)(Y - a) \tag{2.3}$$
$$\text{and } X + Yi - a - bi = X - a + i(Y - b) \tag{2.4}$$

whence $J \subset I$.

$$X^2 + Y^2 - 1 - (X - Yi)(X + Yi - a - bi) =$$

$$(X - Yi)(X + Yi) - (a - bi)(a + bi) - (X - Yi)(X + Yi - a - bi)$$
$$-(a - bi)(a + bi) + (X - Yi)(a + bi)$$
$$= (X - Yi - (a - bi))(a + bi).$$

So J contains $X - a + (Y - b)i$ and $X - a - (Y - b)i$ as well. By that one easily gets $I \subset J$. Therefore in $A = \mathbb{C}[X, Y]/(X^2 + Y^2 - 1)$ the ideal $\mathfrak{a} = I/(X^2 + Y^2 - 1)$ is generated by the residue class of $X + Yi - a - bi$.

c) Let $A = R[X]$ with $R = k[y, z]_{(y,z)}/(y^3 - z^2) \cong k[t^2, t^3]_{(t^2, t^3)}$, where k is any field. Then $Q(R) = k(t)$. Now $t^n \in R$ for all $n \geq 2$. By Proposition 2.4.18 – see below – we conclude that there is a projective A-module P of rank 1, which is not free.

Lemma 2.4.16 *Let R be a domain. Then* $R = \bigcap_{\mathfrak{p} \text{ prime}} R_{\mathfrak{p}}$.

Proof. Let $a \in \bigcap_{\mathfrak{p}} R_{\mathfrak{p}}$ and consider the ideal $I := \{s \in R \mid sa \in R\}$. For every prime ideal $\mathfrak{p}$ there is an $s \in I \setminus \mathfrak{p}$, since $a \in R_{\mathfrak{p}}$. So I is not contained in any prime ideal, hence $I = R$. Therefore $Ra \subset R$, i.e. $a \in R$. $\qquad\square$

Corollary 2.4.17 *Let R be an integral domain with quotient field K and let $a \in K \setminus R$. Then there is a prime ideal $\mathfrak{p}$ of R with $a \notin R_{\mathfrak{p}}$.*

Proposition 2.4.18 *Let R be an integral domain with quotient field K. Suppose that there is an element $a \in K \setminus R$ such that $a^n \in R$ for all sufficiently large n. Then there is a rank 1 projective $R[X]$-module which is not extended from R, i.e. not of the form $P[X]$ for a projective R-module P.*

Proof. Replace a by a suitable power of it and assume that $a^n \in R$ for all $n \geq 2$. Let $f = aX \in K[X]$. Let $P = (1+f)R[X] + (1+f+f^2)R[X] \subset K[X]$. We show that P is an invertible ideal, hence a projective $R[X]$-module of rank 1:

Let $Q = (1 - f)R[X] + (1 - f + f^2)R[X]$. Then

$$PQ = (1-f^2)R[X] + (1-f^3)R[X] + (1+f^3)R[X] + (1+f^2+f^4)R[X] \subset R[X]$$

as $f^2, f^3, f^4 \in R[X]$. We show $PQ = R[X]$. If the overbar is meant modulo PQ, then $\overline{2} = \overline{(1 - f^3)} + \overline{(1 + f^3)} = \overline{0}$, $\overline{f}^4 = \overline{1 - f^2} + \overline{1 + f^2 + f^4} = \overline{0}$, and so $\overline{1} = \overline{1} \cdot \overline{1} = (\overline{f}^2)^2 = \overline{f}^4 = \overline{0}$. Thus, $PQ = R[X]$. Then P is a projective module of rank 1 by Proposition 2.4.9.

If P were extended from R, the $P_{\mathfrak{p}}$ would be extended from $R_{\mathfrak{p}}$ whence $P_{\mathfrak{p}}$ would be free of rank 1, i.e. $P_{\mathfrak{p}} = R_{\mathfrak{p}}[X]c$ for some $c \in K[X]$. Now $1 = (1 + f + f^2) - f(1 + f) \in K[X]P_{\mathfrak{p}} = cK[X]$. Hence $c \in K^{\times}$. Then $P_{\mathfrak{p}}$ is generated by $1 + aX, 1 + aX + a^2X^2$, whence it is generated by 1 modulo X. This means $c \in R_{\mathfrak{p}}^{\times}$. Therefore, $1 + aX \in P_{\mathfrak{p}} = cR_{\mathfrak{p}}[X] = R_{\mathfrak{p}}[X]$ whence $a \in R_{\mathfrak{p}}$. A contradiction. $\qquad\square$

d) Let $\mathbb{H} = \{a + bi + cj + dk \mid a, b, c, d \in \mathbb{R}\}$ be the division ring of the real quaternions. It can be shown that all the projective $\mathbb{H}[x]$-modules are free. However, there are nonfree projective $\mathbb{H}[x, y]$-modules.

This is a special case of the 1971 theorem of M. Ojanguren-R.Sridharan [69] that there are nonfree rank one projective modules over a polynomial extension $D[x_1, \dots, x_n]$, $n > 1$, of a division ring D, which is not a field. The reader can also find a proof of this theorem in ([44], Chapter 2, §3). Here we give the example in the special case when D is the division ring $\mathbb{H}$ of the real quaternions, our argument expands on the one found in [85].

History is witness to the fact that in the decade after this counter-example there were a spate of papers on the Serre's conjecture in the early to mid-seventies, with several partially successful results reconfirming that all should be well in the case of the base ring being a field! In 1980, four years after Serre's conjecture was proved correct by Quillen-Suslin, J. Stafford proved in [90] that every projective module of rank > 2 over $D[x_1, \ldots, x_n]$, where D is a division ring, is free. (In [90] the reader can also find a simple proof in the case when D is finite dimensional over its center.)

We now construct the counter-example mentioned above.

Let $F = \mathbb{H}[x, y] \oplus H[x, y]$ be the free left $\mathbb{H}[x, y]$-module of rank 2, with $e_1 = (1, 0)$, $e_2 = (0, 1)$ as basis. Define a $\mathbb{H}[x, y]$-module homomorphism $\varphi : F \to \mathbb{H}[x, y]$ by sending e_1 to $x + i$, and e_2 to $y + j$. Note that φ is a surjective map as $\varphi((y + j)e_1 - (x + i)e_2) = -2k$. Hence, $\ker(\varphi) = P$ will be a left projective $\mathbb{H}[x, y]$-module of rank 1. We show that P is not free.

Let $t_1 = (y + j)(x - i)e_1 - (x^2 + 1)e_2$, $t_2 = -(y^2 + 1)e_1 + (x + i)(y - j)e_2$. It is easily verified that $\varphi(t_1) = 0 = \varphi(t_2)$, i.e. t_1, $t_2 \in \ker(\varphi) = P$.

If P were free of rank 1, then it would be generated by a single element. It is easy to check that if $P = \mathbb{H}[x, y]t_1$, then $t_2 \notin \mathbb{H}[x, y]t_1$, and if $P = \mathbb{H}[x, y]t_2$, then $t_1 \notin \mathbb{H}[x, y]t_2$. Thus, t_1, t_2 will not generate P. For instance, if $t_2 \in \mathbb{H}[x, y]t_1$, say $t_2 = p(x, y)\{(y + j)(x - i)e_1 - (x^2 + 1)e_2\}$, for some $p(x, y) \in \mathbb{H}[x, y]$. Let deg denote the total degree. Then, since $\deg(fg) = \deg f + \deg g$, for $f, g \in \mathbb{H}[x, y]$, we conclude that $\deg p(x, y) = 0$, i.e. $p(x, y) = p(0, 0) = k$. But this cannot be. Similarly, show that $t_1 \notin \mathbb{H}[x, y]t_2$.

So let $ae_1 + be_2 \in P$ be a generator, with a, b non-zero. Since $ae_1 + be_2 \in \ker(\varphi)$, $a(x + i) + b(y + j) = 0$. Since $t_1 \in P$, there exists $p(= p(x, y))$, $q(= q(x, y)) \in \mathbb{H}[x, y]$ such that $t_1 = p(ae_1 + be_2)$ $t_2 = q(ae_1 + be_2)$. These equations show that $\deg p + \deg a = 2$, $\deg + \deg b = 2$, $\deg q + \deg a = 2$, etc. They also allow us to conclude that $a \in \mathbb{H}[y]$, $b \in \mathbb{H}[x]$, $p(x, y) \in \mathbb{H}[x]$, $q \in \mathbb{H}[y]$.

Consequently, $\deg p$, $\deg q \leq 2$, $\deg a = \deg b$. But $\deg p \neq 2$, as then $\deg a = 0$, which is is not possible as $a(x + i) + b(y + j) = 0$. Similarly $\deg q \neq 2$. Again $\deg p \neq 0$, as then $\deg a = 2$, $\deg b = 2$, and one cannot get $a(x + i) + b(y + j) = 0$. (How can the degree two term corresponding to y^2 in $a(x + i)$ be cancelled by terms of $b(y + j)$? Don't forget that $b \in \mathbb{H}[x]$.)

The only case that remains is that $\deg p = 1 = \deg a = \deg b$. So let $a = a_{00} + a_{01}y$, $b = b_{00} + b_{10}x$, and study the relation $(a_{00} + a_{01}y)(x + i) + (b_{00} + b_{10}x)(y + j) = 0$. By comparing the xy, constant, y, and x coefficients, we get $a_{01} = -b_{10}$, $a_{00}i = -b_{00}j$, $a_{01}i = -b_{00}$, $a_{00} = -b_{10}j$.

Therefore, $a_{00}i = -b_{00}j = a_{01}ij = a_{01}k$. However, $a_{00} = -b_{10}j = a_{01}j$. Hence, $a_{00}i = a_{01}ji = -a_{01}k$!! Hence, $a_{01} = 0$. But then $b = 0$, a contradiction.   □

3

Stably Free Modules

A module P is projective, if and only if there is a module Q such that $P \oplus Q$ is free. We know that this does not imply that P itself is free. It is a fascinating fact that even if Q is finitely generated and free and $P \oplus Q$ is free, P need not be free.

3.1 Generalities

In this section R will denote an arbitrary ring, unless otherwise specified. We are considering right R-modules instead of left R-modules.

A finitely generated R-module P is projective if and only if there exists a finitely generated R-module Q such that $P \oplus Q$ is free. Here we introduce a smaller class of finitely generated modules, the so called stably free modules P, which have the property that there exists a finitely generated free module F such that $P \oplus F$ is free.

Historically it was so that one knew 'from the beginning' that projective modules over a polynomial ring $k[x_1, \ldots, x_n]$, where k is a field, are stably free. This will be shown here in Theorem 3.7.6. And in Proposition 3.1.10 we will give an example of a nonfree, but stably free module, which 'in geometric terms' is the tangent bundle on the 2 dimensional sphere.

In Chapter 4 we will give several proofs of the fact that projective modules over $k[x_1, \ldots, x_n]$ are actually free. Three of them use the fact that these modules are already known to be stably free.

Definition 3.1.1 *A (right) R-module P is said to be* **stably free of type** *m with $(0 \le m < \infty)$ if $P \oplus R^m$ is free. A module is said to be* **stably free** *if it is stably free of type m for some $m \in \mathbb{N}$. (Such a module is of course projective.)*

A projective module P is stably free of type m if and only if

$$P \cong \ker(R^n \xrightarrow{f} R^m)$$

for a suitable epimorphism f, which automatically splits. If M is the $m \times n$ matrix associated with f, then M is right invertible, i.e, there exists an $n \times m$-matrix N such that $MN = I_m$. (By I_m we denote the $m \times m$ unit matrix.) Conversely any right invertible $m \times n$-matrix M defines a finitely generated stably free (right R-module) P of type m, namely the solution space of M,

$$P = \left\{ \alpha = \begin{pmatrix} a_1 \\ \vdots \\ a_n \end{pmatrix} \,\middle|\, M\alpha = 0 \right\}.$$

In this way the *study of finitely generated stably free right R modules is equivalent to the study of right invertible rectangular matrices over R.*

Proposition 3.1.2 *The kernel P of an epimorphism $f : R^n \to R^m$ is free if and only if f can be lifted to an isomorphism $f' : R^n \to R^m \oplus R^r$ for some r, such that $\mathrm{pr}_1 \circ f' = f$.*

Proof. Suppose P is free, then there exists an r such that $P \xrightarrow{g} R^r$ is an isomorphism. One may write $R^n = Q \oplus P$ in such a way that the restriction of f to Q gives an isomorphism $f_0 : Q \to R^m$. Then $f_0 \oplus g : R^n \to R^m \oplus R^r$ clearly gives the desired isomorphism.

Conversely suppose that an isomorphism f' with $\mathrm{pr}_1 \circ f' = f$ exists. Then $P = \ker f \cong \ker(\mathrm{pr}_1) = R^r$ is free.   □

Note that in the situation above $R^m \oplus R^r \cong R^n$ need not imply $m + r = n$, in general. A well-known example of P.M. Cohn [14] is: Let V be an infinite dimensional vector space over a field k, and $R = \mathrm{End}_k(V)$ the endomorphism ring of R. Then $R \cong R \oplus R$. Namely, since V is infinite dimensional, $V \oplus V \cong V$ and so

$$R \cong \mathrm{Hom}(V, V) \cong \mathrm{Hom}(V \oplus V, V) \cong \mathrm{Hom}(V, V) \oplus \mathrm{Hom}(V, V) \cong R \oplus R.$$

Definition 3.1.3 *We say that a ring R satisfies the* **(right) invariant basis property (IBP)** *if for any $s, t \geq 0$, $R^s \cong R^t \Rightarrow s = t$.*

Examples 3.1.4 a) A division ring (= skew field) satisfies IBP.

b) A commutative ring $\neq (0)$ satisfies IBP; see Lemma 2.2.9.

c) Any (right) Noetherian ring satisfies (right) IBP. Otherwise an isomorphism $R^m \xrightarrow{\sim} R^n$ with $m < n$, composed with the canonical projection $R^n \to R^m$

would give a surjective endomorphism $f : R^n \to R^n$ with nontrivial kernel. Then the series

$$\ker(f) \subsetneq \ker(f^2) \subsetneq \ker(f^3) \subsetneq \cdots$$

would be an infinite strictly increasing series of submodules of a finitely generated module over a Noetherian ring. This is impossible as we will see in Chapter 6.

Remark 3.1.5 Consider f, f' in Proposition 3.1.2. Let M denote the $m \times n$ matrix corresponding to f and let N denote the $(m+r) \times n$ matrix corresponding to f', if f' exists. The condition " $\mathrm{pr}_1 \circ f' = f$ " says that M is a submatrix of N, consisting of its first n rows. The condition "f is an isomorphism" says that N is a (not necessarily square) invertible matrix. i.e. there exists another matrix N' of size $n \times (m+r)$ such that $NN' = I_{m+r}$, $N'N = I_n$. The following is the matrix theoretic version of Proposition 3.1.2.

Proposition 3.1.6 *For any right invertible $m \times n$ matrix M, $m < n$, the (stably free) solution space of M is free if and only if M can be completed to an invertible matrix by adding a suitable number of new rows.*

Definition 3.1.7 *We say that $v = (v_1, \ldots, v_n)$ is a (right)* **unimodular row** *if $\sum_{i=1}^n v_i R = R$.*

The set of all (right) unimodular rows of length n with entries in R is denoted by $\mathrm{Um}_n(R)$.

Remarks 3.1.8 a) A row $(v_1, \ldots, v_n)$ is unimodular, if and only if there are $w_i \in R$ such that $\sum v_i w_i = 1$. Therefore, if $w = (w_1, \ldots, w_n)$, then $vw^t = 1$. Thus, $v \in \mathrm{Um}_n(R)$ if and only if there is a row w of length n such that $\langle v, w \rangle = vw^t = 1$. For example, the ith co-ordinate vector

$$e_i = (0, \ldots, 0, 1, 0, \ldots, 0)$$

is a unimodular vector.

b) A row, considered as a $1 \times n$-matrix, describes a linear map of right modules $R^n \to R$, which is surjective if and only if the row is unimodular. It follows that there is a surjective map from the set of unimodular rows of length n to the set of isomorphism classes of stably free modules of rank $n - 1$ and type 1.

Example 3.1.9 Let $A = \mathbb{R}[x_0, \ldots, x_n]/(x_0^2 + \cdots + x_n^2 - 1)$, $n \geq 1$. Let P be the projective module corresponding to the unimodular row $(x_0, \ldots, x_n) \in \mathrm{Um}_{n+1}(A)$. Then P is free for $n = 1, 3, 7$.

Proof. Let $e_0, \ldots, e_n$ be a basis for the complex numbers ($n = 1$), quaternions ($n = 3$), and Cayley numbers ($n = 7$) over $\mathbb{R}$ such that $e_0 = 1$ and the norm of $x = \sum\limits_{i=0}^{n} x_i e_i$ is $\sum\limits_{i=0}^{n} x_i^2$ ($= N(x)$).

Consider the matrix H_x of the linear transformation $y \mapsto yx$. If $x \neq 0$ then $H_x \in \mathrm{GL}_{n+1}(A)$. Moreover, $e_1 H_x = (x_0, \ldots, x_n)$ and so H_x is a completion of $(x_0, \ldots, x_n)$ in the sense of Proposition 3.1.6. $\square$

Here is an interesting example of a stably free projective module which is not free.

Proposition 3.1.10 *Let*

$$A = \mathbb{R}[x_0, x_1, x_2]/(x_0^2 + x_1^2 + x_2^2 - 1)$$

Then the projective A-module P, corresponding to the unimodular row $(\overline{x}_0, \overline{x}_1, \overline{x}_2)$, (where $\overline{x}_i$ denotes the residue class of x_i) is not free.

Proof. If P where free then by Proposition 3.1.6 the row $(\overline{x}_0, \overline{x}_1, \overline{x}_2)$ can be completed to an invertible matrix σ with $e_1 \sigma = (\overline{x}_0, \overline{x}_1, \overline{x}_2)$. We shall think of σ_{ij}, $\det(\sigma)$ as "functions" on the real 2-sphere

$$S^2 = \{(v_0, v_1, v_2) \in \mathbb{R}^3 : v_0^2 + v_1^2 + v_2^2 = 1\}$$

below.

For any point on S^2 we can define a tangent vector as follows: If $v \in S^2$ consider

$$\varphi(v) = (\sigma_{21}^{-1\,t}(v), \sigma_{22}^{-1\,t}(v), \sigma_{23}^{-1\,t}(v)) \in \mathbb{R}^3.$$

Clearly, $\langle v, \varphi(v) \rangle = 0$ and so $\varphi(v)$ is a tangent vector to S^2 at the point v. Since $\sigma_{2i}^{-1\,t}$ are polynomials, the map $\varphi : S^2 \to \mathbb{R}^3$ is a differentiable function. Since $\sigma^{-1\,t} \in \mathrm{GL}_3(A)$ the vector $\varphi(v)$ can never be the zero vector. Thus φ is a nowhere zero vector field on S^2. This is well-known to be impossible by elementary topology (cf. [56]). (We will give a short sketch of the argument in Proposition 5.4.4.) Thus, P cannot be a free module. $\square$

Proposition 3.1.11 *For any ring R, the following statements are equivalent.*

(1) *Any finitely generated stably free right R-module is free.*

(2) *Any finitely generated stably free right R-module of type 1 is free.*

(3) *Any right unimodular row over R can be completed to an invertible matrix (by adding a suitable number of new rows).*

Proof. (2) ⇔ (3) follows from Proposition 3.1.6.

(1) ⇒ (2) is obvious.

(2)⇒ (1). We prove the result by induction on m. This is clear for $m = 1$ by (2). Assume the result for $m - 1$, and let P be finitely generated stably free of type m. Then $P \oplus R^m \cong R^n$, i.e. $(P \oplus R^{m-1}) \oplus R \cong R^n$. This implies $P \oplus R^{m-1}$ is stably free of type 1. By (2) it is free. Thus $P \oplus R^{m-1} \cong R^k$ for some k. By induction P is free.   □

Definitions 3.1.12 a) *A ring which satisfies the above properties is called a (right)* **Hermite ring.**

b) *A (right unimodular) row which satisfies condition (iii) above is called* **completable.**

More generally, the reader can establish

Corollary 3.1.13 *The following properties of a ring R are equivalent:*

(1) *Any finitely generated stably free right R-module of rank $\geq t$ is free.*

(2) *Any finitely generated stably free right R-module of rank $\geq t$ and of type 1 is free.*

(3) *Any unimodular row of length $\geq t+1$ is completable.*

Lemma 3.1.14 *Let R be a commutative ring. Suppose $P \oplus R^{n-1} = R^n$, i.e. P is stably free of rank 1. Then $P \cong R$.*

Proof. Represent P as the solution space of a right invertible $(n-1 \times n)$-matrix M. We show that the maximal minors $b_1, \ldots, b_n$ of M generate the unit ideal. Suppose not; then there exists a maximal ideal m containing $b_1, \ldots, b_n$. Since $\overline{M} \in M_{n-1,n}(R/m)$ is right invertible, it has rank n-1. By linear algebra it can be completed to an invertible matrix $M^* \in GL_n(R/m)$. But by Laplace expansion of $\det(M^*)$ w.r.t. the last row we get $\det(M^*) = 0$ as all the maximal minors are in m. A contradiction.

By the above there are $a_1, \ldots, a_n \in R$ such that $a_1 b_1 + \ldots + a_n b_n = 1$. Hence we can complete M to a matrix of determinant 1 by adding a last row $a_1, \ldots, a_n$ with appropriate signs. Then by Proposition 3.1.6 this implies that P is free. Hence $P \cong R$.   □

Alternatively the above argument can be written neatly using exterior powers as follows:

Proof. $\bigwedge^n(P \oplus R^{n-1}) = \sum_{i+j=n}(\bigwedge^i P \otimes \bigwedge^j R^{n-1}) = \bigwedge^1 P \otimes \bigwedge^{n-1} R^{n-1}$ as $\bigwedge^i P = 0$ if $i > 1$, and $\bigwedge^j R^{n-1} = 0$ if $j \neq n-1$. Therefore, as $\bigwedge^{n-1} R^{n-1} \cong R$, $\bigwedge^n(P \oplus R^{n-1}) \cong \bigwedge^1 P \cong P$. But $P \oplus R^{n-1} = R^n$ and so $\bigwedge^n(P \oplus R^{n-1}) = \bigwedge^n R^n \cong R$. Hence $P \xrightarrow{\sim} R$.   □

3.2 Non- Free over Localized Polynomial Rings

Here we give a general method to find polynomials $f \in k[X_1, \ldots, X_n]$ and nonfree, stably free modules over $k[X_1, \ldots, X_n]_f$.

If I is an ideal, contained in the nilradical of a ring R, then by Proposition 2.3.5 every projective module over R/I can be lifted to one over R. This does not generally hold if I is only assumed to be contained in the Jacobson radical of R. But for stably free modules the situation is different.

Proposition 3.2.1 *Let A be a not necessarily commutative ring and $I \subset$ Jac(A) a two-sided ideal. Every (finitely generated) stably free A/I-module can be lifted to a stably free A-module.*

For simplicity we give here a short proof of this special case of Theorem 5.7.12

Proof. If Q is stably free over A/I, there is a split exact sequence

$$0 \longrightarrow Q \longrightarrow (A/I)^m \xrightarrow{\beta} (A/I)^n \longrightarrow 0$$

with a matrix β over A/I. Let $\alpha : A^m \to A^n$ be any lift of β. Since $I \subset$ Jac(A) and $\alpha(A^m) + IA^n = A^n$, by Nakayama's Lemma α is surjective as well. So with $P := \ker(\alpha)$ we get an exact sequence

$$0 \longrightarrow P \longrightarrow A^m \xrightarrow{\alpha} A^n \longrightarrow 0,$$

which splits, because A^n is free. So

$$0 \longrightarrow P \otimes A/I \longrightarrow (A/I)^m \xrightarrow{\beta} (A/I)^n \longrightarrow 0$$

is exact as well, which proves $Q \cong P \otimes A/I$. □

Proposition 3.2.2 *Let A be a (commutative) ring and I an ideal such that there exists a non-free, stably free A/I-module. Then there is an $f \in 1 + I$ and a non-free stably free A_f-module.*

Proof. Let $S := 1 + I$. then $S^{-1}I \subset$ Jac($S^{-1}A$) and $S^{-1}A/S^{-1}I \cong A/I$. So a stably free, nonfree A/I-module lifts to one, say P', over $S^{-1}A$ by the above proposition. Since the description of P' as a stably free module can be done by finitely many relations, there is already a stably free module P over A_f for some $f \in S$ with $S^{-1}P \cong P'$. □

One may apply this to $A = k[X_1, \ldots, X_n]$ to get a lot of examples of non-free projective modules over some A_f which are not extended from A, since every projective A-module is free, as we will see in the next chapter.

3.3 Action of $\mathbf{GL}_n(R)$ on $\mathbf{Um}_n(R)$

The group $\mathrm{GL}_n(R)$ of invertible $n \times n$ square matrices over R acts on the set $\mathrm{Um}_n(R)$ of unimodular rows in the following natural manner: If $v \in \mathrm{Um}_n(R)$, $\sigma \in \mathrm{GL}_n(R)$, then

$$v \mapsto v\sigma.$$

Note that if $w \in M_{1,r}(R)$ is such that $vw^t = 1$, then $v\sigma(w\sigma^{-1}{}^t)^t = 1$, and so $v\sigma \in \mathrm{Um}_n(R)$. Thus, the above map defines an action of $\mathrm{GL}_n(R)$ on $\mathrm{Um}_n(R)$. If $v' = v\sigma$ for some $\sigma \in \mathrm{GL}_n(R)$, then we write this as $v \sim v'$ or $v \sim_{\mathrm{GL}_n(R)} v'$. We will use this notation sometimes also if v, v' are not necessarily unimodular.

Let G be a subgroup of $\mathrm{GL}_n(R)$. We write $v \sim_G v'$, if there is a $\sigma \in G$ with $v\sigma = v'$.

Proposition 3.3.1 *The orbits of $\mathrm{Um}_n(R)$ under the $\mathrm{GL}_n(R)$-action are in one to one corespocorrespondencehe isomorphism classes of right R modules P for which $P \oplus R \cong R^n$. Under this correspondence orbit of $(1, 0, \dots, 0)$ corresponds to the free module R^{n-1}.*

Proof. To any $(b_1, \dots, b_n) \in \mathrm{Um}_n(R)$ we can associate $P = P(b_1, \dots, b_n)$, the solution space (i.e. kernel) of $(b_1, \dots, b_n) : R^n \to R$. Such a P is a typical module for which $P \oplus R \cong R^n$. Suppose

$$P(b_1, \dots, b_n) \cong P(c_1, \dots, c_n)$$

for another $(c_1, \dots, c_n) \in \mathrm{Um}_n(R)$. Then we can complete the following commutative diagram,

$$
\begin{array}{ccccccccc}
0 & \longrightarrow & P(b_1, \dots, b_n) & \longrightarrow & R^n & \xrightarrow{(b_1, \dots, b_n)} & R^1 & \longrightarrow & 0 \\
 & & \downarrow{\scriptstyle \beta} & & \downarrow{\scriptstyle \exists \alpha} & & \| & & \\
0 & \longrightarrow & P(c_1, \dots, c_n) & \longrightarrow & R^n & \xrightarrow{(c_1, \dots, c_n)} & R^1 & \longrightarrow & 0
\end{array}
$$

with a suitable isomorphism $R^n \xrightarrow{\alpha} R^n$. (Note that the rows are split exact). If $M \in \mathrm{GL}_n(R)$ denote the matrix of this isomorphism α we will have

$$(b_1, \dots, b_n) = (c_1, \dots, c_n)M.$$

Conversely suppose $(b_1, \dots, b_n) = (c_1, \dots, c_n)M$ for some $M \in \mathrm{GL}_n(R)$. Then the automorphism defined by M induces an isomorphism of the two kernels

$$P(b_1, \dots, b_n) \cong P(c_1, \dots, c_n).$$

$\square$

Corollary 3.3.2 *Let* $(b_1, \ldots, b_n) \in \mathrm{Um}_n(R)$. *The following statements are equivalent:*

(1) $(b_1, \ldots, b_n)$ *is completable;*

(2) $P(b_1, \ldots, b_n) \cong R^{n-1}$.

(3) $(b_1, \ldots, b_n) \sim (1, 0, \ldots, 0)$.

Proof. (2) $\Leftrightarrow$ (3) follows from Proposition 3.3.1.

(1) $\Rightarrow$ (3). $(b_1, \ldots, b_n) \in \mathrm{Um}_n(R)$ is completable to an invertible matrix $M' \in \mathrm{GL}_n(R)$. If $M'M = I_n$, then $e_1 M'M = (b_1, \ldots, b_n)M = e_1 I_n = e_1$, i.e. $(b_1, \ldots, b_n) \sim e_1$.

(3) $\Rightarrow$ (1): Suppose $(b_1, \ldots, b_n) = (1, 0, \ldots, 0)M$. Then M is a completion of $(b_1, \ldots, b_n)$ to a square invertible matrix. $\qquad\square$

3.4 Elementary Action on Unimodular Rows

Definitions 3.4.1 *Let A be a ring, $a \in A$ and $r > 0$ an integer.*

a) *For $i, j \leq r$ let e_{ij} denote the $r \times r$-matrix whose only non-zero entry is 1 on the (i, j)-th place, and define $E_{ij}^a := I_r + a e_{ij}$ for $a \in A$ and $i \neq j$. Such E_{ij}^a are called* **elementary matrices.** *They are clearly invertible, since $E_{ij}^a E_{ij}^{-a} = I_r$. (The exponential place of the a is justified by the rule $E_{ij}^a E_{ij}^b = E_{ij}^{a+b}$.)*

b) *Let $\mathrm{E}_r(A)$ denote the subgroup of $\mathrm{GL}_r(A)$, generated by all E_{ij}^a (where $i \neq j$ and $a \in A$). $\mathrm{E}_r(A)$ is called the* **elementary group.**

c) *By $\mathrm{diag}(a_1, \ldots, a_r)$ we denote the diagonal matrix with entries $a_1, \ldots, a_r$ along the diagonal.*

Remarks 3.4.2 a) Clearly any surjective ring homomorphism $A \to B$ induces a surjective group homomorphism $\mathrm{E}_r(A) \to \mathrm{E}_r(B)$. This so called 'lifting property' will be extremely useful for us. (Note that it does not hold generally for GL_r or SL_r.)

b) If A is commutative – such that $\mathrm{SL}_r(A)$ is defined – $\mathrm{E}_r(A) \subset \mathrm{SL}_r(A)$. These two groups do not always coincide, even not for all principal domains. Nevertheless they are equal in important cases, so for semilocal or Euclidean domains, as we will see below.

c) Multiplying a matrix on the right (resp. left) side by E_{ij}^a means to change its j-th column (resp. i-th row) by adding the i-th column times a (resp. a times the j-th row) to it.

We call these transformations **elementary (column, resp. row) transformations.**

d) By three elementary column transformations one can interchange two columns – up to a change of a sign. Namely let v, w denote the two columns, forget about the others and carry out the following elementary transformations

$$(v, w) \mapsto (v, w + v) \mapsto (-w, w + v) \mapsto (-w, v).$$

The same holds for rows.

This clearly implies e.g.

$$\begin{pmatrix} 0 & 1 \\ -1 & 0 \end{pmatrix} \in E_2(A), \text{ moreover } \begin{pmatrix} 0 & I_n \\ -I_n & 0 \end{pmatrix} \in E_{2n}(A).$$

Proposition 3.4.3 *If R is commutative, any diagonal matrix in $\mathrm{SL}_n(R)$ belongs to $\mathrm{E}_n(R)$.*

Proof. If $D = \mathrm{diag}(d_1, \dots, d_n)$ where d_i's are unit, we can factorize

$$D = \mathrm{diag}(d_1, d_1{}^{-1}, 1, \dots, 1) \cdot \mathrm{diag}(1, d_1, d_2, d_3, \dots, d_n).$$

By induction on n we are reduced to proving that $\mathrm{diag}(d, d^{-1}) \in \mathrm{E}_n(R)$. This is shown by bringing $\mathrm{diag}(d, d^{-1})$ to the identity matrix by elementary row and column transformations.

$$\begin{pmatrix} d & 0 \\ 0 & d^{-1} \end{pmatrix} \mapsto \begin{pmatrix} 0 & d^{-1} \\ -d & 0 \end{pmatrix} \mapsto \begin{pmatrix} 1 & d^{-1} \\ -d & 0 \end{pmatrix} \to \begin{pmatrix} 1 & d^{-1} \\ 0 & 1 \end{pmatrix} \mapsto I_2$$

$\square$

Remark 3.4.4 The above is a special case of Whitehead's Lemma:

If $\alpha \in \mathrm{GL}_n(R)$ then $\begin{pmatrix} \alpha & 0 \\ 0 & \alpha^{-1} \end{pmatrix} \in \mathrm{E}_{2n}(R)$. This is proved analogously:

$$\begin{pmatrix} \alpha & 0 \\ 0 & \alpha^{-1} \end{pmatrix} = \begin{pmatrix} 0 & -I_n \\ I_n & 0 \end{pmatrix} \begin{pmatrix} I_n & 0 \\ -\alpha & I_n \end{pmatrix} \begin{pmatrix} I_n & \alpha^{-1} \\ 0 & I_n \end{pmatrix} \begin{pmatrix} I_n & 0 \\ -\alpha & I_n \end{pmatrix}$$

Proposition 3.4.5 *If $(b_1, \dots, b_n) \in \mathrm{Um}_n(R)$ contains a right unimodular sub-row of shorter length (especially if one of the b_i is a unit or if one is 0), then $(b_1, \dots, b_n) \sim_{\mathrm{E}_n(R)} e_1$.*

Proof. Let $(b_{i_1}, \dots, b_{i_m}) \in \mathrm{Um}_m(R)$ be contained in $(b_1, \dots, b_n) \in \mathrm{Um}_n(R)$. Let $i \notin \{i_1, \dots, i_m\}$. Write $b_{i_1} a_{i_1} + \dots + b_{i_m} a_{i_m} = 1$ for some $b_{ij} \in R$ $1 \leq j \leq m$. After a sequence of elementary transformations we may change b_i to $b_i - (b_{i_1} a_{i_1} + \dots + b_{i_m} a_{i_m})(b_i - 1) = 1$. Further elementary transformations change the other b_j's to 0. Then we have $(b_1, \dots, b_n) \sim_{\mathrm{E}_n(R)} e_i$. Since $\begin{pmatrix} 0 & 1 \\ -1 & 0 \end{pmatrix} \in \mathrm{E}_2(R)$, we can show that $e_i \sim_{\mathrm{E}_n(R)} e_1$ and so $(b_1, \dots, b_n) \sim_{\mathrm{E}_n(R)} e_1$. $\square$

Proposition 3.4.6 *Let R be a Euclidean domain and $(a_1,\ldots,a_n) \in \mathrm{Um}_n(R)$ with $n > 1$. Then*

a) $(a_1,\ldots,a_n)$ *is elementarily equivalent to* $e_1 = (1,0,\ldots,0)$, *i.e.* $(a_1,\ldots,a_n) \sim_{\mathrm{E}_n(R)} e_1$.

b) *Consequently* $\mathrm{SL}_n(R) = \mathrm{E}_n(R)$.

Proof. a) Let $\delta : R \to \mathbb{N}$ be the Euclidean norm (also called Euclidean degree.) In the set of all $(a_1',\ldots,a_n')$, elementarily equivalent to $(a_1,\ldots,a_n)$, choose one with minimal $\delta(a_1')$. If $a_1' = 0$, we are through by Proposition 3.4.5. Otherwise $a_1'|a_j'$ for $j > 1$. Namely if, say, $a_1' \nmid a_2'$, write $a_2' = a_1'q + r$ with $\delta(r) < \delta(a_1')$. Then $(a_1',a_2',\ldots,a_n') \sim_{\mathrm{E}_n(R)} (a_1',r,a_3',\ldots,a_n') \sim_{\mathrm{E}_n(R)} (r,-a_1',a_3',\ldots,a_n')$, which contradicts the minimality of $\delta(a_1')$. But if $a_1'|a_j'$, then clearly $(a_1',\ldots,a_n') \sim_{\mathrm{E}_n(R)} (a_1',0,\ldots,0)$. Again we use Proposition 3.4.5.

b) As R is commutative ring $\mathrm{E}_n(R) \subset \mathrm{SL}_n(R)$. For the converse inclusion above let $M \in \mathrm{SL}_n(R)$. By (i) we can perform suitable elementary transformations to bring M to M_1 with first row $(1,0,\ldots,0)$. Now a sequence of row transformations brings M_1 to

$$M_2 = \begin{pmatrix} 1 & 0 \\ 0 & M' \end{pmatrix}$$

where $M' \in \mathrm{SL}_{n-1}(R)$. The proof now proceeds by induction on n. □

(Note that a field is a special case of a Euclidean domain. But of course, if R is a field, then the proposition nearly is trivial. Namely $(b_1,\ldots,b_n)$ over a field is unimodular, if and only if one of the b_i is non-zero, i.e. a unit. To transform $(b_1,\ldots,b_n)$ to $e_1 = (1,0,\ldots,0)$ in this case by elementary transformations, is the classical 'Gauss Algorithm'.)

Corollary 3.4.7 *Let $R = \prod_{i=1}^r k_i$ be a finite direct product of fields (or more generally Euclidean domains) k_i. Then the statements of the Proposition 3.4.6) also hold.*

Proof. Clearly one can identify: $\mathrm{GL}_n(R) = \prod_{i=1}^r \mathrm{GL}_n(k_i)$, i.e. one can regard an $\alpha \in \mathrm{GL}_n(R)$ as an r-tuple $(\alpha_1,\ldots,\alpha_r)$ with $\alpha_i \in \mathrm{GL}_n(k_i)$. So let $\alpha = (\alpha_1,\ldots,\alpha_r) \in \mathrm{GL}_n(R) = \prod_{i=1}^r \mathrm{GL}_n(k_i)$.

CLAIM: $\alpha \in \mathrm{E}_n(R) \iff \alpha_i \in \mathrm{E}_n(k_i)$ for $i = 1,\ldots,r$.

The implication '$\Longrightarrow$'being clear, let us show '$\Longleftarrow$'. So assume that every α_i is a product of elementary matrices, say

$$\alpha_i = \varepsilon_{i1} \cdots \varepsilon_{im_i}.$$

Every r-tuple $\varepsilon'_{ij} := (I_n, \dots, I_n, \varepsilon_{ij}, I_n, \dots, I_n)$ (with ε_{ij} in the i-th place) is an elementary matrix over $\prod_{i=1}^{r} k_i$. Then

$$\alpha = \varepsilon'_{11} \cdots \varepsilon'_{1m_1} \cdots \varepsilon'_{r1} \cdots \varepsilon'_{rm_r} \in \mathrm{E}_n(R).$$

Now let $(b_1, \dots, b_n)$ be unimodular over R and regard it as an r-tuple $\left((b_1, \dots, b_n)_1, \dots, (b_1, \dots, b_n)_r\right)$, where $(b_1, \dots, b_n)_i$ is unimodular over k_i. By Proposition 3.4.6 there are $\alpha_i \in \mathrm{E}_n(k_i)$ with $(b_1, \dots, b_n)_i \alpha_i = (1, 0, \dots, 0)_i$. For $\alpha := (\alpha_1, \dots, \alpha_r)$ we get $(b_1, \dots, b_n)\alpha = e_1$ and $\alpha \in \mathrm{E}_n(R)$ by the claim. $\qquad\square$

Corollary 3.4.8 *Let R be a semilocal ring and $v \in \mathrm{Um}_n(R)$ be a unimodular vector. Then $v \sim_{\mathrm{E}_n(R)} e_1$.*

Proof. Let $\mathfrak{m}_1, \dots, \mathfrak{m}_r$ be the maximal ideals of R and $J = \bigcap_{i=1}^{r} \mathfrak{m}_i$ its Jacobson radical. By the Chinese Remainder Theorem 1.5.13

$$R/J \simeq \prod_{i=1}^{r} R/\mathfrak{m}_i.$$

is a finite direct product of fields. Let 'overbar' denote 'modulo J'. By Corollary 3.4.7 there is an $\overline{\varepsilon} \in \mathrm{E}_n(\overline{R}$ with $\overline{v}\overline{\varepsilon} = \overline{e_1}$. We can lift $\overline{\varepsilon}$ to an $\varepsilon \in \mathrm{E}_n(R)$, and we get $v\varepsilon = (a_1, \dots, a_n)$ with $a_1 \equiv 1 (\mathrm{mod} J)$. So $a_1 \in R^\times$ and therefore $(a_1, \dots, a_n) \sim_{\mathrm{E}_n(R)} e_1$, which implies the corollary. $\qquad\square$

Clearly the same holds, if $R/\mathrm{Jac}(R)$ is a finite product of Euclidean domains.

Proposition 3.4.9 *Let R be a commutative ring with $(a_1, \dots, a_r, b_1, \dots, b_s) \in \mathrm{Um}_{r+s}(R)$; $s \geq 1$. Let $I = \sum Ra_i$ and $\overline{R} = R/I$. If*

$$(\overline{b}_1, \dots, \overline{b}_s) \sim_{\mathrm{E}_s(\overline{R})} (\overline{1}, \overline{0}, \dots, \overline{0}),$$

then $(a_1, \dots, a_r, b_1, \dots, b_s) \sim_{\mathrm{E}_{r+s}(R)} (1, 0, \dots, 0)$.

Proof. Lifting the elementary transformations which bring $(\overline{b}_1, \dots, \overline{b}_s)$ to $(\overline{1}, \overline{0} \dots, \overline{0})$ we get

$$(a_1, \dots, a_r, b_1, \dots, b_s) \sim_{\mathrm{E}_{r+s}(R)} (a_1, \dots, a_r, b_1', \dots, b_s')$$

where $(b_1', \dots, b_s') \equiv (1, 0, \dots, 0) (\mathrm{mod}\ I)$. Performing further elementary transformations we can bring $(a_1, \dots, a_r, b_1', \dots, b_s')$ to $(a_1, \dots, a_r, 1, 0, \dots, 0)$. So the row $(a_1, \dots, a_r, b_1, \dots, b_s)$ contains a right unimodular subrow of shorter length. Hence by Proposition 3.4.5 we get

$$(a_1, \dots, a_r, b_1, \dots, b_s) \sim_{\mathrm{E}_{r+s}(R)} e_1.$$

$\qquad\square$

3.5 Interesting Examples of Completable Rows

R. G. Swan and J. Towber in [101], while studying "cancellative" properties of projective modules over affine algebras, stumbled upon the following remarkable fact:

Given any unimodular 3-vector $(a, b, c) \in \mathrm{Um}_3(R)$ over a commutative ring R, then the unimodular 3-vector (a^2, b, c) can always be completed to an invertible matrix!

Later on in [41] M. Krusemeyer gave a delightful explanation as to why $(a^2, b, c) \in \mathrm{Um}_3(R)$ is always completable, and exhibited the identity

$$\det \begin{pmatrix} a^2 & b & c \\ -b - 2ac' & c'^2 & a' - b'c' \\ -c + 2ab' & -a' - b'c' & b'^2 \end{pmatrix} = (aa' + bb' + cc')^2 = 1$$

(Recall that the determinant of an alternating matrix is a square - now figure out M. Krusemeyer's explanation!)

The above theorem of Swan-Towber was also explained and dramatically generalized by A. Suslin in his doctoral thesis – see [92] – wherein he shows that if $(a_0, a_1, \ldots, a_r) \in \mathrm{Um}_{r+1}(R)$, then the unimodular vector $(a_0, a_1, a_2^2, \ldots, a_r^r)$ can always be completed to an invertible matrix! Before proving this we need the following

Lemma 3.5.1 *Let R be a commutative ring and let $a, b \in R$ such that $Ra + Rb = R$. Then*

$$\begin{pmatrix} a^n & 0 \\ 0 & I_{n-1} \end{pmatrix} = a\varepsilon + b\mu,$$

with some $\varepsilon \in E_n(R)$ and $\mu \in M_n(R)$.

Proof. Let $\overline{R} = R/Rb$, then $\overline{a} := (a \bmod b) \in \overline{R}^\times$. By Whitehead's Lemma 3.4.4

$$\begin{pmatrix} \overline{a}^n & 0 \\ 0 & I_{n-1} \end{pmatrix} = \begin{pmatrix} \overline{a} & & 0 \\ & \ddots & \\ 0 & & \overline{a} \end{pmatrix} \overline{\varepsilon} = \overline{a}\,\overline{\varepsilon},$$

where $\overline{\varepsilon} \in E_n(\overline{R})$. Let $\varepsilon \in E_n(R)$ be a lift of $\overline{\varepsilon}$. Then

$$\begin{pmatrix} a^n & 0 \\ 0 & I_{n-1} \end{pmatrix} = a\varepsilon + b\mu,$$

for some $\mu \in M_n(R)$. □

Proposition 3.5.2 *(M. P. Murthy) [63]*
Let R be a commutative ring and $(x_0, x_1, \dots, x_n) \in \mathrm{Um}_{n+1}(R)$. Suppose $(\overline{x}_0, \overline{x}_1, \dots, \overline{x}_{n-1})$ is completable over $\overline{R} = R/Rx_n$. Then $(x_0, \dots, x_{n-1}, x_n^n)$ is completable.

Proof. Since $(\overline{x}_0, \dots, \overline{x}_{n-1})$ is completable over $\overline{R} = R/Rx_n$, we have a $\overline{\varphi} \in GL_n(\overline{R})$ such that the first column of $\overline{\varphi}$ is $(\overline{x}_0, \dots, \overline{x}_{n-1})^t$. Let φ be a lift of $\overline{\varphi}$ in $M_n(R)$, whose first column is $(x_0, \dots, x_n)^t$. Since $b := \det(\varphi)$ is a unit modulo x_n, there are $a, c \in R$ with $ab - cx_n = 1$. Let φ' be the $n \times n$-matrix which arises from φ by multiplying its second column by a. Then for the adjoint matrix $\mathrm{adj}(\varphi')$ of φ' we get $\varphi' \cdot \mathrm{adj}(\varphi') = \det(\varphi')I_n = a\det(\varphi)I_n = abI_n$. So

$$\begin{pmatrix} \varphi' & cI_n \\ x_nI_n & \mathrm{adj}\ \varphi' \end{pmatrix} \begin{pmatrix} \mathrm{adj}\ \varphi' & -cI_n \\ -x_nI_n & \varphi' \end{pmatrix} = I_{2n}.$$

Hence $\begin{pmatrix} \varphi' & cI_n \\ x_nI_n & \mathrm{adj}\ \varphi' \end{pmatrix} \in GL_{2n}(R)$. Now by Lemma 3.5.1, there exist $\varepsilon \in E_n(R)$, $\mu \in M_n(R)$ such that

$$\begin{pmatrix} x_n^n & 0 \\ 0 & I_{n-1} \end{pmatrix} = x_n\varepsilon + ab\mu.$$

Let $\gamma = \mu\mathrm{adj}(\varphi')$. Then $\gamma\varphi' = \mu\mathrm{adj}(\varphi')\varphi' = ab\mu$. Now consider the matrix $\begin{pmatrix} I_n & 0 \\ \gamma & \varepsilon \end{pmatrix} \in GL_{2n}(R)$. We have

$$\begin{pmatrix} I_n & 0 \\ \gamma & \varepsilon \end{pmatrix} \begin{pmatrix} \varphi' & cI_n \\ x_nI_n & \mathrm{adj}\ \varphi' \end{pmatrix} = \begin{pmatrix} \varphi' & cI_n \\ \begin{bmatrix} x_n^n & 0 \\ 0 & I_{n-1} \end{bmatrix} & * \end{pmatrix}$$

It is now easy to make elementary row and column operations and transform this matrix to a matrix of the form $\begin{pmatrix} \alpha & 0 \\ 0 & I_{n-1} \end{pmatrix}$ with $\alpha \in GL_{n+1}(R)$ and $\alpha e_1^t = (x_0, \dots, x_{n-1}, x_n^n)^t$. $\qquad\square$

Theorem 3.5.3 *(A. Suslin) [92]*
Let R be a commutative ring and let $(x_0, \dots, x_n) \in \mathrm{Um}_{n+1}(R)$. Then $(x_0, x_1, x_2^2, \dots, x_n^n)$ is completable.

Proof. It follows by induction on n. If $n = 1$ and $x_0y_0 + x_1y_1 = 1$, then

$$\begin{pmatrix} x_0 & x_1 \\ -y_1 & y_0 \end{pmatrix}$$

is a completion.

If $n > 1$, by induction assumption $(x_0, x_1, x_2^2, \dots, x_{n-1}^{n-1})$ is completable modulo x_n. So Proposition 3.5.2 performs the induction step. $\qquad\square$

Note that this construction of A.A. Suslin leads to the same matrix as M. Krusemeyer's in the case $n = 2$.

3.6 Direct sums of a stably free module

In this section we shall prove an interesting theorem of M. Gabel, which was sharpened by T. Y. Lam.

Lemma 3.6.1 *[26] (M. Gabel)*
Let $\varphi : R^n \to R^m$ be an onto homomorphism. Let $P = \ker \varphi$. Assume that there is a basis $\{d_1, \dots, d_n\}$ on R^n such that, for some $k \leq n - m$, the images $\{\varphi(d_1), \dots, \varphi(d_k)\}$ generate R^m. Then $P \overset{\sim}{\to} R^{n-m}$.

Proof. Let $F = \Sigma_{i=1}^{k} Rd_i \overset{\sim}{\to} R^k$. Since $\{\varphi(d_1), \dots, \varphi(d_k)\}$ generate R^m, $P + F = R^m$. If K denotes the kernel of $\varphi \mid P : P \to R^n$, then $K = P \cap F$. Let $n - m = k + s$. We have two exact sequences.

$$0 \longrightarrow K \longrightarrow F \overset{\varphi}{\longrightarrow} R^m \longrightarrow 0$$

$$0 \longrightarrow K \longrightarrow P \longrightarrow P/K \longrightarrow 0$$

where $P/K \simeq R^n/F \simeq R^{n-k} \simeq R^{m+s}$. Thus,

$$P \simeq K \oplus R^{m+s} \overset{\sim}{\to} K \oplus R^m \oplus R^s \simeq F \oplus R^s \simeq R^{k+s} \simeq R,^{n-m}$$

as claimed.   □

Lemma 3.6.2 *(Whitehead's Lemma for Rectangular Matrices)*
Let $A \in M_{m,n}(R), B \in M_{s,t}(R)$. Assume that B has a right inverse. Let $A = (M, V)$ with $M \in M_{m \times s}(R), V \in M_{m,n-s}(R)$. Let P, Q denote the solution spaces of A, B. Then $P \oplus Q$ is isomorphic to the solution space of (MB, V).

Proof. Let $S \in M_{t,s}(R)$ with $BC = I_s$. $Q \oplus P$ is the solution space of $\begin{pmatrix} B & 0 \\ 0 & A \end{pmatrix}$
By elementary transformations:

$$\begin{pmatrix} B & 0 \\ 0 & A \end{pmatrix} = \begin{pmatrix} B & 0 & 0 \\ 0 & M & V \end{pmatrix} \mapsto \begin{pmatrix} B & -BC & 0 \\ 0 & M & V \end{pmatrix} = \begin{pmatrix} B & -I_s & 0 \\ 0 & M & V \end{pmatrix}$$

$$\mapsto \begin{pmatrix} B & -I_s & 0 \\ MB & 0 & V \end{pmatrix} \mapsto \begin{pmatrix} 0 & -I_s & 0 \\ MB & 0 & V \end{pmatrix}$$

The last matrix clearly has solution space isomorphic to that of (M, B, V). □

Corollary 3.6.3 *Let $A \in M_{m,n}(R)$, $B \in M_{n,t}(R)$, and B have a right inverse. Then the solution space of AB is isomorphic to the direct sum of the solution spaces of A and B.*

Theorem 3.6.4 *(T.Y.Lam)*
Let R be a commutative ring, and P be a stably free R-module with $P \oplus R^m \simeq R^n$, with $n > m$. Then, the r-fold sum $rP = P \oplus \ldots \oplus P$ (r times) is free for $r \geq m + m/(n - m)$.

Proof. Let $P = \ker(\varphi)$, where $\varphi : R^n \to R^m \to 0$ is an epimorphism. Let $A \in M_{m,n}(R)$ be the matrix of φ.

Let $A = (M, V)$, $M \in M_{m,m}(R)$, $V \in M_{m,n-m}(R)$. By Whitehead's Lemma 3.6.2

(1) $P \oplus P$ corresponds to $(M(M, V), V) = (M^2, MV, V)$.
(2) $P \oplus P \oplus P$ corresponds to (M^3, M^2V, MV, V), etc.

Let N_i denote the submodule of R^m generated by the column vectors of the matrix $(M^{i-1}, M^{i-2}V_1, \ldots, MV, V)$. By Cayley-Hamilton theorem M satisfies its characteristic polynomial of degree m. Consequently, $N_i = N_{m-1}$ for all $i \geq m - 1$.

Let $N = r(n - m) + m$ for some $r \geq m + m/(n - m)$. Let $\alpha = R^N \to R^m$ be given by the matrix $(M^r, M^{r-1}V, \ldots, MV, V) \in M_{m,N}(R)$, i.e. α is given by $\alpha e_j = j$-th column of this matrix. Clearly, $\ker(\alpha) \simeq rP$. Now,

$$
\begin{aligned}
R^m = \mathrm{Im}(\alpha) &= \sum_{i=1}^{m} \alpha(e_i)R + N_r \\
&= \sum_{i=1}^{m} \alpha(e_i)R + N_{m-1} \\
&= \sum_{i=1}^{m} \alpha(e_i)R + \sum_{j=p+1}^{N} \alpha(e_j)R
\end{aligned}
$$

where $N - p = (m-1)(n-m)$. Thus R^m is generated by images of $m + (N - p) \leq N - m$ basis vectors of R^n. By Gabel's lemma, $\ker \alpha \simeq rP$ is free. □

3.7 Projectives over $k[x_1, \ldots, x_n]$ are Stably Free

In the remaining part of this chapter we will show that finitely generated projective modules over the polynomial ring $k[x_1, \ldots, x_n]$, where k is a field, are stably free. This is a preparation for some of the proofs of Serre's Conjecture, which says that they are indeed free.

3.7.1 Schanuel's Lemma

We begin with Schanuel's Lemma:

Lemma 3.7.1 (Schanuel) *If P, P' are projective A-modules and we have exact sequences of A-modules,*

$$0 \longrightarrow K \longrightarrow P \overset{\alpha}{\longrightarrow} M \longrightarrow 0$$

$$0 \longrightarrow K' \longrightarrow P' \overset{\beta}{\longrightarrow} M \longrightarrow 0$$

then $P' \oplus K \cong P \oplus K'$.

Proof. Since P, P' are projective one has R-linear maps f, g so that

$$0 \longrightarrow K \longrightarrow P \overset{\alpha}{\longrightarrow} M \longrightarrow 0$$

$$g \Big\Updownarrow f$$

$$0 \longrightarrow K' \longrightarrow P' \overset{\beta}{\longrightarrow} M \longrightarrow 0$$

$\beta \circ f = \alpha$, $\alpha \circ g = \beta$.

Let $X \subset P \oplus P'$ be the pull-back of α, β, i.e.

$$X = \{(p, p') \in P \oplus P' \mid \alpha(p) = \beta(p')\}.$$

One has the diagram

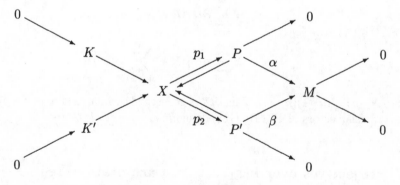

Here p_1, p_2 are the natural projections which are surjective as α, β are so. Their kernels are K', K respectively. Since P, P' are projective the crossed exact sequences splits. Hence $P \oplus K' \cong X \cong P' \oplus K$.   □

3.7.2 Proof of the Stable Freeness

Lemma 3.7.2 (R.G. Swan) *Let R be a subring of A. Let P, Q be projective A-modules and let $\alpha : Q \to P$, $\beta : P \to Q$ be monomorphisms. If*

A, $P/\alpha\beta P$, $Q/\beta\alpha Q$ are projective over R, then $P/\alpha Q$ and $Q/\beta P$ are also R-projective.

Proof. Clearly, P and Q are R-projective as they are direct summands of free A-modules and A is a direct summand of a free R-module.

We have exact sequences

$$
\begin{array}{ccccccccc}
0 & \longrightarrow & Q & \stackrel{\alpha}{\longrightarrow} & P & \longrightarrow & P/\alpha Q & \longrightarrow & 0 \\
 & & & & & & \| & & \\
0 & \longrightarrow & \alpha Q/\alpha\beta P & \longrightarrow & P/\alpha\beta P & \longrightarrow & P/\alpha Q & \longrightarrow & 0 \\
 & & \cong \uparrow \alpha & & & & & & \\
 & & Q/\beta P & & & & & &
\end{array}
$$

By Schanuel's Lemma, $P/\alpha\beta P \oplus Q \cong Q/\beta P \oplus P$. Hence, $Q/\beta P$ is R-projective. Similarly, $P/\alpha Q$ is R-projective. $\qquad\square$

Corollary 3.7.3 *Let R be a subring of A and $s \in A$ be a non-zero-divisor. Let P, Q be projective A-modules with $sP \subset Q \subset P$. If A and A/sA are R-projective then so is P/Q.*

Proof. Take α to be the inclusion map $Q \hookrightarrow P$ and β to be multiplication by s from $P \to Q$. Then $P/\alpha\beta P = P/sP$ and $Q/\beta\alpha Q = Q/sQ$ are A/sA-projective, and so R-projective. By Lemma 3.7.2 the module $P/Q (= P/\alpha Q)$ is R-projective. $\qquad\square$

Definition 3.7.4 *R-modules P, Q are called stably isomorphic, if $P \oplus R^m \cong Q \oplus R^m$ for some m.*

Theorem 3.7.5 (M. P. Murthy - C. Pedrini) *Let P, Q be finitely generated projective $R[x]$-modules. Suppose that $fP \subset Q \subset P$ for some monic polynomial $f \in R[x]$. then P and Q are stably isomorphic. In particular, if $P_f \cong Q_f$, then P and Q are stably isomorphic.*

Proof. Let $M = P/Q$. Since f is monic, $R[x]/(f)$ is free as an R-module. Therefore P/fP is R-projective, whence M is R-projective by Corollary 3.7.3. We have the exact sequences of $R[x]$-modules

$$0 \to Q \to P \to M \to 0$$

$$0 \to M[x] \xrightarrow{\cdot x} M[x] \to M \to 0,$$

where $M[x] = M \otimes_R R[x]$ is defined as in Example 1.2.3 h) and $\cdot x$ denotes the homothesy of x. By Schanuel's Lemma, $Q \oplus M[x] \cong P \oplus M[x]$. Since M is R-projective, $M[x]$ is $R[x]$-projective, and so the above isomorphism enables us to conclude that Q, P are stably isomorphic. $\qquad\qquad\square$

Theorem 3.7.6 (D. Hilbert) *Let P be a finitely generated projective $k[x_1, \ldots, x_n]$-module, where k is a field. Then P is stably free.*

Proof. We prove the result by induction on n, it being clear for $n = 0$. Let $S = k[x_n] \setminus \{0\}$. Then $S^{-1}k[x_1, \ldots, x_n] = k(x_n)[x_1, \ldots, x_{n-1}]$, where $k(x_n)$ is the quotient field of $k[x_n]$. By the inductive hypothesis $S^{-1}P$ is stably free. Therefore, $P_{f(x_n)}$ is stably free for some monic $f(x_n) \in k[x_n]$. By Remark 3.7.5 the module P is stably free. $\qquad\qquad\square$

Remark 3.7.7 Let $A = k[x_1, \ldots, x_n]$, k a field and M be a finitely generated A-module. D.Hilbert showed that every finitely generated A-module M has a finite free resolution of length $\leq n$

$$0 \to A^{\beta_n} \to \cdots \to A^{\beta_1} \to A^{\beta_0} \to M \to 0.$$

For projective M this implies

$$M \oplus \bigoplus_{i \text{ odd}} A^{\beta_i} \cong \bigoplus_{i \text{ even}} A^{\beta_i}.$$

Also it implies that for every multiplicative subset S of A every finitely generated projective $S^{-1}A$-module is stably free. We will show this. For a short and independent proof we refer the reader to Zariski and Samuel's book on Commutative Algebra [110].

4

Serre's Conjecture

In his paper [84] J-P. Serre posed the question whether any projective modules over the polynomial ring $k[x_1, \ldots, x_n]$ in several variables over a field is actually free. Later this was called Serre's conjecture. It was proved in 1976 independently by D. Quillen and A. Suslin by different methods. In this chapter we present four proofs, and present one more proof in Chapter 7. The reader can also find other interesting proofs in [94], [90].

In the whole chapter all rings are supposed to be *commutative*.

4.1 Elementary Divisors over Principal Domains

Theorem 4.1.1 *Let A be a principal domain and $\alpha \in M_{r \times s}(A)$. Then there are $\varepsilon \in \mathrm{SL}_r(A)$, $\varepsilon' \in \mathrm{SL}_s(A)$) such that $\varepsilon \alpha \varepsilon' = \mathrm{diag}(a_1, \ldots, a_t)$ with $a_i | a_{i+1}$, for $1 \leq i \leq t-1$, where $t := \mathrm{Min}\{r, s\}$.*

(We do not exclude the case that there is a $j \geq 1$ with $a_j = a_{j+1} = \cdots = a_t = 0$. It should be clear what we mean by $\mathrm{diag}(a_1, \ldots, a_t)$ also if $r \neq s$. It's not the geometric diagonal.)

The elements a_i are called the **elementary divisors** of α.

We will apply this theorem primarily in the case of a Euclidean ring A, where $E_n(A) = \mathrm{SL}_n(A)$ by Proposition 3.4.6. The proof of the theorem in the case of a Euclidean A is simpler and shows directly that one can choose $\varepsilon, \varepsilon'$ in the elementary groups then. Therefore we will give this proof first.

Proof. a) Let first A be Euclidean and δ the Euclidean norm. The claim is clear, if $\alpha = 0$ or $r = s = 1$. Otherwise consider the set $S := \{\sigma \alpha \tau \mid \sigma \in E_r(A), \tau \in E_s(A)\}$. In S choose a matrix $\alpha' = (a'_{ij})$ such that $a'_{11} \neq 0$ and $\delta(a'_{11}) \leq \delta(a''_{11})$ for all $\alpha'' = (a''_{ij}) \in S$ with $a''_{11} \neq 0$.

Then $a'_{11}|a'_{1j}$ and $a'_{11}|a'_{i1}$ for all i,j. For otherwise, using division with remainder, by an elementary transformation we could replace say a'_{1j} by $r = a'_{1j} - qa'_{11}$ with a suitable $q \in A$, such that $r \neq 0$ and $\delta(r) < \delta(a'_{11})$ and again – as shown in Remark 3.4.2 d) – by three elementary transformations we could replace a'_{11} by r. This would contradict the minimality of $\delta(a'_{11})$.

Now, using at most $r + s - 2$ elementary transformations, we may assume $a'_{1j} = a'_{i1} = 0$ for $i, j > 1$, i.e. that α' is of the form

$$\alpha' = \begin{pmatrix} a_1 & 0 \\ 0 & \beta \end{pmatrix}, \qquad a_1 = a'_{11}, \ \beta \in \mathrm{M}_{(r-1)\times(s-1)}(A).$$

By induction on $\mathrm{Max}\{r,s\}$ we see that β can be transformed to the form $\mathrm{diag}(a_2,\dots,a_t)$. It only remains to show that $a_1|a_2$. By an elementary transformation we get – looking only at the top left 2×2-matrix

$$\begin{pmatrix} a_1 & 0 \\ 0 & a_2 \end{pmatrix} \to \begin{pmatrix} a_1 & a_2 \\ 0 & a_2 \end{pmatrix}$$

Now we see $a_1|a_2$ by the same argument as above.

So we are done for Euclidean rings.

b) Now let A be a general principal domain. The proof in this general case is similar. Consider the set $S := \{\sigma\alpha\tau \mid \sigma \in \mathrm{SL}_r(A),\ \tau \in \mathrm{SL}_s(A)\}$. In S choose a matrix $\alpha' = (a'_{ij})$ such that Aa''_{11} is maximal among all ideals Aa''_{11} for $\alpha'' = (a''_{ij}) \in S$.

Again we claim that a'_{11} divides all entries in the first row and first column of α'. Assume e.g., it does not divide a'_{12}, so that $Ad \supsetneq Aa'_{11}$, if d is a g.c.d. of a'_{11} and a'_{12}. Let $d = b_1 a'_{11} + b_2 a'_{12}$. Then for

$$\tau := \begin{pmatrix} b_2 & -a'_{11}/d & 0 \\ b_1 & a'_{12}/d & 0 \\ 0 & 0 & I_{s-2} \end{pmatrix} \in \mathrm{SL}_s(A) \quad \text{we get} \quad \alpha'\tau = \begin{pmatrix} d & * \\ * & * \end{pmatrix}.$$

This contradicts the maximality condition on Aa'_{11}.

Then we proceed as done previously. $\square$

Corollary 4.1.2 *Let A be a principal domain.*

a) *Every submodule U of A^r is free of some rank $m \leq r$, and there are bases $x_1,\dots,x_r$ of A^r and $y_1,\dots,y_m$ of U, such that $y_i = a_i x_i$ for suitable $a_i \in A$, $i \leq m$.*

If A is even Euclidean, one may pass by elementary transformations from a given pair of bases of A^r and U to one which fulfills the above property.

b) *Every finitely generated A-module is of the form*

$$A/Aa_1 \oplus A/Aa_2 \oplus \cdots \oplus A/Aa_r$$

for some $a_i \in A$. ($a_i = 0$ is not excluded.)

Proof. Choosing a finite generating system of U one gets a linear map

$$A^s \to A^r \quad \text{given by a matrix } \alpha.$$

Changing bases means multiplying α on both sides by invertible matrices. The rest is clear. □

Remark 4.1.3 Things become much more difficult for A-modules which are not finitely generated For example $\mathbb{Q}$ is not a direct sum of cyclic groups.

Namely show $\text{Hom}_{\mathbb{Z}}(\mathbb{Q}, \mathbb{Z}/m) = 0$ for every $m \in \mathbb{Z}$. If $m \neq 0$, the group Hom is as well a $\mathbb{Q}$-vector space, as annihilated by m. For $m = 0$, if n were the smallest positive integer in the image of some non-zero map $f : \mathbb{Q} \to \mathbb{Z}$, say $n = f(a)$, what should be the image of $a/2$?

4.1.4 In Section 8.8 we will prove an analogous structure theorem for finitely generated modules over so called Dedekind rings, which were studied by Dedekind in connection with Fermat's Last Theorem and other Problems of Number Theory.

4.2 Horrocks' Theorem

We first prove some preparatory lemmas, which show that if $(R, \mathfrak{m})$ is a local ring then $\text{Spec}(R[x])$ can be written as a union of two principal open sets:

$$\text{Spec}(R[x]) = D(f) \cup D(g),$$

where $f \in R[x]$ is an arbitrary monic polynomial and $g \in R[x]$ is any polynomial all of whose coefficients, except for its constant term, are lying in $\mathfrak{m}$.

Lemma 4.2.1 *Let R be a ring, $a \in R$ and I an ideal of the polynomial ring $R[x]$. Assume that I contains a monic polynomial as well as an element of the form $1 + ah$ with $h \in R[x]$. Then I contains an element of the form $1 + ar$, for some $r \in R$.*

Proof. The ring $R[x]/I$ is integral over $R/(I \cap R)$, since I contains a monic polynomial. And $1 + ah \in I$ means that the residue class of a is invertible in $R[x]/I$, hence by Proposition 1.4.12 it is also invertible in $R/I \cap R$, which proves the statement. □

Corollary 4.2.2 *Let* $(R, \mathfrak{n})$ *be a local ring and let* $\mathfrak{m}$ *be a maximal ideal of* $R[x]$ *possessing a monic polynomial. Then* $\mathfrak{m} \cap R = \mathfrak{n}$. *Consequently,* $A = R[x]/(f)$ *is semilocal for any monic* f.

Proof. Assume, there were an $a \in \mathfrak{n} \backslash (\mathfrak{m} \cap R)$. Since $\mathfrak{m}$ is maximal $(\mathfrak{m}, a) = R[x]$, and so $1 + ah \in \mathfrak{m}$ for some $h \in R[x]$. By Lemma 4.2.1 we have $1 + a\lambda \in \mathfrak{m}$ for some $\lambda \in R$. But $1 + a\lambda$ is a unit as R is local. Therefore, $\mathfrak{m} \cap R = \mathfrak{n}$.

We note that $(R/\mathfrak{n})[x]/(\overline{f})$ has only finitely many prime ideals corresponding to the irreducible factors of $\overline{f} \in (R/\mathfrak{n})[x]$. Since every maximal ideal of $R[x]/(f)$ contracts to $\mathfrak{n}$ in R there is a bijective correspondence between maximal ideals of $R[x]/(f)$ and maximal ideals of $(R/\mathfrak{n})[x]/(\overline{f})$. The latter correspond to the irreducible factors of $\overline{f}$. This settles the rest. $\qquad\square$

Corollary 4.2.3 *Let* $(R, \mathfrak{n})$ *be a local ring and let* $f \in R[x]$ *be a monic polynomial. If* $h \in 1 + \mathfrak{n}[x]$ *then* f, h *are comaximal.*

Proof. Let $\mathfrak{m}$ be a maximal ideal of $R[x]$ containing f. Then by Corollary 4.2.2 $\mathfrak{m} \cap R = \mathfrak{n}$, i.e. $\mathfrak{n}[x] \subset \mathfrak{m}$. Therefore, any $h \in 1 + \mathfrak{n}[x]$ cannot lie in $\mathfrak{m}$. Thus, f, h are comaximal. $\qquad\square$

We now prove the main theorem of this section, due to G. Horrocks – see [33]. The monic inversion principle is one of the chief techniques in this subject.

Theorem 4.2.4 (Horrocks) *Let* R *be a local ring,* $A = R[x]$ *and* S *be the set of monic polynomials of* A. *If* P *is a finitely generated projective* A-*module such that* P_S *is free over* A_S, *then* P *is a free* A-*module.*

Proof. Since A has no non-trivial idempotents, P has a constant rank; say rk $P = n$. First we show by induction on n that P splits as a direct sum $L \oplus R^{n-1}$ for some rank-1-projective A-module L. This is clear for $n = 1$, so let $n \geq 2$.

Let $\mathfrak{m}$ be the maximal ideal of R and $k = R/\mathfrak{m}$. Let $p_1, \cdots, p_n \in P$ make up an A_S-basis of P_S. (Since S consists of non-zero-divisors of A we may regard P as a submodule of P_S and further may multiply any basis of P_S by a common denominator, thereby pushing it into P.) Now $\overline{P} = P/\mathfrak{m}[x]P = P \otimes_A k[x]$ is a projective, hence a free $k[x]$-module of rank n. So let $q_1, \ldots, q_n \in P$ be chosen in such a way that their residue classes $\overline{q_1}, \ldots, \overline{q_n}$ form a basis of $\overline{P} = P/\mathfrak{m}P = P \otimes_A k[x]$ over $k[x]$. By the Elementary Divisors Theorem 4.1.1 we can change $\overline{p_1}, \ldots, \overline{p_n}$ and $\overline{q_1}, \ldots, \overline{q_n}$ by elementary transformations such that – giving the new elements the old names – $\overline{p_1} = \overline{\alpha}\,\overline{q_2}$, for some $\alpha \in R[x]$. Since elementary transformations can be lifted, the new $\overline{p_1}, \ldots, \overline{p_n}$ have representatives $p_1, \ldots, p_n$ which form again a base of P_S.

Set $p = q_1 + x^r p_1$, with r to be specified later. Then $\bar{p} = \bar{q}_1 + x^r \bar{\alpha} \, \bar{q}_2$. So $\bar{p}, \bar{q}_2, \cdots, \bar{q}_n$ is a basis of $\bar{P}$. We have $sq_1 = \sum_{1 \leq i \leq n} a_i p_i$ for some $s \in S$, $a_1, \cdots, a_n \in A$. Thus $sp = (a_1 + sx^r)p_1 + \sum_{2 \leq i \leq n} a_i p_i$. Choose r large enough, such that $a_1 + sx^r$ is monic. (Recall that s is a monic polynomial.) Then $p, p_2, \cdots, p_n$ is an A_S-basis of P_S.

Let $T = 1 + \mathfrak{m}[x]$. Then $\mathfrak{m}$ is contained in $\mathrm{Jac}(A_T)$. Note that $\bar{P} = \bar{P}_T$. Since $\bar{p}, \bar{q}_2, \cdots, \bar{q}_n$ generate $\bar{P} = \bar{P}_T$, it follows from Nakayama's Lemma that $p, q_2, \cdots, q_n$ generate P_T. Since P_T is projective of rank n, by Proposition 2.2.15 we conclude that P_T is actually free with basis $p, q_2, \cdots, q_n$.

Now consider $P' = P/Ap$. Note that P'_S, P'_T are free of rank $n-1$. By Corollary 4.2.3 any maximal ideal $\mathfrak{n}$ of A either avoids S or avoids T. Thus $P'_{\mathfrak{n}}$ is free of rank $n-1$. By Proposition 2.2.8 the module P' is projective. Hence the exact sequence

$$0 \to Ap \to P \to P' \to 0$$

splits, i.e. $P \cong P' \oplus A$. By induction $P' \cong L \oplus A^{n-2}$ for some rank 1 projective A-module L. Hence $P \cong L \oplus A^{n-1}$.

It remains to show that L is free. We know that after inverting some monic f we have

$$R_f^{n-1} \oplus L_f \cong P_f \cong R_f^n,$$

i.e. L_f is stably free hence free by Lemma 3.1.14. The rest is done by the following lemma. $\qquad \square$

Lemma 4.2.5 *Let L be a rank-1-projective $R[x]$-module such that L_f is stably free for some monic polynomial $f \in R[x]$. Then L is free.*

Proof. L is stably free by Theorem 3.7.5. Since $\mathrm{rk} \, L = 1$ we know by Lemma 3.1.14 that L is free. $\qquad \square$

4.3 Quillen's Local Global Principle

This principle is the crux of the matter.

Definition 4.3.1 *Let $R \to A$ be a ring homomorphism. An A-module M is called **extended from** R, if there is an R-module N with $M \cong N \otimes_R A$.*

Especially, if $R \to R[x]$ is the canonical injection, an $R[x]$-module M is extended from R, if and only if $M \cong N[x]$ for some R-module N.

In the latter case $N \cong M/xM$. In general such a simple description of N by M is not possible.

Lemma 4.3.2 *Let A be a ring, $s, t \in A$ be comaximal, i.e. $As + At = A$, and M an A-module. Then the sequence*

$$0 \to M \xrightarrow{(i_{M,s}, \, i_{M,t})} M_s \oplus M_t \xrightarrow{(i_{M_s,t}, \, -i_{M_t,s})} M_{st}$$

is exact.

Note that module homomorphisms $M \to N_1 \oplus N_1$, resp. $M_1 \oplus M_2 \to N$ are in one to one correspondence with pairs (f_1, f_2) where $f_i : M \to N_i$, resp. $f_i : M_i \to N$ are homomorphisms.

Proof. The equality $As + At = A$ means that for every maximal ideal $\mathfrak{m}$ of A one has $s \notin \mathfrak{m}$ or $t \notin \mathfrak{m}$. Therefore for every $\mathfrak{m}$ both of $(i_{M,s})_\mathfrak{m}$, $(i_{M_t,s})_\mathfrak{m}$ or both of $(i_{M,t})_\mathfrak{m}$, $(i_{M_s,t})_\mathfrak{m}$ are isomorphisms. So one easily checks the exactness locally. □

Corollary 4.3.3 *Let A be a commutative ring and let $s, t \in A$ be comaximal. Let M, M' be two A-modules such that there are isomorphisms $\alpha : M_s \xrightarrow{\sim} M'_s$ and $\beta : M_t \xrightarrow{\sim} M'_t$ with $\alpha_t = \beta_s$ (as maps $M_{st} \to M'_{st}$). Then $M \cong M'$.*

If you wish, you may see D.0.2. □

Corollary 4.3.4 (Patching Local Isomorphisms) *Let A be a ring and $s, t \in A$ comaximal. Let M, M' be A-modules and $\alpha : M_s \xrightarrow{\sim} M'_s$ and $\beta : M_t \xrightarrow{\sim} M'_t$ be isomorphisms. If $(\alpha \circ \beta^{-1})_{st} = (\gamma_1)_t \circ (\gamma_2)_s$ for suitable $\gamma_1 \in \mathrm{Aut}\, M'_s$, $\gamma_2 \in \mathrm{Aut}\, M'_t$, then $M \cong M'$.*

Proof. Let $\alpha' = \gamma_1^{-1} \circ \alpha$, $\beta' = \gamma_2 \circ \beta$. Then $\alpha'_t = \beta'_s$. Now apply Corollary 4.3.3. □

The above corollary leads one to search for automorphism of M_{st} which factorize as a product of two automorphism as above. D. Quillen found an important class of such automorphism of an extended finitely generated module $M_{st}[x]$: Any automorphisms $\alpha(x) \in \mathrm{Hom}(M_{st}, M_{st})[x]$ of $M_{st}[x]$ with $\alpha(0) = \mathrm{id}_{M_{st}}$!

Lemma 4.3.5 *Let P be a finitely generated projective R-module. Then*

$$\mathrm{End}_{R[x]}(P[x]) \cong \mathrm{End}_R(P)[x]$$

and

$$\mathrm{End}_{R_s}(P_S) \cong \mathrm{End}_R(P)_S.$$

This is a special case of Proposition 1.6.11. We will give here an extra proof.

Proof. If $P \cong R^n$ is a free module then

$$\mathrm{End}_{R[x]}(R[x]^n) = \mathrm{M}_n(R[x]) = \mathrm{M}_n(R)[x] = \mathrm{End}_R(R^n)[x].$$

In general one has a natural map

$$\mathrm{End}_R(P)[x] \longrightarrow \mathrm{End}_{R[x]}(P[x]),$$

and this can be checked to be "locally" an isomorphism due to the above case. Hence by Corollary 1.3.16 it is an isomorphism.

Analogously one proves the second statement.   □

Remark 4.3.6 The above lemma also extends to finitely presented R-modules M. Namely for such modules M, by Proposition 1.6.11, we know that

$$\mathrm{Hom}_{R[x]}(M[x], N[x]) \cong \mathrm{Hom}_R(M, N)[x]$$

and

$$\mathrm{Hom}_{R_S}(M_S, N_S) \cong \mathrm{Hom}_R(M, N)_S$$

for all N.

Lemma 4.3.7 *Let S be a multiplicative subset of a ring R, P a finitely generated projective R-module and $\alpha(x) \in \mathrm{Aut}(P_S[x])$ with $\alpha(0) = \mathrm{id}_{P_S}$. Then there is an $s \in S$ such that $\alpha(sx)$ is in the image of the canonical homomorphism $\mathrm{Aut}(P[x]) \to \mathrm{Aut}(P_S[x])$. Consequently $\alpha(bx)$ belongs to this image for every $b \in Rs$.*

Proof. $\mathrm{End}_{R_S[x]}(P_S[x]) = \mathrm{End}_{R_S}(P_S)[x]$ by Lemma 4.3.5. Since $\alpha(x) \in \mathrm{End}_{R_S}(P_S, P_S)[x]$ is an automorphism with $\alpha(0) = id_{P_S}$, we can write

$$\alpha(x) = \mathrm{id}_{P_S} + x\alpha_1 + \cdots + x^n \alpha_n,$$
$$\alpha(x)^{-1} = \mathrm{id}_{P_S} + x\alpha'_1 + \cdots + x^n \alpha'_n,$$

for some $\alpha_i, \alpha'_i \in \mathrm{End}(P_S) = \mathrm{End}(P)_S$. As P is finitely generated, for each α_i, α'_i, there are $s_i, s'_i \in S$ such that $s_i \alpha_i, s'_i \alpha'_i$ are in the image H of $\mathrm{End}(P) \to \mathrm{End}(P_S)$.

Let s' be the product of all s_i, s'_i, then $s'\alpha_i, s'\alpha'_i \in H$ for all i. This implies that $\alpha(s'x), \alpha'(s'x)$ are images of say $\beta(x), \beta'(x)$ under the map $\mathrm{End}(P[x]) \to \mathrm{End}(P_S[x])$, where we may assume $\beta(0) = \beta'(0) = \mathrm{id}_P$. Let $\gamma(x) := \beta(x)\beta'(x)$. Then $\gamma(0) = \mathrm{id}_P$ and $\gamma(x)_S = \mathrm{id}_{P_S[x]}$, i.e. $\gamma(x) = \mathrm{id}_P + \gamma_1 x + \cdots + \gamma_m x^m$ with $\gamma_i \in \mathrm{End}_R(P)$ such that $t_i \gamma_i = 0$ for suitable $t_i \in S$. Therefore $\gamma(tx) = \mathrm{id}_{P[x]}$ for $t = \prod t_i$. Then $s = s't$ fulfills the statement of the lemma.   □

Lemma 4.3.8 (D. Quillen) *(Splitting Lemma)* *Let P be a finitely generated projective R-module and s_1, s_2 be comaximal elements of R. Let $\alpha(x) \in \mathrm{Aut}_{R[x]}(P_{s_1 s_2}[x])$ be an automorphism with $\alpha(0) = \mathrm{id}_{P_{s_1 s_2}}$. Then there are automorphisms $\alpha_1(x) \in \mathrm{Aut}(P_{s_1}[x])$, $\alpha_2 \in \mathrm{Aut}(P_{s_2}[x])$, with $\alpha_1(0) = \mathrm{id}_{P_{s_1}}$, $\alpha_2(0) = \mathrm{id}_{P_{s_2}}$, and with $\alpha(x) = \alpha_1(x)_{s_2}\alpha_2(x)_{s_1}$.*

Proof. Consider $\beta(x, y, z) := \alpha((z+y)x)\alpha(zx)^{-1} \in \mathrm{Aut}(P_{s_1 s_2}[x, y, z])$. Clearly $\beta(x, 0, z) = \mathrm{id}_{P_{s_1 s_2}[x,z]}$. Therefore by Lemma 4.3.7, there exists an $m \in \mathbb{N}$ such that,

(1) $\beta(x, s_2{}^m y, z) = \beta_1(x, y, z)_{s_2}$ for some $\beta_1(x, y, z) \in \mathrm{Aut}(P_{s_1}[x, y, z])$;

(2) there is an $\alpha_2(x) \in \mathrm{Aut}(P_{s_2}[x])$ with $\alpha(\lambda s_1{}^m x) = \alpha_2(x)_{s_1}$, for every $\lambda \in R$.

Note that there are $\lambda, \mu \in R$ with $\lambda s_1{}^m + \mu s_2{}^m = 1$. Also note that $\beta(x, s_2^m y, z) = \alpha((z + s_2^m y)x)\alpha(zx)^{-1}$. Now specialize $z \mapsto \lambda s_1{}^m$, $y \mapsto \mu$ to get

$$\alpha((\lambda s_1{}^m + s_2{}^m \mu)x)\alpha(\lambda s_1{}^m x)^{-1} = \alpha(x)\alpha(\lambda s_1{}^m x)^{-1} = \beta_1(x, \mu, \lambda s_1^m)_{s_2},$$

and let $\alpha_1(x) := \beta_1(x, \mu, \lambda s_1^m) \in \mathrm{Aut}(P_{s_2}[x])$. Then

$$\alpha(x) = \big(\alpha(x)\alpha(\lambda s_1{}^m x)^{-1}\big)\big(\alpha(\lambda s_1{}^m x)\big) = \alpha_1(x)_{s_2}\alpha_2(x)_{s_1}$$

as desired. $\qquad\qquad\qquad\qquad\qquad\qquad\qquad\qquad\qquad\qquad\qquad\qquad\square$

Theorem 4.3.9 (Quillen's Local Global Principle) *Let R be a ring and P a finitely generated projective module over $R[x]$. Suppose that $P_\mathfrak{m}$ is extended from $R_\mathfrak{m}$ for all $\mathfrak{m} \in \mathrm{Spmax}\, R$, then P is extended from R.*

Clearly by $P_\mathfrak{m}$ we mean the $R_\mathfrak{m}[x]$-module $S^{-1}P$, where $S := R \setminus \mathfrak{m}$.

Proof. Define $Q(P) := \{s \in R \mid P_s \text{ is extended from } R_s\}$. The crucial point is:

CLAIM: $Q(P)$ is an ideal of R. (It is called the **Quillen ideal** of P.)

PROOF: Clearly, if $s \in Q(P)$, $\lambda \in R$, then $\lambda s \in Q(P)$. So we need only prove that $s_1, s_2 \in Q(P)$ implies $s_1 + s_2 \in Q(P)$. i.e. we have to show that $P_{s_1 + s_2}$ is extended from $R_{s_1 + s_2}$, provided P_{s_1}, P_{s_2} are extended from R_{s_1}, resp. R_{s_2}. But the latter imply that $P_{s_1(s_1 + s_2)}$ and $P_{s_2(s_1 + s_2)}$ are extended as well. So we may rename $R_{s_1 + s_2}$ by R and $P_{s_1 + s_2}$ by P and we have to show that P is extended, provided P_{s_1} and P_{s_2} are so for comaximal s_1, s_2.

Consider the 'covering' diagram,

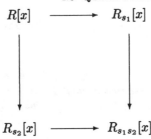

Let $\overline{P} = P/xP$ and generally the overbar mean 'modulo x'. We are given isomorphisms:

$$\varphi_1 : P_{s_1} \xrightarrow{\sim} \overline{P}_{s_1}[x], \qquad \varphi_2 : P_{s_2} \xrightarrow{\sim} \overline{P}_{s_2}[x]$$

We may assume that $\overline{\varphi_1} = \mathrm{id}_{\overline{P}} = \overline{\varphi_2}$ by replacing φ_1 by $\varphi_1 \circ (\overline{\varphi}_1^{-1} \otimes \mathrm{id}_{R_{s_1}[x]})$ and φ_2 by $\varphi_2 \circ (\overline{\varphi}_2^{-1} \otimes \mathrm{id}_{R_{s_2}[x]})$ respectively. $(\overline{\varphi_i} := \varphi_i \otimes_{R[x]} R)$

We are interested in patching these 'local' data to an isomorphism

$$\varphi : P \to \overline{P} \otimes R[x].$$

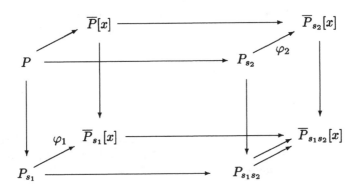

By the patching technique, described in Corollary 4.3.4, this will follow if the automorphism $\vartheta(x) = (\varphi_2)_{s_1} \circ (\varphi_1)_{s_1}^{-1} \in \mathrm{Aut}(\overline{P}_{s_1 s_2}[x])$ splits as a product

$$\vartheta(x) = \vartheta_1(x)_{s_2} \vartheta_2(x)_{s_1},$$

for $\vartheta_i \in \mathrm{Aut}(\overline{P}_{s_i}[x])$, $i = 1, 2$. Since $\vartheta(0) = \mathrm{id}_{\overline{M}_{s_1 s_2}}$, this will be the case by Lemma 4.3.8.

The claim suffices to establish the theorem: Let $\mathfrak{m}$ be a maximal ideal of R. As $P_{\mathfrak{m}}$ is extended from $R_{\mathfrak{m}}$, there is an $s \notin \mathfrak{m}$ such that P_s is extended from R_s. Hence $s \in Q(P)$. Thus $Q(P) \not\subset \mathfrak{m}$ for all $\mathfrak{m} \in \mathrm{Spmax}(R)$. Therefore $Q(P)$, being an ideal, must be the whole ring, i.e. $1 \in Q(P)$, i.e. P is extended. $\square$

Remark 4.3.10 The theorem also holds for any finitely presented $R[x]$-module P, since Lemma 4.3.5 does so.

Theorem 4.3.11 *Let* $S = S_0 \oplus S_1 \oplus S_2 \oplus \cdots$ *be a commutative graded ring and let* M *be a finitely presented S-module. Assume that for every maximal ideal* $\mathfrak{m}$ *of* S_0. $M_\mathfrak{m}$ *is extended from* $(S_0)_\mathfrak{m}$. *Then* M *is extended from* S_0.

Proof. We recall C. Weibel's homotopy trick: One has a ring homomorphism $S \to S[T]$ defined by

$$\varphi(S_0 + S_1 + S_2 + \cdots) = S_0 + S_1 T + S_2 T^2 + \cdots$$

and let $P = S[T] \otimes_\phi M$ be extension of M and P via φ. By assumption, for every maximal ideal $\mathfrak{m}$ of S_0, we have $M_\mathfrak{m} = S \otimes_{S_0} N_\mathfrak{m}$, where, obviously, $N = M/S_+ M$ with $S_+ = S_1 \oplus S_2 \oplus \cdots$. This implies that

$$P_\mathfrak{m} = S[T] \otimes_\varphi S \otimes_{S_0} N_\mathfrak{m} = S[T] \otimes_{S_0} N_\mathfrak{m} = S[T] \otimes_S (S \otimes_{S_0} N)_\mathfrak{m}.$$

Thus P is locally extended fom S via the canonical inclusions $S_\mathfrak{m} \to S_\mathfrak{m}[T]$. By Quillen's Local Global Principle, $P = S[T] \otimes_S Q$ where $Q \simeq P/(T-a)P$, for any $a \in S$. Since the composite homomorphism $S \to S[T] \to S[T]/(T-1) = S$ is the identity of S, we have $P/(T-1)P = S[T]/(T-1) \otimes_S S[T] \otimes_\varphi M = S \otimes_S M = M$. On the other hand, the composite homomorphism $S \overset{\varphi}{\to} S[T] \to S[T]/(T) = S$ is the same as $S \to S/S_+ = S_0 \to S$, hence

$$P/TP = S[T]/(T) \otimes_S P = S[T]/(T) \otimes_\varphi M = S \otimes_{S_0} S/S_+ \otimes_S M = S \otimes_{S_0} N$$

and this shows that $M \simeq S \otimes_{S_0} N$. $\square$

Theorem 4.3.12 (Quillen-Suslin Monic Inversion Principle) *Let* R *be a ring, P a finitely generated projective $R[x]$-module such that* P_f *is free over* $R[x]_f$ *for some monic polynomial* f *in* $R[x]$. *Then* P *is extended from* R.

Proof. Let $\mathfrak{m} \in \mathrm{Spmax}\, R$. Then $(P_\mathfrak{m})_f$ is a free $R_\mathfrak{m}[x]_f$-module. By Horrocks' Theorem 4.2.4, the $R_\mathfrak{p}$-module $P_\mathfrak{m}$ is free, hence extended from $R_\mathfrak{m}$. Therefore P is extended from R by Quillen's Local-Global Theorem 4.3.9. $\square$

Lemma 4.3.13 (Nagata Transformation) *Let* k *be a field and* f *a polynomial in* $k[x_1, \ldots, x_d]$. *There exists a change of variables*

$$t_1 \mapsto t_1, \quad t_i \mapsto t_i + t_1^{r_i} \quad (2 \leq i \leq d),$$

such that

$$f(t_1, t_2 + t_1^{r_2}, \ldots) = c \cdot h(t_1, \ldots, t_d)$$

where $c \in k^\times$ *and* h *is monic as a polynomial in* t_1 *over* $k[t_2, \ldots, t_d]$.

Proof. Let $f(t_1, \ldots, t_d) = \sum_{i \in \mathbb{N}^d} a_i {t_1}^{i_1} \cdots {t_d}^{i_d}$, so

$$f(t_1, t_2 + {t_1}^{r_2}, \ldots) = \sum_i a_i {t_1}^{i_1} (t_2 + {t_1}^{r_2})^{i_2} \cdots (t_d + {t_1}^{r_d})^{i_d}$$

$$= \sum_i a_i {t_1}^{i_1 + r_2 i_2 + \cdots + r_d i_d}$$

$$+ \text{ terms of lower degree in } t_1$$

We can choose $r_2, \ldots, r_d$ such that the integers $i_1 + r_2 i_2 + \cdots + r_d i_d$ are distinct for all the intervening d-tuples $i = (i_1, \ldots, i_d)$. In fact if m is an integer greater than all involved i_j, we may choose $r_j = m^{j-1}$, since then the integers $i_1 + r_2 i_2 + \cdots + r_d i_d$ will have different m-adic expansions. Having thus chosen the $r_2, \ldots, r_d$, the monomials $a_i {t_1}^{i_1 + r_2 i_2 + \cdots + r_d i_d}$ in $f(t_1, t_2 + {t_1}^{r_2}, \ldots)$ will not cancel out each other, and the one with the highest degree and with $a_i \neq 0$ will emerge as the leading term in $f(t_1, t_2 + {t_1}^{r_2}, \ldots)$, regarded as a polynomial in t_1. $\qquad\square$

We now give two proofs of Serre's Conjecture on the freeness of projective modules over a polynomial extension of a field.

Theorem 4.3.14 (Quillen - Suslin) *If k is a field, every finitely generated projective $k[x_1, \ldots, x_n]$-module is free.*

Proof. (Via the Monic Inversion Principle)

We prove the result by induction on the number n of variables. When $n = 1$ the result is known by Corollary 4.1.2. So let $n > 1$.

Assume that all finitely generated projective $k[x_1, \ldots, x_{n-1}]$-modules are free and let P be a finitely generated projective module over $k[x_1, \ldots, x_n]$. Let $S := k[x_n] \setminus (0)$.

Then $S^{-1} k[x_1, \ldots, x_n] = k(x_n)[x_1, \ldots, x_{n-1}]$, where $k(x_n) = S^{-1} k[x_n]$ is the quotient field of $k[x_n]$. Now $S^{-1} P$ is a projective module over a polynomial extension of $n - 1$ variables over the field $k(x_n)$. By induction $S^{-1} P$ is free. Since P is finitely generated there exists a monic $f \in S = k[x_n] \setminus (0)$, such that P_f is free. Therefore P is extended from $k[x_1, \ldots, x_{n-1}]$ by the Monic Inversion Principle 4.3.12, i.e. there exists a $P' \in \mathbb{P}(k[x_1, \ldots, x_{n-1}])$ such that $P \cong P' \otimes k[x_1, \ldots, x_n]$. By induction P' is free. Hence P is free. $\qquad\square$

Proof. (Via Quillen's Local-Global Principle)

By Theorem 3.7.6 we know that P is stably free. It therefore suffices to show that $k[x_1, \ldots, x_n]$ is an Hermite ring.

Let $v(x_n) = (v_1(x_n), \dots, v_r(x_n)) \in \mathrm{Um}_r(k[x_1, \dots, x_n])$. We can make a Nagata transformation of variables, fixing x_n, by Lemma 4.3.13 so that $v_1(x_n)$ is a monic polynomial in x_n with coefficients in $B := k[x'_1, \dots, x'_{n-1}]$, for some new variables x'_i, $1 \le i \le n - 1$.

We show that $v(x_n) \sim_{\mathrm{SL}_r(B[x_n])} v(0)$. By Quillen's Local-Global Principle 4.3.9 it suffices to show that $v(x_n)_{\mathfrak{m}} \sim_{\mathrm{SL}_r(B_{\mathfrak{m}}[x_n])} v(0)_{\mathfrak{m}}$, for every maximal ideal $\mathfrak{m}$ of B. This will follow from:

Lemma 4.3.15 *Let $(R, \mathfrak{m})$ be a local ring and $v(x) = (v_1(x), \dots, v_r(x)) \in \mathrm{Um}_r(R[x])$, $r \ge 3$. If $v_1(x)$ is a monic polynomial, then $v(x)$ can be completed to an elementary matrix.*

Proof: Write $S = R[x]$, let the 'overbar' denote 'modulo $(v_1(x))$' and $e_1 := (1, 0, \dots, 0)$.

$\overline{S}_{\mathfrak{m}}$ is a semilocal ring by Corollary 4.2.2. And $(\overline{v_2(x)}, \dots, \overline{v_r(x)}) \sim_{\mathrm{E}_{r-1}(\overline{S}_{\mathfrak{m}})} \overline{e}_1$ by Corollary 3.4.8. Let $\overline{\varepsilon} \in \mathrm{E}_{r-1}(\overline{S})$ be such that $(\overline{v_2(x)}, \dots, \overline{v_n(x)})\overline{\varepsilon} = \overline{e_1}$, and let $\varepsilon \in \mathrm{E}_{r-1}(S)$ be a lift of $\overline{\varepsilon}$. Write $(v_2(x), \dots, v_r(x))\varepsilon = (v'_2(x), \dots, v'_r(x))$. Now

$$v(x) \sim_{\mathrm{E}_r(R[x])} (v_1(x), v'_2(x), \dots, v'_r(x)) \sim_{\mathrm{E}_r(R[x])} e_1$$

the last assertion following as $v'_2(x) \equiv 1 \pmod{v_1(x)}$.   □

4.4 Suslin's Proof of Serre's Conjecture

He finds a fascinating method to analyse the action of invertible matrices on unimodular polynomial rows with a monic entry.

Lemma 4.4.1 *Let R be a ring, I an ideal of $R[t]$ which contains a monic polynomial and J an ideal of R such that $I + J[t] = R[t]$. Then $(R \cap I) + J = R$.*

Proof. Write $\overline{R} := R/(R \cap I)$ and $\overline{J}$ the image of J in $\overline{R}$. The ring extension $\overline{R} \hookrightarrow R[t]/I$ is integral, since I contains a monic polynomial.

If $(R \cap I) + J \ne R$, there would be a maximal ideal $\mathfrak{m}$ of $\overline{R}$ containing $\overline{J}$. By the Lying Over Theorem 1.4.18 and Proposition 1.4.16 there were a maximal ideal $\mathfrak{m}'$ of $R[t]/I$ over $\mathfrak{m}$. Its preimage in $R[t]$ would contain $I + J[t]$, contradicting the assumption.   □

Lemma 4.4.2 *Let R be a ring and $f = (f_1, f_2) \in R[t]^2$. Let $c \in R \cap (f_1 R[t] + f_2 R[t])$. Then for any commutative R-algebra A for which c is a non-zero-divisor and any $b, b' \in A$,*

$$b \equiv b' \pmod{cA} \implies f(b) \sim_{\mathrm{SL}_2(A)} f(b').$$

Proof. We can write $c = f_1 g_1 + f_2 g_2$ with $g_1, g_2 \in R[t]$. Over A_c define the matrix

$$M = \frac{1}{c} \begin{pmatrix} g_1(b) & -f_2(b) \\ g_2(b) & f_1(b) \end{pmatrix} \cdot \begin{pmatrix} f_1(b') & f_2(b') \\ -g_2(b') & g_1(b') \end{pmatrix}$$

It belongs to $\mathrm{SL}_2(A)$. Namely first $\det(M) = (1/c^2) \cdot c \cdot c = 1$. Then modulo cA, the product of the two matrices on the right hand side is congruent to 0. So the entries of M belong to A.

Now we have

$$\begin{aligned} f(b) \cdot M &= \frac{1}{c}(f_1(b), f_2(b)) \begin{pmatrix} g_1(b) & -f_2(b) \\ g_2(b) & f_1(b) \end{pmatrix} \cdot \begin{pmatrix} f_1(b') & f_2(b') \\ -g_2(b') & g_1(b') \end{pmatrix} \\ &= \frac{1}{c}(c, 0) \begin{pmatrix} f_1(b') & f_2(b') \\ -g_2(b') & g_1(b') \end{pmatrix} \\ &= (1, 0) \begin{pmatrix} f_1(b') & f_2(b') \\ -g_2(b') & g_1(b') \end{pmatrix} \\ &= (f_1(b'), f_2(b')) = f(b'). \end{aligned}$$

$\square$

Remark 4.4.3 The condition that c is a non-zero-divisor for A – the only case we need here – may be dropped. For this aim one can carry out a formal consideration, which is not quite obvious. (See Exercise 2.)

Lemma 4.4.4 *Let R be a ring and $f \in R[t]^n$. Then for any commutative R-algebra A and any subgroup $G \subset \mathrm{GL}_n(A)$*

$$I = I_{f,A,G} := \{c \in R \mid b \equiv b' \pmod{Ac} \Rightarrow f(b) \sim_G f(b')\}$$

is always an ideal in R.

Proof. Let $c, c' \in I$, $r, r' \in R$. To see that $rc + r'c' \in I$, let $b - b' = a(rc + r'c')$ where $b, b', a \in A$. Since $b - arc = b' + ar'c'$, we have $f(b) \sim_G f(b - (ar)c) = f(b' + (ar')c') \sim_G f(b')$. So $I_{f,A,G}$ is an ideal in R. $\square$

Theorem 4.4.5 (A. Suslin) *Let R be a ring and $f = (f_1, \ldots, f_n) \in \mathrm{Um}_n(R[t])$, $n \geq 2$, with f_1 monic. Then for any commutative R-algebra A which is a domain, and any $b, b' \in A$ we have $f(b) \sim_G f(b')$ where G is the subgroup of $\mathrm{GL}_n(A)$ generated by $\mathrm{E}_n(A)$ and $\mathrm{SL}_2(A)$.*

Proof. For f and G as in the statement of the theorem, we want to prove that the ideal $I_{f,A,G}$ in Lemma 4.4.4 is the unit ideal in R. So for any given maximal ideal $\mathfrak{m} \subset R$, it suffices to find an element $c \in I \setminus \mathfrak{m}$.

Note that $(\overline{f}_2, \ldots, \overline{f}_n) \in \mathrm{Um}_{n-1}(\overline{R[t]})$ where $\overline{R[t]} = (R/\mathfrak{m})[t]/(\overline{f}_1)$, which is a semilocal ring, since it is a residue class ring of a principal domain by a nonzero ideal. Thus by Corollary 3.4.8 there is an $\overline{M} \in \mathrm{E}_{n-1}(R[t]/(f_1) + \mathfrak{m}[t])$ such that $(\overline{f}_2, \ldots, \overline{f}_n)\overline{M} = (\overline{1}, \overline{0}, \ldots, \overline{0})$. Lift $\overline{M}$ to $M \in \mathrm{E}_{n-1}(R[t])$ and let

$$(f_2, \ldots, f_n)M = (g_2, \ldots, g_n) \in R[t]^{n-1}.$$

Then $g_2 \equiv 1 \,(\mathrm{mod}\ (f_1) + \mathfrak{m}[t])$, whence $f_1 R[t] + g_2 R[t] + \mathfrak{m}[t] = R[t]$. Since f_1 is monic, we can infer from Lemma 4.4.1 that $R \cap (f_1, g_2) + \mathfrak{m} = R$. In particular there exists an element $c \in R \cap (f_1, g_2)$, $c \notin \mathfrak{m}$. We will be done if we can show that $c \in I_{f,A,G}$. To check this let $b \equiv b' (\mathrm{mod}\ cA)$. For $i \geq 2$ we have

$$g_i(b) - g_i(b') \in (b - b')A \subset cA \subset f_1(b)A + g_2(b)A.$$

Thus via Lemma 4.4.2

$$f(b) = (f_1(b), \ldots, f_n(b)) \sim_{\mathrm{E}_n} (f_1(b), g_2(b), \ldots, g_n(b))$$
$$\sim_{\mathrm{E}_n} (f_1(b), g_2(b), g_3(b'), \ldots, g_n(b')) \sim_{\mathrm{SL}_2} (f_1(b'), g_2(b'), \ldots, g_n(b'))$$
$$\sim_{\mathrm{E}_n} (f_1(b'), f_2(b'), \ldots, f_n(b')) = f(b').$$

$\square$

Corollary 4.4.6 *Let R be a ring, $f(t) = (f_1(t), \ldots, f_n(t))$, $(n \geq 2)$ a unimodular row over $R[t]$ and $f_1(t)$ be monic. Then $f(t) \sim_G f(0)$ where G is the subgroup of $\mathrm{GL}_n(R[t])$ generated by $\mathrm{E}_n(R[t])$ and $\mathrm{SL}_2(R[t])$.*

Proof. Apply Theorem 4.4.5 to $A = R[t]$ and $b = t$, $b' = 0$. $\square$

Corollary 4.4.7 *If k is a field then every finitely generated projective module P over the polynomial ring $A := k[t_1, \ldots, t_d]$ is free.*

Proof. Since P is stably free by Theorem 3.7.6, it is enough to show that every unimodular row $(f_1, \ldots, f_n)$ over $k[t_1, \ldots, t_d]$ can be completed to an invertible matrix. (See Corollary 3.1.11.) This is clear, if $f_1 = 0$, by Proposition 3.4.5. If $f_1 \neq 0$, using some Nagata transformation we may assume that f_1 is monic in t_1 (upto a factor in $k^\times$). Then by Corollary 4.4.6

$$(f_1, \ldots, f_n) \sim_{\mathrm{SL}_n(A)} (f_1(0, t_2, \ldots, t_n) \ \cdots \ , f_n(0, t_2, \ldots, t_n)).$$

We finish by induction on d. $\square$

In Appendix A we give a proof of a deep theorem of A. Suslin that the elementary subgroup $\mathrm{E}_n(A)$ of $\mathrm{GL}_n(A)$ is a *normal* subgroup when $n \geq 3$. We use this fact to deduce sharper results.

Corollary 4.4.8 (R.A. Rao) *Let R be a ring, $f(x) = (f_1(x), \cdots, f_n(x)) \in$ $\mathrm{Um}_n(R[x])$ with $n \geq 3$ and $f_1(x)$ monic. Then $f(x)$ can be completed to an elementary matrix.*

We use a trick of S. Mandal [53]. (See the last Exercise on Appendix A for a Horrock's like argument.)

Proof. Consider $f^*(x, t) = (t^{d_i} f(x + t - t^{-1})) \in \mathrm{Um}_n(R[x,t])$, where $d_i = \deg f_i(x)$. (See Lemma 4.5.2). By Corollary 4.4.6 there is a $\sigma(x,t)$ in the group generated by $\mathrm{SL}_2(R[x,t])$ and $\mathrm{E}_n(R[x,t])$ such that $f^*(x,t)\sigma = e_1$. Since $\mathrm{E}_n(R[x,t])$ is a normal subgroup of $\mathrm{GL}_n(R[x,t])$ we may write $\sigma = \varepsilon(x,t)(I_{n-2} \perp \delta(x,t))$ for some $\varepsilon(x,t) \in \mathrm{E}_n(R[x,t])$, $\delta(x,t) \in \mathrm{SL}_2(R[x,t])$. Hence $f^*(x,t)\varepsilon(x,t)(I_{n-2} \perp \delta(x,t)) = e_1$, whence

$$f^*(x,t)\varepsilon(x,t) = e_1(I_{n-2} \perp \delta(x,t)^{-1}) = e_1.$$

Now put $t = 1$ to recover $f(x)\varepsilon(x,1) = e_1$. $\square$

Theorem 4.4.9 (Quillen - Suslin) *If k is a field and $A = k[t_1, \ldots, t_d]$ then any $f = (f_1, \ldots, f_n) \in \mathrm{Um}_n(A)$ with $n \geq 3$ can be completed to an elementary matrix.*

The proof is analogous to that of Corollary 4.4.7. We only use Corollary 4.4.8 to achieve the stronger result.

4.5 Vaserstein's Proof of Serre's Conjecture

Quillen's proof of Serre's conjecture does not use the fact that projective modules over $K[x_1, \ldots, x_n]$ are already stably free. But using this fact and Quillen's ideas, one obtains a remarkably short proof of Serre's conjecture, as L.N. Vaserstein wrote in a letter to H. Bass.

The next lemma is a special case of Lemma 4.3.7 when $P = R[x]^n$. Also the proof is the same as there. The only difference is that is is written in terms of matrices. Note that there is an obvious ring isomorphism $\mathrm{M}_n(R[x]) \xrightarrow{\sim} \mathrm{M}_n(R)[x]$.

Lemma 4.5.1 *Let R be a commutative ring and S be a multiplicatively closed subset of R. Let $\tau(x) \in \mathrm{GL}_n(R_S[x])$ be such that $\tau(0) = I_n$. Then there exists a matrix $\hat{\tau}(x) \in \mathrm{GL}_n(R[x])$ such that $\hat{\tau}(x)$ localizes to $\tau(sx)$ for some $s \in S$, (i.e. $\hat{\tau}(x)_s = \tau(sx)$) and $\hat{\tau}(0) = I_n$.*

Proof. Since $\tau(x) \in \mathrm{GL}_n(R_S[x])$, there exists $\mu(x) \in \mathrm{GL}_n(R_S[x])$ such that $\tau(x)\mu(x) = I_n$. Also as $\tau(0) = I_n$, we have $\mu(0) = I_n$.

Thus $\tau(x) = (\delta_{ij} + x f_{ij}(x))$ and $\mu(x) = (\delta_{ij} + x g_{ij}(x))$ where $f_{ij}(x), g_{ij}(x) \in R_S[x]$. Since there are only a finite number of denominators, we can find an $s_1 \in S$ such that $\tau(s_1 x)$ and $\mu(s_1 x)$ are images of matrices $\tau_1(x)$, resp. $\mu_1(x)$ over $R[x]$ with $\tau_1(0) = I_n = \mu_1(0)$.

Let $\tau_1(x)\mu_1(x) = (\delta_{ij} + x h_{ij}(x))$, then, since $(\tau_1 \mu_1)_S = I_n$, there is an $s_2 \in S$ with $s_2 h_{ij} = 0$ for all i, j. So $\hat{\tau}(x) := \tau_1(s_2 x) \in \mathrm{GL}_n(R)$ and localizes to $\tau(s_1 s_2 x)$. $\qquad\square$

Lemma 4.5.2 (S. Mandal) *a) Let $f(x) \in R[x]$ be a monic polynomial of degree d. Then $f^*(x, t) = t^d f(x + t - t^{-1}) \in (R[x])[t]$ is a bimonic polynomial in t (i.e. the leading coefficient and the constant term of $f^*(x, t)$, regarded as a polynomial in t over $R[x]$ are 1).*

b) Let $f(x) = (f_1(x), \dots, f_r(x)) \in \mathrm{Um}_r(R[x])$ with $f_1(x)$ monic. Then $f^*(x, t) = (t^{d_1} f_1(x + t - t^{-1}), \dots, t^{d_r} f_r(x + t - t^{-1}) \in \mathrm{Um}_r(R[x, t])$, where $d_i = \deg f_i$. Moreover, clearly $f^*(x, 1) = f(x)$.

Proof. a) Let $f(x) = x^d + a_1 x^{d-1} + \dots + a_d$, for $a_i \in R$, $1 \le i \le d$. From

$$f(x + t - t^{-1}) = (x + t - t^{-1})^d + a_1(x + t - t^{-1})^{d-1} + \dots + a_d.$$

it is clear that $t^d f(x + t - t^{-1}) = t^{2d} + \dots + 1$.

b) For this part note that $f^*(x) \in \mathrm{Um}_r(R[x, t, t^{-1}])$. Hence $t^l \in \sum_{i=1}^r t^{d_i} f_i^*(x, t)) R[x, t]$ for some $l \ge 0$. But $t^{d_1} f_1^*(x, t) = 1 \mod t R[x, t]$, whence $t^l R[x, t] + t^{d_1} f_1^*(x, t) R[x, t] = R[x, t]$. $\qquad\square$

Theorem 4.5.3 *If k is a field, any finitely generated projective module over $A = k[x_1, \dots, x_n]$ is free.*

Proof. (L.N. Vaserstein) We know by Theorem 3.7.6 that every projective module over $k[x_1, \dots, x_n]$ is stably free. By Lemma 3.1.11, it is enough to show that every unimodular row of length $r \ge 3$ over $k[x_1, \dots, x_n]$ is completable. Let $R = k[x_1, \dots, x_{n-1}]$ and $f(x_n) = (f_1(x_n), \dots, f_r(x_n)) \in \mathrm{Um}_r(R[x_n])$. After making a Nagata transformation of variables (see Lemma 4.3.13), fixing x_n, the polynomial $f_1(x_n)$ becomes monic. We claim that $f(x_n) \sim_{\mathrm{GL}_n(R[x_n])} f(0)$. Consider $f^*(x_n, t) \in \mathrm{Um}_r(R[x_n, t])$ defined in Lemma 4.5.2, Since $f^*(x_n, 1) = f(x_n)$ it suffices to show that $f^*(x_n, t)$ is completable. In particular we may assume (after renaming $f^*(x_n, t)$ as $f(t) \in \mathrm{Um}_r(A[t])$) that we have a unimodular row $f(t) = (f_1(t), \dots, f_r(t))$ with $f_1(t)$ monic and $f_1(0) = 1$. Using elementary transformations we may further ensure that $f_i(0) = 0$ for $i > 1$. By Lemma 4.3.15, if $\mathfrak{m} \in \mathrm{Spmax}(A)$ then $f(t)_{\mathfrak{m}} \sim_{\mathrm{E}_r(A_{\mathfrak{m}}[t])} e_1$. Therefore, there is an $s \in A \setminus \mathfrak{m}$ such that $f(t)_s \sim_{\mathrm{E}_r(A_s[t])} e_1$.

Let $Q(f(t)) = \{s \in A \mid f(t)_s \sim_{\mathrm{SL}_r(A_s[t])} e_1\}$ be the "Quillen ideal" of $f(t)$. As in Quillen's proof it is enough to show show that $Q(f(t))$ is indeed an ideal: For then it will follow from above that $Q(f(t)) = A[t]$, i.e. $1 \in Q(f(t))$, i.e. $f(t) \sim_{\mathrm{SL}_r(A[t])} e_1$ as required.

Clearly we need only show that $s_1, s_2 \in Q(f(t))$ implies $s_1 + s_2 \in Q(f(t))$. Further, after inverting $s_1 + s_2$, we may assume $s_1 + s_2 = 1$, and also $s_1, s_2 \in Q(f(t)) \setminus (0)$. Let $\alpha(t) \in \mathrm{SL}_r(A_{s_1}[t])$ with $f(t)\alpha(t) = e_1$. Since $f(0) = e_1$ we may modify $\alpha(t)$ and assume $\alpha(0) = I_r$. Let $\gamma(x, y, t) = \alpha((x+y)t)\alpha(yt)^{-1} \in \mathrm{SL}_r(A_{s_1}[x, y, t])$ and $\beta(t) \in \mathrm{SL}_r(A_{s_2}[t])$ such that $f(t)\beta(t) = e_1$ and $\beta(0) = I_r$. By Lemma 4.5.1 there is an $m > 0$ such that for all $b \in (s_1^m)$, $c \in (s_2^m)$,

$$\gamma(bx, y, t) \in \mathrm{SL}_r(A[x, y, t]), \qquad \beta(ct) \in \mathrm{SL}_r(A[x, y, t]).$$

(We consider A_{s_i} as subrings of the quotient field $Q(A)$, the s_i being nonzero in a domain.)

Note that

$$f((x+y)t)\gamma(x, y, t) = f((x+y)t)\alpha((x+y)t)\alpha(yt)^{-1}$$
$$= e_1\alpha(yt)^{-1} = f(yt).$$

Since $(s_1, s_2) = A$ there are $b \in (s_1^m)$, $c \in (s_2^m)$ with $b + c = 1$. Now, put $x = b$, $y = c$, then $\gamma(b, c, t) \in \mathrm{SL}_r(A[t])$, $\beta(ct) \in \mathrm{SL}_r(A[t])$, and

$$f(t)\gamma(b, c, t) = f(ct)\beta(ct) = e_1,$$

as required.   □

Continuous Vector Bundles

There are a large number of useful analogies and relations between algebra and topology. In this chapter we will describe one such relationship: that existing between projective modules and vector bundles. The main application of this so far have been to the construction of non-trivial examples of projective modules, the non-triviality being proved by passing to the associated vector bundle and using topological methods.

5.1 Categories and Functors

To give an adequate formulation of the statements in this chapter, we introduce here the language of categories and functors, which is extremely important in many fields in modern mathematics.

5.1.1 We begin with examples, one of a category and one of a functor.

The class of all R-modules over a fixed ring R together with the sets $\operatorname{Hom}_R(M, N)$ for all pairs of R-modules M, N and the compositions

$$\operatorname{Hom}_R(M, N) \times \operatorname{Hom}_R(L, M) \longrightarrow \operatorname{Hom}_R(L, N), \qquad (g, f) \mapsto g \circ f$$

make up a *category*.

Let $R \to S$ be a ring homomorphism. The map which assigns to every R-module M the S-module $S \otimes_R M$ and to every R-linear map $f : M \to N$ the S-linear map $\operatorname{id}_S \otimes_R f : S \otimes_R M \to S \otimes_R N$ is a *functor*.

Definitions 5.1.2 a) *A* **category** **C** *consists of*

(1) *a class of* **objects** *(sometimes denoted by* $\operatorname{Ob}(\mathbf{C})$ *),*

(2) *for every pair M, N of objects a set of* **morphisms** $\mathrm{Hom}_{\mathbf{C}}(M, N)$ *(sometimes one writes* $\mathrm{Mor}_{\mathbf{C}}(M, N)$ *or* $\mathbf{C}(M, N)$ *for this set),*

(3) *a specified element* $\mathrm{id}_M \in \mathrm{Hom}_{\mathbf{C}}(M, M)$ *for every object M,*

(4) *for every triple L, M, N of objects a map*

$$\mathrm{Hom}_{\mathbf{C}}(M, N) \times \mathrm{Hom}_{\mathbf{C}}(L, M) \longrightarrow \mathrm{Hom}_{\mathbf{C}}(L, N) \quad (g, f) \mapsto g \circ f,$$

such that
$$h \circ (g \circ f) = (h \circ g) \circ f, \ \text{and} \ \ \mathrm{id}_N \circ f = f, \quad f \circ \mathrm{id}_M = f,$$
whenever the left hand sides are defined.

b) *An* **isomorphism** *in a category is a morphism $f \in \mathrm{Hom}(M, N)$ for which there is a morphism $g \in \mathrm{Hom}(N, M)$ with $g \circ f = \mathrm{id}_M$ and $f \circ g = \mathrm{id}_N$. Objects M, N are called isomorphic if there exists an isomorphism in $\mathrm{Hom}(M, N)$. We write $M \cong N$ in this case. Clearly the relation '$\cong$' is an equivalence relation in the class of objects.*

Examples 5.1.3 a) Sets and maps. Note that Lord Russel's famous antinomy tells us that we cannot speak about the set of all sets. But to speak of the *class* of all sets does not lead to a contradiction. The inconsistency in the definition of the barber of the village, to be the man who shaves every man who doesn't shave himself, totally disappears, if the barber is a woman.
b) Topological spaces and continuous maps.
c) Rings (resp. groups, resp. R-modules) and ring (resp. group, resp. R-module) homomorphisms.
d) Vector bundles over a topological space X and bundle homomorphisms over X (which will be defined in the next section).
And so on.
e) If G is a group (or more generally a monoid) one can form a category with exactly one object, say X, and $\mathrm{Hom}(X, X) = G$, '$\circ$' being the product in G.

Definition 5.1.4 *Let* $\mathbf{C}$, $\mathbf{D}$ *be categories. A* **functor** $F : \mathbf{C} \longrightarrow \mathbf{D}$ *is a map which associates to every object M in $\mathbf{C}$ an object $F(M)$ in $\mathbf{D}$ and to every morphism $f \in \mathrm{Hom}_{\mathbf{C}}(M, N)$ in $\mathbf{C}$ a morphism $F(f) \in \mathrm{Hom}_{\mathbf{D}}(F(M), F(N))$ in $\mathbf{D}$, such that*

$$F(\mathrm{id}_M) = \mathrm{id}_{F(M)} \ \text{and} \ F(g \circ f) = F(g) \circ F(f) \ \text{if } g \circ f \text{ is defined.}$$

Note 5.1.5 One often writes $f : M \to N$ for $f \in \mathrm{Hom}_{\mathbf{C}}(M, N)$. The meaning of a commutative diagram in a category then is clear. A functor transforms isomorphisms into isomorphisms and commutative diagrams into commutative diagrams.

Example 5.1.6 Let A be a ring and E an A-module. Then the assignments $M \mapsto \mathrm{Hom}_A(E, M)$, $f \mapsto f_*$, (resp. $M \mapsto M \otimes_A E, f \mapsto f \otimes E$) make up a functor from the category of A-modules to that of abelian groups. (If A is commutative, these are also functors from the category of A-modules to itself.)

Definitions 5.1.7 *Let* $F : \mathbf{C} \longrightarrow \mathbf{D}$ *be a functor.*

a) F *is called* **faithful** *if* F *maps* $\mathrm{Hom}_{\mathbf{C}}(M, N)$ *injectively to* $\mathrm{Hom}_{\mathbf{D}}(F(M), F(N))$ *for every pair* M, N *of objects in* $\mathbf{C}$.

b) F *is called* **full** *if* F *maps* $\mathrm{Hom}_{\mathbf{C}}(M, N)$ *surjectively to* $\mathrm{Hom}_{\mathbf{D}}(F(M), F(N))$ *for every pair* M, N *of objects in* $\mathbf{C}$.

c) F *is called an* **equivalence** *if it is full and faithful and for every object* P *in* $\mathbf{D}$ *there is an object* M *in* $\mathbf{C}$ *with* $P \cong F(M)$.

Remark 5.1.8 If F is full and faithful and M, N are objects in $\mathbf{C}$, then $F(M) \cong F(N)$ implies $M \cong N$. So a full and faithful functor gives an injective map of the class of isomorphism classes of objects in $\mathbf{C}$ into that in $\mathbf{D}$.

An equivalence gives a bijective map of the class of isomorphism classes of objects in $\mathbf{C}$ to that in $\mathbf{D}$. Further, using the Axiom of Choice for classes, to an equivalence $F : \mathbf{C} \longrightarrow \mathbf{D}$ one can construct a functor $G : \mathbf{D} \longrightarrow \mathbf{C}$ such that $G(F(M)) \cong M$ and $F(G(P)) \cong P$ for every object M in $\mathbf{C}$ and P in $\mathbf{D}$.

Moreover, these isomorphisms can be chosen as so called **natural** isomorphisms. This means that for morphisms $f : M \to N$, $g : P \to Q$ in $\mathbf{C}$, resp. $\mathbf{D}$ one gets commutative diagrams

$$
\begin{array}{ccc}
G(F(M)) & \xrightarrow{\quad \sim \quad} & M \\
{\scriptstyle G(F(f))}\downarrow & & \downarrow{\scriptstyle f} \\
G(F(N)) & \xrightarrow{\quad \sim \quad} & N
\end{array}
\qquad
\begin{array}{ccc}
F(G(P)) & \xrightarrow{\quad \sim \quad} & P \\
{\scriptstyle F(G(g))}\downarrow & & \downarrow{\scriptstyle g} \\
F(G(Q)) & \xrightarrow{\quad \sim \quad} & Q
\end{array}
$$

Remark 5.1.9 A functor as defined above is often called **covariant**, since it preserves the "direction of arrows". A **contravariant functor** reverses it. This means it gives maps $\mathrm{Hom}_{\mathbf{C}}(M, N) \to \mathrm{Hom}_{\mathbf{D}}(F(N), F(M))$. Clearly one must require $F(g \circ f) = F(f) \circ F(G)$.

An example is the following: Fix an R-module E. Then one gets a contravariant functor F from the category of R-modules to that of abelian groups by $F(M) := \mathrm{Hom}_R(M, E)$ and $F(f) := f^*$, where f^* is defined as in Chapter 1 Section 4.

If $\mathbf{C}$ is a category one defines the **opposite** category $\mathbf{C}^{\mathrm{op}}$ as follows: $\mathbf{C}^{\mathrm{op}}$ has the same objects as $\mathbf{C}$ and $\mathrm{Hom}_{\mathbf{C}^{\mathrm{op}}}(M, N) := \mathrm{Hom}_{\mathbf{C}}(N, M)$. A contravariant functor $\mathbf{C} \longrightarrow \mathbf{D}$ is the same as a covariant functor $\mathbf{C}^{op} \longrightarrow \mathbf{D}$ or $\mathbf{C} \longrightarrow \mathbf{D}^{op}$. So in abstract category theory one may restrict oneself to covariant functors, and call them simply 'functors'.

5.2 Vector Bundles

5.2.1 We asked DALE HUSEMÖLLER to give us some information on the historical development of vector bundles theory. He wrote us a short note, which we – thankfully – reproduce here without change.

"The general concept of fibre bundles has its origins already in the 1930's with Seifert, Ehresmann and Whitney. Then in the 1940's came the book of Steenrod, and in 1949-1950 the Seminar of Henri Cartan containing a general introduction to fibre bundles through principal bundles.

In the 1950's differential topology, that is the topology of smooth, piecewise linear, and topological manifolds, had a great development. The first impulse came from Thom's 1953 thesis which reconsidered the theory of characteristic classes and his 1954 article on cobordism and transversality and the second impulse came from Milnor's 1956 discovery of exotic differential structures on the 7-sphere followed by his work on the Hauptvermutung.

Fibre bundles and especially vector bundles which are fibre bundles with a vector space V as fibre and structure group $GL(V)$ played a role everywhere in this development. Milnor gave courses at Princeton and wrote up notes, which were mimeographed and distributed in a small circle of people – sometimes supported by the Government Grants of the time. It was just before the Xerox machine and long before electronic servers.

There were two sets of notes of Milnor which had a wider influence, especially when the new Xerox machine made their reproduction easier. They were *Differential topology* and *Characteristic classes*. The second were notes of Milnor's lectures which were taken by Stasheff and which were eventually published with additions by Milnor and Stasheff in the Annals of Mathematics series. In *Differential topology* Milnor studies vector bundles directly without a reference to a general theory of fibre bundles, and in these and another set of notes he proved the homotopy classification theorem.

Then in 1957 came the K-theory in Grothendieck's general Riemann-Roch theorem. This was followed a year later by Bott's periodicity theorem which was very soon after interpreted as giving a new cohomology theory, called topological K-theory. This was introduced by Atiyah and Hirzebruch. The analytic implications were soon after discovered by Atiyah and Singer in the general index theorem.

The affine version of Grothendieck's K-theory lead to the study of the class groups of projective modules over a ring. Both Serre and Swan showed that finitely generated projective modules over the ring $C(X)$ of continuous complex (or real) valued functions on a compact space X came as modules of cross sections of vector bundles. So the vector bundles were related to finitely generated projective modules by an equivalence of categories."

In [103] L.N. Vaserstein gave a new very general version as well of this correspondence as well as that of a correspondence to homotopy classes of maps to Grassmannians, which we will present here.

5.2.2 We denote by $\mathbb{F}$ one of the skew-fields $\mathbb{R}$, $\mathbb{C}$ or $\mathbb{H}$ (of the real or complex numbers or quaternions). The reader may restrict his attention to the fields $\mathbb{R}$ and $\mathbb{C}$.

Recall that a finite dimensional $\mathbb{F}$-vector space V possesses a canonical topology. ($\mathbb{F}^n$ has the product topology, and since the automorphisms of $\mathbb{F}^n$ clearly are continuous all isomorphisms $\mathbb{F}^n \overset{\sim}{\to} V$ induce the same toplogy on V.)

Definitions 5.2.3 *Let X be a topological space.*

a) *A* **quasi vector bundle** *ξ over X consists of a topological space $E(\xi)$, a continuous map $p_\xi : E(\xi) \to X$, and the structure of a finite dimensional $\mathbb{F}$-vector space on every* **fibre** *$F_x(\xi) := p_\xi^{-1}(x)$ (where $x \in X$), in such a way, that the two topologies on every $F_x(\xi)$, one coming from the vector space structure, one induced by the topology of $E(\xi)$, coincide.*

$E(\xi)$ is called the **total space** *of ξ.*

The quasi vector bundle, given by $p : E \to X$ will sometimes be denoted by $[p : E \to X]$, sometimes, if possible, by E

b) *A* **homomorphism** *$f : \xi \to \xi'$ of quasi vector bundles over X is a continuous map (denoted by the same letter) $f : E(\xi) \to E(\xi')$, such that*

$$
\begin{array}{ccc}
E(\xi) & \overset{f}{\longrightarrow} & E(\xi') \\
\downarrow{\scriptstyle p_\xi} & & \downarrow{\scriptstyle p_{\xi'}} \\
X & \overset{\mathrm{id}}{\longrightarrow} & X
\end{array}
$$

commutes, and for every $x \in X$ the induced map $F_x(\xi) \to F_x(\xi')$ is $\mathbb{F}$-linear. Consequently an **isomorphism** *is a homomorphism which has an inverse, or equivalently which is bijective. It is then clear, what we mean by the phrase "ξ and η are isomorphic" or "ξ is isomorphic to η".*

c) *A quasi vector bundle over X is called a* **trivial** *vector bundle, if it is isomorphic to the quasi vector bundle $\mathrm{pr}_2 : V \times X \to X$, where V is a finite dimensional $\mathbb{F}$-vector space. (The vector space structure on the fibres of pr_2 should be clear.)*

d) *Let U be a subspace of X. The **restriction** $\xi|_U$ of a quasi vector bundle ξ over X to U is obviously defined, namely by the restriction of p_ξ to $p_\xi^{-1}(U) \to U$.*

e) *A **vector bundle** over X is a quasi vector bundle ξ which is locally trivial. By this we mean that every $x \in X$ has a neighbourhood U, such that the restriction $\xi|_U$ is a trivial vector bundle. **Homomorphisms** of vector bundles are those of quasi vector bundles.*

Remarks 5.2.4 a) Let X be a topological space. The vector bundles over X together with the vector bundle homomorphisms make up a category.

b) The **rank** or **dimension** $\mathrm{rk} : X \to \mathbb{N}$, $x \mapsto \mathrm{rk}_\mathbb{F} F_x(\xi) = \dim_\mathbb{F} F_x(\xi)$ clearly is locally constant for a vector bundle. (This does not hold for quasi vector bundles.)

Definition 5.2.5 *A vector bundle of constant rank 1 is called a **line bundle**.*

5.2.6 A vector bundle gives us data of the following kind. There is a covering $(U_i)_{i \in I}$ of X, such that $\xi_i := \xi|_{U_i}$ is trivial for every i, i.e. we may assume $E(\xi_i) = V_i \times U_i$, where V_i is an $\mathbb{F}$-vector space and $p_{\xi_i} = \mathrm{pr}_2$. If $U_i \cap U_j \neq O$ clearly $V_i \cong V_j$. For every $x \in U_i \cap U_j$ we get an isomorphism $\alpha_x : V_i \to V_j$. Choosing bases of every V_i we may assume $\alpha_x \in \mathrm{Gl}_n(\mathbb{F})$. So we have a map $U_i \cap U_j \to \mathrm{Gl}_n(\mathbb{F})$, $x \mapsto \alpha_x$.

CLAIM: This map is continuous.

To see this, it is enough to show, that every column of α_x depends continuously on x. Fix a basis in V_i, take its r-th vector e_r and consider

$$V_i \times (U_i \cap U_j) \longrightarrow p_\xi^{-1}(U_i \cap U_j) \longrightarrow V_j \times (U_i \cap U_j)$$

$$U_i \cap U_j \xrightarrow{\ \mathrm{id}\ } U_i \cap U_j \xrightarrow{\ \mathrm{id}\ } U_i \cap U_j$$

The maps of the first line are the the isomorphisms according to (5.2.3) b) and c). They are continuous. Therefore $(e_r, x) \mapsto (\alpha_x(e_r), x)$ is continuous, hence $x \mapsto \alpha_x(e_r)$ is so too.

5.2.7 On the other hand let X be a topological space, $(U_i)_{i \in I}$ a covering of it, V_i an $\mathbb{F}$-vector space for every $i \in I$ and continuous maps

$$\alpha^{ij} : U_i \cap U_j \to \mathrm{Isom}(V_i, V_j)$$

with $\alpha^{ii}(x) = \mathrm{id}_{V_i}$ for $x \in U_i$ and $\alpha^{jk}(x) \circ \alpha^{ij}(x) = \alpha^{ik}(x)$ for $x \in U_i \cap U_j \cap U_k$. (Here we wrote $\alpha_x^{ij} := \alpha^{ij}(x)$.)

From such data we construct a vector bundle ξ in the following way.

On the 'disjoint union' (or 'coproduct') $\bigsqcup_{i \in I}(V_i \times U_i)$ we define the equivalence relation "$\sim$" by $(v, x) \sim (w, y) \iff$ there are i, j such that $x = y \in U_i \cap U_j$, $v \in V_i$, $w \in V_j$, $\alpha^{ij}(x)(v) = w$.

Then $E(\xi) := \bigsqcup(V_i \times U_i)/\sim$ and the projection $p_\xi : E(\xi) \to X$ is defined to be the obvious one.

Clearly every fibre has a well defined vector space structure.

Further every map

$$V_i \times (U_i \cap U_j) \to V_j \times (U_i \cap U_j) \,, \ (v, x) \mapsto (\alpha^{ij}(x)(v), x)$$

is a homeomorphism. Namely it factorizes through $V_i \times \mathrm{Isom}(V_i, V_j) \times (U_i \cap U_j)$ in the following way:

$$(v, x) \mapsto (v, \alpha(x), x), \quad (v, \beta, x) \mapsto (\beta(v), x).$$

And both maps are continuous. Since $\alpha^{ii}(x) = \mathrm{id}_{V_i}$ and $\alpha^{ji}(x) \circ \alpha^{ij}(x) = \alpha^{ii}(x)$, the inverse is continuous too.

So, since the glueing of the $V_i \times U_i$ is done by homeomorphism, one sees, that the canonical maps $V_i \times U_i \to E(\xi)$ are embeddings. By this one easily derives the property of being a vector bundle.

Definition 5.2.8 *Kernel and image of a vector bundle homomorphism* $f : \xi \to \eta$ *are sensibly defined. Namely the* **kernel** $\ker(f)$ *is the union of the kernels* $\ker_x(f)$ *of the induced maps* $F_x(\xi) \to F_x(\eta)$ *together with the restricted map* $p_\xi : \bigcup_{x \in X} \ker_x(f) \to X$.

The **image** *is the image of the map* $f : E(\xi) \to E(\eta)$ *together with the restriction of* p_η *to* $f(E(\xi))$.

Clearly kernel and image are quasi vector bundles, but in general they need not be vector bundles. Consider e.g. the (trivial) bundle $\xi := [\mathrm{pr}_2 : \mathbb{F} \times \mathbb{F} \to \mathbb{F}]$ and the bundle homomorphism $\xi \to \xi$, given by $\mathbb{F} \times \mathbb{F} \to \mathbb{F} \times \mathbb{F}$, $(y, x) \mapsto (xy, x)$ – which is a notorious counterexample in mathematics.

But in important cases kernel and image are vector bundles, as the following lemma shows.

Lemma 5.2.9 *Let* $f : \xi \to \eta$ *be a vector bundle homomorphism (over a topological space X). The following assertions are equivalent:*

(1) *The rank of* $\ker(f)$ *is locally constant;*

(2) *the rank of* $\mathrm{im}(f)$ *is locally constant;*

(3) $\ker(f)$ *is a vector bundle;*

(4) $\operatorname{im}(f)$ *is a vector bundle;*

Proof. (1) $\Longleftrightarrow$ (2) is trivial, and (3) $\Rightarrow$ (1) and (4) $\Rightarrow$ (2) are so as well.

(1), (2) $\Rightarrow$ (4) and (3). Let $x \in X$ and U be a neighbourhood of x over which ξ and η are trivial, say $\xi|_U = [V \times U \to U]$ and $\eta|_U = [W \times U \to U]$. Let V' be a sub vector space – say of dimension r – of V which over x is mapped isomorphically to the image in W. Then V' is mapped injectively over some neighbourhood $U' \subset U$ of x. Namely, consider the map $f|V' \times U$ as a matrix, depending continuously on $y \in U$. By construction, its $r \times r$-minors over x do not all vanish. So they don't over a neighbourhood of x. (An $\mathbb{H}$-vector space is to be regarded as an $\mathbb{R}$-vector space to make this argument work.) Since the image has constant rank near x, it is isomorphic to $[\mathrm{pr}_2 : V' \times U' \to U']$. Hence (4) is established.

Now let $s_1, \ldots, s_n$ be a basis of V, such that $s_1, \ldots, s_k$ generate V'. Then for $i > k$ and $y \in U'$ near x we can write $f(s_i, y)$ in the form

$$f(s_i, y) = (\sum_{j=1}^{k} a_{ij}(y)s_j \, , \, y).$$

So $s_i'(y) := s_i - \sum_{j=1}^{k} a_{ij}(y)s_j$ for $i > k$ belongs to the kernel of f for every $y \in U'$. The $s_{k+1}'(y), \ldots, s_n'(y)$ are linearly independent and are in the right number. So they generate the kernel for $y \in U'$. $\qquad\square$

We will use this especially in the case when f is surjective hence (2) is fulfilled.

Definitions 5.2.10 a) *The ring of continuous functions* $f : X \to \mathbb{F}$ *will be denoted by* $\mathcal{C}(X)$.

b) *Let* ξ *be a vector bundle over the topological space* X. *A* **section** s *of* ξ *is defined to be a continuous map* $X \to E(\xi)$ *with* $p_\xi \circ s = \mathrm{id}_X$. *(This means* $s(x) \in F_x(\xi)$ *for every* $x \in X$.)

c) *The set of sections of* ξ *is denoted by* $\Gamma(\xi)$. *It is a* $\mathcal{C}(X)$-*module in a canonical way.*

d) *Let* $f : \xi \to \eta$ *be a vector bundle homomorphism over* X. *This is given by a 'fibre preserving' map* $f : E(\xi) \to E(\eta)$. *Therefore* $f \circ s \in \Gamma(\eta)$ *if* $s \in \Gamma(\xi)$. *Define* $\Gamma(f)(s) := f \circ s$. *Since* f *is fibrewise linear,* $\Gamma(f)$ *is a* $\mathcal{C}(X)$-*module homomorphism.*

e) *Let* $U \subset X$. *By a* **section over** U *we mean a section of the restricted bundle* $\xi|_U$. *We use the notion '* **global section** *', if we want to stress that we mean a section over all of* X.

Remarks 5.2.11 a) A vector bundle ξ over X is trivial, if and only if there are $s_1, \dots, s_n \in \Gamma(\xi)$ such that for every $x \in X$ the n-tuple $s_1(x), \dots, s_n(x)$ is a basis of $F_x(\xi)$.

b) Clearly Γ is a functor from the category of vector bundles over X to the category of $\mathcal{C}(X)$-modules.

c) If $\lambda = [\mathrm{pr}_2 : \mathbb{F}^n \times X \to X]$ is a trivial vector bundle, then $\Gamma(\lambda) \cong \mathcal{C}(X)^n$. And a global basis of ξ is a basis of $\Gamma(\xi)$ as a $\mathcal{C}(X)$-module.

Definition 5.2.12 *Let $\xi_1, \dots, \xi_n$ be finitely many vector bundles over the same space X. Their **direct sum** $\bigoplus_{j=1}^{n} \xi_j = \xi_1 \oplus \cdots \oplus \xi_n$ is defined by the total space $E(\bigoplus_{j=1}^{n} \xi_j) :=$*

$$\{(a_1, \dots, a_n) \in E(\xi_1) \times \cdots \times E(\xi_n) \mid p_{\xi_i}(a_i) = p_{\xi_j}(a_j) \text{ for all } i, j\}$$

and the obvious projection: $p_{\xi_1 \oplus \cdots \oplus \xi_n}(a_1, \dots, a_n) := p_{\xi_1}(a_1) \ (= p_{\xi_j}(a_j)$ for any j). For every fibre we have $F_x(\xi_1 \oplus \cdots \oplus \xi_n) = F_x(\xi) \times \cdots \times F_x(\xi_n)$. Define a quasi vector bundle structure by equipping the latter cartesian product with the direct sum structure in the usual way.

Remarks 5.2.13 a) If $\xi_1, \dots, \xi_n$ are all trivial vector bundles, then clearly $\bigoplus_{j=1}^{n} \xi_j$ is one.

b) If $(U_i)_{i \in I_1}, \dots, (U_i)_{i \in I_n}$ (where the I_j are assumed to be pairwise disjoint) be trivializing open coverings of ξ, resp. η. Then all ξ_j are trivial over every of the following open sets: $U_{i_1} \cap \dots \cap U_{i_n}$, $(i_1, \dots, i_n) \in I_1 \times, \dots \times I_n$. These sets make up a covering of X. This means that $\bigoplus_{j=1}^{n} \xi_j$ is a vector bundle.

c) If $\xi_1, \dots, \xi_n$ are subbundles of a vector bundle ζ such that over every $x \in X$ one has $F_x(\zeta) = F_x(\xi_1) \oplus \cdots \oplus F_x(\xi_n)$ then $\zeta \cong \xi_1 \oplus \cdots \oplus \xi_n$ in a canonical way.

d) A trivial vector bundle of rank n is a direct sum of n trivial vector bundles of rank 1.

e) $\Gamma(\bigoplus_{j=1}^{n} \xi_j) \cong \bigoplus_{j=1}^{n} \Gamma(\xi_j)$ in a canonical way.

5.2.14 The direct sum of vector bundles has some formal properties in common with the direct sum of modules.

For every $i = 1, \dots, n$ there is a canonical projection $p_i : \bigoplus_{j=1}^{n} \xi_j \to \xi_i$. There is a bijective correspondence between the vector bundle homomorphisms

$$f : \eta \to \bigoplus_{j=1}^{n} \xi_j,$$

with the n-tuples of homomorphisms

$$f_i : \eta \to \xi_i, \ i = 1, \ldots, n$$

given by $f \to p_i \circ f$

For every $i = 1, \ldots, n$ there is a canonical embedding $e_i : \xi_i \to \bigoplus_{j=1}^{n} \xi_j$. There is a canonical bijective correspondence of the vector bundle homomorphisms

$$f : \bigoplus_{j=1}^{n} \xi_j \to \eta$$

with the n-tuples of homomorphisms

$$f_i : \xi_i \to \eta, \ i = 1, \ldots, n$$

given by $f_i = f \circ e_i$.

5.2.15 Therefore every vector bundle homomorphism

$$f : \bigoplus_{j=1}^{n} \xi_j \to \bigoplus_{i=1}^{m} \eta_i$$

is uniquely described by an $m \times n$-matrix (f_{ij}) whose entries are homomorphisms

$$f_{ij} : \xi_j \to \eta_i.$$

And it is clear that

$$\Gamma(f) : \bigoplus_{j=1}^{n} \Gamma(\xi_j) \to \bigoplus_{i=1}^{m} \Gamma(\eta_i)$$

is uniquely described by the $m \times n$-matrix $(\Gamma(f_{ij}))$ whose entries are the homomorphisms

$$\Gamma(f_{ij}) : \Gamma(\xi_j) \to \Gamma(\eta_i).$$

A special case is that all ξ_j, η_i are trivial bundles of rank 1. Every vector bundle homomorphism

$$g : [\mathbb{F}^n \times X \overset{\mathrm{pr}_2}{\to} X] \to [\mathbb{F}^m \times X \overset{\mathrm{pr}_2}{\to} X]$$

is uniquely described by an $m \times n$-matrix whose entries are elements of $\mathcal{C}(X)$. And $\Gamma(g)$ is described by the same matrix.

5.3 Vector Bundles and Projective Modules

Definitions 5.3.1 a) *A finite partition of unity on a topological space X is a finite family (or set) of continuous functions $f_i : X \to [0,1]$ with $\sum_i f_i(x) = 1$ for every $x \in X$. (By $[0,1]$ we denote the closed unit interval.)*

b) **finite envelope of unity** *on a topological space X is a finite family of continuous functions $f_i : X \to [0,1]$ with $\mathrm{Max}_i\{f_i(x)\} = 1$ for every $x \in X$.*

c) *The* **support** $\mathrm{Supp}(f)$ *of a function $f : X \to \mathbb{R}$ is the closure of the set $\{x \in X \mid f(x) \neq 0\}$.*

d) *Let $\mathcal{U} = (U_i)_{1 \leq i \leq n}$ be a finite open covering of X. A partition of unity, resp. envelope of unity is called* **subordinate** *to $\mathcal{U}$, if and only if it is of the form $(f_i)_{1 \leq i \leq n}$ with $\mathrm{Supp}(f_i) \subset U_i$.*

e) *A* **locally finite partition of unity** *on X is a family of continuous functions $(f_i : X \to [0,1])_{i \in I}$ such that for every $x \in X$ there are only finitely many $i \in I$ with $f_i(x) \neq 0$ and $\sum_{i \in I} f_i(x) = 1$.*

Remark 5.3.2 Let $g_1, \dots, g_n$ be finitely many nonnegative continuous real functions on a topological space X such that

$$\text{for every } x \in X \text{ there is an } i \text{ with } g_i(x) > 0. \tag{5.1}$$

Define $s := g_1 + \cdots + g_n$, which clearly has no zero. Then apparently $\{g_1/s, \dots, g_n/s\}$ is a partition of unity on X.

Analogously define $s'(x) := \mathrm{Max}_i\{g_i(x)\}$ for every $x \in X$. Then $\mathrm{Max}\{g_1/s', \dots, g_n/s'\}$ is an envelope of unity.

Clearly any partition and any envelope of unity fulfil Equation (5.1). So the three concepts of a partition of unity, an envelope of unity and a family of nonnegative functions, fulfilling Equation (5.1) are equivalent in some sense.

Lemma 5.3.3 *Let $(f_i)_{1 \leq i \leq n}$ be a finite partition of unity on a topological space X. Then $\mathcal{U} := (U_i)_{1 \leq i \leq n}$ with $U_i := \{x \in X \mid f_i(x) \neq 0\}$ is an open covering of X and there is a new partition (and also a new envelope) of unity $(f'_i)_{1 \leq i \leq n}$, which is subordinate to $\mathcal{U}$.*

Proof. For every $x \in X$ there is at least one i with $f_i(x) > 0$. Therefore $\mathcal{U}$ is a covering. But note, if $U_i \neq (\overline{U_i})$ for some i, then $(f_i)_{i=1,\dots,n}$ is not subordinate to $[\mathcal{U}]$.

Define

$$h(t) := \mathrm{Max}\left\{0,\ t - \frac{1}{n+1}\right\}$$

(Remember: n is the number of the f_i.) Then $h(t) > 0$ for $t \geq 1/n$. So the $g_i := h \circ f_i$ have the property (5.1) since for every $x \in X$ at least for one $i \in \{1, \dots, n\}$ we have $f_i(x) \geq 1/n$, whence $g_i(x) > 0$. Further $\mathrm{Supp}(g_i) \subset \{x \in X \mid f_i \geq 1/(n+1)\} \subset U_i$. Remark 5.3.2 then proves the lemma. $\square$

Proposition 5.3.4 *Let ξ be a vector bundle on a topological space X. The following two properties are equivalent:*

(i) *There is a finite partition of unity $(f_i)_{1 \leq i \leq n}$ on X such that ξ, restricted to any $V_i = \{x \in X \mid f_i(x) > 0\}$, is trivial.*

(ii) *There is a finite open covering $\mathcal{U} = (U_i)_{1 \leq i \leq n}$ of X, such that ξ restricted to any U_i is trivial and there exists a partition of unity subordinate to $\mathcal{U}$.*

Proof. (i) $\Longrightarrow$ (ii) is Lemma 5.3.3.

(ii) $\Longrightarrow$ (i). Let $(f_i)_{1 \leq i \leq n}$ be a partition of unity, subordinate to $\mathcal{U}$. Define V_i as in (i). Then $(V_i)_{1 \leq i \leq n}$ is a finite open covering of X, as we know. Since $V_i \subset \overline{V_i} \subset U_i$, the bundle ξ restricted to any V_i is trivial. $\qquad\square$

Definition 5.3.5 *A vector bundle ξ on a topological space X is called* **soft** *(strongly of finite type, if and only if it has the properties* (i), (ii) *of the Proposition 5.3.4.*

Remarks 5.3.6 a) If X is a normal space and $\mathcal{U} = (U_i)_i$ any *finite* open covering of X, then there exists a partition of unity, subordinate to $\mathcal{U}$. See [38] Chapter 5, Problem W. (In our notion of 'normal' the Hausdorff condition is included.)

Therefore, if ξ is a vector bundle on a normal space X which is of **finite type**, i.e. if there is a finite open covering (U_i) of X, such that $\xi|_{U_i}$ is trivial for every i, then it is soft.

b) If X is a compact space, then every vector bundle over X is soft. (In our notion of 'compact' the Hausdorff condition is included.)

5.3.7 In this chapter we will use several times the possibility to equip $\mathbb{F}^n$ with an inner product $\langle \ , \ \rangle : \mathbb{F}^n \times \mathbb{F}^n \to \mathbb{F}$ by

$$\langle (x_1, \dots, x_n), (y_1, \dots, y_n) \rangle := \sum_{i=1}^{n} x_i \overline{y_i} \qquad (5.2)$$

where $\overline{y}$ equals y if $\mathbb{F} = \mathbb{R}$ and is the complex, resp. quaternionic conjugate of y if $\mathbb{F} = \mathbb{C}$, resp. $\mathbb{H}$. (Note that in the last case $\overline{yz} = \overline{z} \cdot \overline{y}$.) It should be clear what we mean by the conjugate of a matrix.

As usual, vectors $v, w \in \mathbb{F}^n$ are called **orthogonal** , if $\langle v, w \rangle = 0$. In this case we write $v \perp w$

We define the norm $|v|$ of any $v \in \mathbb{F}^n$ by $|v| := \sqrt{\langle v, v \rangle}$. (This is the usual Euclidean 'length' of v, if one identifies $\mathbb{F}^n$ with $\mathbb{R}^{kn}$, with $k = 1, 2$ or 4.) It is well known and easy to show that the topology, defined by this norm, on $\mathbb{F}^n$ coincides with the product topology.

An **orthonormal basis** of $\mathbb{F}^n$ is an n-tuple of vectors $v_1, \ldots, v_n$ in $\mathbb{F}^n$, such that $\langle v_i, v_j \rangle = \delta_{ij}$. (Kronecker's δ.) It is automatically a basis. The canonical basis of $\mathbb{F}^n$ is an orthonormal basis. If $w_1, \ldots, w_n$ is any basis of $\mathbb{F}^n$, then there is a canonical way to transform this into a basis, the **Gram - Schmidt orthogonalization**. Namely set first $v_1 := |w_1|^{-1}w_1$. Then assume that $v_1, \ldots, v_r$ is already an orthonormal basis of the space, generated by $w_1, \ldots, w_r$. If $r < n$, set first $v'_{r+1} := w_{r+1} - \sum_{i=1}^{r} \langle w_{r+1}, v_i \rangle v_i$. Then v'_{r+1} is orthogonal to all the $v_1, \ldots, v_r$. So, setting $v_{r+1} := |v'_{r+1}|^{-1}v'_{r+1}$, we get the orthonormal basis $v_1, \ldots, v_{r+1}$ of the space, generated by $w_1, \ldots, w_r, w_{r+1}$. We will use the fact that the Gram - Schmidt orthogonalization is a *continuous* map from the set of all bases of $\mathbb{F}^n$ to itself. Here the topology on the set of bases is given by considering a base as an n^2-tuple over $\mathbb{F}$. Check this!

5.3.8 To every endomorphism α of $\mathbb{F}^n$ there is a unique **adjoint** α^*, defined by the property $\langle \alpha v, w \rangle = \langle v, \alpha^* w \rangle$ for all $v, w \in \mathbb{F}^n$. With respect to the canonical basis and the above described canonical inner product, α^* as a matrix is the transposed conjugate of α. One has $(\beta \circ \alpha)^* = \alpha^* \circ \beta^*$.

Following a suggestion of S. Lang ([45] VIII, 7 'Terminology') we will call an $n \times n$-matrix σ over $\mathbb{F}$ **unitary**, if it describes an automorphism of $\mathbb{F}^n$ as an inner product space, i.e. if besides linearity it fulfills $\langle \sigma(x), \sigma(y) \rangle = \langle x, y \rangle$ for all $x, y \in \mathbb{F}^n$ in any of the three cases $\mathbb{F} = \mathbb{R}, \mathbb{C}$ or $\mathbb{H}$. A matrix σ is unitary if and only if it is invertible and $\sigma^{-1} = \sigma^*$.

One calls α **hermitian**, if and only if $\alpha = \alpha^*$. By the Spectral Theorem ([45] XV Theorem 6.7) an endomorphism α is hermitian if and only if there is a unitary σ such that

$$\sigma \alpha \sigma^* = \text{diag}(\lambda_1, \ldots, \lambda_n) \quad \text{with} \quad \lambda_j \in \mathbb{R} \tag{5.3}$$

5.3.9 The space $\text{Hom}(\mathbb{F}^m, \mathbb{F}^n)$ of all homomorphisms $\mathbb{F}^m \to \mathbb{F}^n$ carries the so called **operator norm** $| \ |$, defined by

$$|\alpha| := \text{Max}\left\{ |\alpha(v)| \ \Big| \ |v| = 1 \right\} = \text{Max}\left\{ |\alpha(v)|/|v| \ \Big| \ v \neq 0 \right\}$$

It defines a topology on $\text{Hom}(\mathbb{F}^m, \mathbb{F}^n)$ which coincides with the product topology if $\text{Hom}(\mathbb{F}^m, \mathbb{F}^n)$ is interpreted as the set of $n \times m$-matrices. It fulfills $|\alpha \circ \beta| \leq |\alpha||\beta|$. In case of a hermitian endomorphism, $|\alpha|$ equals the maximum of the absolute values of the eigenvalues of α.

If α is unitary, $|\alpha| = 1$.

5.3.10 Every linear subspace U of $\mathbb{F}^n$ is the image of a unique orthogonal projection $\varepsilon \in M_n(\mathbb{F})$.

CLAIM: $\varepsilon \in M_n(\mathbb{F})$ is an orthogonal projection if and only if $\varepsilon = \varepsilon^2 = \varepsilon^*$.

A **projection** ε is an endomorphism of $\mathbb{F}^n$ such that there is a splitting $\mathbb{F}^n = U \oplus U'$ with $\varepsilon = \mathrm{id}_U \oplus 0_{U'}$. We know by Paragraph 1.5.17 and Proposition 1.5.18 that ε is a projection, if and only if $\varepsilon = \varepsilon^2$.

An **orthogonal projection** is a projection such that for the U, U' as above the condition $U' = U^\perp$ holds. ($U^\perp := \{v \in \mathbb{F}^n \mid \langle v, u \rangle = 0 \text{ for all } u \in U\}$ means the orthogonal complement of U.)

If ε is an orthogonal projection with image U, then orthogonal bases of U and $U^\perp$ together make up an orthogonal base of $\mathbb{F}^n$. With respect to such a base, ε is described by the matrix $\mathrm{diag}(1, \dots, 1, 0, \dots, 0)$ and so clearly fulfills $\varepsilon = \varepsilon^*$.

Conversely from $\varepsilon = \varepsilon^*$ we derive $\langle \varepsilon(x), y \rangle = \langle x, \varepsilon(y) \rangle$ and hence have the equivalences

$$x \in \ker(\varepsilon) \Leftrightarrow \langle \varepsilon(x), y \rangle = 0 \text{ for all } y \Leftrightarrow \tag{5.4}$$

$$\langle x, \varepsilon(y) \rangle \text{ for all } y \Leftrightarrow x \in \mathrm{im}(\varepsilon)^\perp \tag{5.5}$$

So the claim holds.

Therefore the set $\mathrm{G}_k^n(\mathbb{F})$ of linear subspaces of dimension k of $\mathbb{F}^n$, the so called **Grassmannian** is in bijective correspondence to the set of idempotent hermitian $n \times n$-matrices over $\mathbb{F}$ of rank k.

We equip $\mathrm{G}_k^n(\mathbb{F})$ with the topology given by the operator norm on $M_n(\mathbb{F})$. And the set $\mathrm{G}^n(\mathbb{F})$ of all sub vector spaces of $\mathbb{F}^n$, i.e. the disjoint union of the $\mathrm{G}_k^n(\mathbb{F})$, gets the disjoint union topology.

Note that $|\varepsilon| = 1$ if $\varepsilon = \varepsilon^2 = \varepsilon^*$ and $\varepsilon \neq 0$. Hence $\mathrm{G}^n(\mathbb{F})$ is compact. (It is a Hausdorff space since $M_n(\mathbb{F})$ is one.)

Lemma 5.3.11 *Let ξ be a subbundle of a trivial vector bundle $\lambda = [\mathbb{F}^n \times X \overset{\mathrm{pr}_2}{\to} X]$. The orthogonal projection $\varepsilon(x) : \mathbb{F}^n \to F_x(\xi)$ is continuosly depending on x.*

Proof. Fix $x \in X$. There is an open neighbourhood U' of x such that $\xi_{|U'}$ is trivial. Let $t_1, \dots, t_k$ be a basis of ξ over U' and $s_1, \dots, s_n$ be the canonical basis of λ over X, hence over U'. There are $n-k$ sections under the $s_1, \dots, s_n$, say $s_{k+1}, \dots, s_n$ such that $t_1(x), \dots, t_k(x), s_{k+1}(x), \dots, s_n(x)$ is a basis of $F_x(\lambda)$. By the determinant argument of the proof of Lemma 5.2.9 we see that there is a neighbourhood $U \subset U'$ of x over which $t_1, \dots, t_k, s_{k+1}, \dots, s_n$ is a basis. The Gram-Schmidt orthogonalization, which depends continuously on $y \in U$, transforms the latter basis into a basis $t_1', \dots, t_k', s_{k+1}', \dots, s_n'$, which is an orthonormal basis of $\mathbb{F}^n$ over every $y \in U$. With respect to this basis the orthogonal projection to ξ is given by $\mathrm{diag}(1, \dots, 1, 0, \dots, 0)$ where k is the number of the 1's. With respect to the canonical basis this projection is given by $\sigma \mathrm{diag}(1, \dots, 1, 0, \dots, 0)\sigma^{-1}$ where the columns of σ are the $t_1', \dots, t_k', s_{k+1}', \dots, s_n'$ which are continuously dependent on $y \in U$. $\square$

Theorem 5.3.12 *If ξ is a soft vector bundle over the topological space X, then there are a trivial vector bundle λ and a vector bundle η over X with $\lambda \cong \xi \oplus \eta$.*

Proof. Let $(U_i)_{i=1,\ldots,n}$ be a finite covering of X together with a subordinate partition of unity $(f_i)_{i=1,\ldots,n}$ such that $\xi|_{U_i}$ is trivial and isomorphic to

$$[\mathrm{pr}_2 : V_i \times U_i \to U_i]$$

with vector spaces V_i. For any i we define a map from the trivial bundle $V_i \times X$ over X to ξ by

$$(v,x) \mapsto \begin{cases} (f_i(x)v, x) & \text{for } x \in U_i \\ (0,x) & \text{for } x \notin U_i \end{cases}$$

Together these define a map $\pi : (\bigoplus_{i=1}^n V_i) \times X \to E(\xi)$, which clearly is a surjective vector bundle homomorphism.

Its kernel η is a vector bundle since its image is a vector bundle, namely ξ. By Lemma 5.3.11 the orthogonal projection $\varepsilon : \lambda \to \eta$ is a vector bundle homomorphism. Since $\mathrm{im}(\varepsilon) = \eta$ is a vector bundle, its kernel is so, too. And we have $\lambda = \ker(\varepsilon) \oplus \eta$. Finally $\ker(\varepsilon)$ clearly is mapped isomorphically to ξ by π. □

Corollary 5.3.13 *If X is a topological space and ξ a soft vector bundle over X, then $\Gamma(\xi)$ is a finitely generated projective $\mathcal{C}(X)$-module .*

We will show the converse:

Proposition 5.3.14 *For every finitely generated projective $\mathcal{C}(X)$-module P there is a soft vector bundle ξ over X with $\Gamma(\xi) \cong P$*

Proof. There is an n and an idempotent endomorphism α of $\mathcal{C}(X)^n$ with $\alpha(\mathcal{C}(X)^n) \cong P$. (Compare this with Paragraph 1.5.17 and Proposition 1.5.18.)

We interprete α as a matrix whose entries are continuous $\mathbb{F}$-valued functions on X. So α may be considered as a vector bundle endomorphism of the n-dimensional trivial vector bundle $\lambda := [\mathrm{pr}_2 : \mathbb{F}^n \times X \to X]$. Let ξ denote its image (which *a priori* only is a quasi vector bundle).

CLAIM: ξ is a vector bundle.

We have to show its local triviality.

Let $x, y \in X$ and set $\gamma(x,y) := \alpha(x)\alpha(y) + (I - \alpha(x))(I - \alpha(y))$, where I denotes the identity. Since $\alpha(x)$, $\alpha(y)$ are idempotent, we see immediately

$$\alpha(x)\gamma(x,y) = \alpha(x)\alpha(y) = \gamma(x,y)\alpha(y)$$

Now fix x. Since $\gamma(x,x) = I$, there is an open neighbourhood U of x such that $\gamma(x,y)$ is invertible for every $y \in U$. So for $y \in U$ we have $\alpha(y) = \gamma(x,y)^{-1}\alpha(x)\gamma(x,y)$. So over U we get the vector bundle isomorphism

$$F_x(\xi) \times U \to \xi_{|U} \,, \quad (v,y) \mapsto (\ \gamma(x,y)^{-1}v \,,\ y\ )$$

(If $v \in \alpha(x)(\mathbb{F}^n)$, then $\gamma(x,y)^{-1}(v) \in \gamma(x,y)^{-1}\alpha(x)\gamma(x,y)(\mathbb{F}^n) = \alpha(y)(\mathbb{F}^n)$ and conversely.)

CLAIM: ξ is a soft vector bundle.

By Lemma 5.3.11, ξ is the image of an orthogonal projection, i.e. we have an $\varepsilon \in M_n(\mathcal{C}(X))$ with $\varepsilon = \varepsilon^2 = \varepsilon^*$ and $\varepsilon(\lambda) = \xi$, i.e. $\varepsilon(x) = F_x(\xi)$, where $\varepsilon(x) = \varepsilon(x)^2 = \varepsilon(x)^* \in M_n(\mathbb{F})$ depends continuously on x.

The set $G^n := \{\alpha \in M_n(\mathbb{F}) \mid \alpha = \alpha^2 = \alpha^*\}$ is compact. Clearly it is closed and $|\alpha| = 1$ or 0. So on G^n there is a finite partition of unity $f_1, \ldots, f_m$ such that $|\alpha - \beta| < 1/3$ whenever $f_i(\alpha)f_i(\beta) \neq 0$ for some i.

Considering ε as a continuous map $X \to G^n$, the $f_i \circ \varepsilon$ make up a finite partition of unity on X.

We will show that ξ is trivial on every $U_i := \{x \in X \mid f_i \circ \varepsilon(x) \neq 0\}$.

For $x,y \in X$ set as above $\gamma(x,y) := \varepsilon(x)\varepsilon(y) + (I - \varepsilon(x))(I - \varepsilon(y))$. We compute $\gamma(x,y) = 2\varepsilon(x)\varepsilon(y) + I - \varepsilon(x) - \varepsilon(y) = I + (\varepsilon(y) - \varepsilon(x))(1 - 2\varepsilon(y))$.

Now, if $x,y \in U_i$ by the definition of f_i and U_i we derive: $|\gamma(x,y) - I| \leq |\varepsilon(y) - \varepsilon(x)| \cdot |I - 2\varepsilon(y)| < \frac{1}{3} \cdot 3 = 1$ From this we conclude that $\gamma(x,y)$ is invertible for $x,y \in U_i$.

Now, for $x,y \in U_i$ we have $\varepsilon(y) = \gamma(x,y)^{-1}\varepsilon(x)\gamma(x,y)$. As above we have the triviality on U_i. $\square$

Theorem 5.3.15 *The functor Γ from the category of soft vector bundles over X and the category of finitely generated projective $\mathcal{C}(X)$-modules is an equivalence.*

Proof. Since we know already that Γ induces a bijective map of the isomorphism classes of the objects in these categories, it will be enough to show that it is full and faithful, i.e. Γ induces an isomorphism

$$\mathrm{Hom}_X(\xi,\eta) \xrightarrow{\sim} \mathrm{Hom}_{\mathcal{C}(X)}(\Gamma(\xi), \Gamma(\eta)),$$

where $\mathrm{Hom}_X(\xi,\eta)$ denotes the group of vector bundle homomorphisms $\xi \to \eta$.

CLAIM: Let λ_1, λ_2 be trivial vector bundles. Then the canonical map $\mathrm{Hom}_X(\lambda_1, \lambda_2) \to \mathrm{Hom}_{\mathcal{C}(X)}(\Gamma(\lambda_1), \Gamma(\lambda_2))$ is bijective.

Indeed homomorphisms $\lambda_1 \to \lambda_2$ and $\Gamma(\lambda_1) \to \Gamma(\lambda_2)$ both are described by matrices of the same shape whose entries are in $\mathcal{C}(X)$.

Then let $f : \xi \to \eta$ be any vector bundle homomorphism and $\xi \oplus \xi' = \lambda_1$, $\eta \oplus \eta' = \lambda_2$ be trivial bundles.

We may describe f as the composition

$$\xi \to \xi \oplus \xi' \xrightarrow{\begin{pmatrix} f & 0 \\ 0 & 0 \end{pmatrix}} \eta \oplus \eta' \to \eta$$

where the first and the last homomorphism are the canonical ones. Applying Γ we get

$$\Gamma(\xi) \to \Gamma(\xi) \oplus \Gamma(\xi') \xrightarrow{\begin{pmatrix} \Gamma(f) & 0 \\ 0 & 0 \end{pmatrix}} \Gamma(\eta) \oplus \Gamma(\eta') \to \Gamma(\eta) \qquad (5.6)$$

By the claim $\Gamma(f) \neq 0$, if $f \neq 0$. Also by the claim we see that to every module homomorphism $g : \Gamma(\xi) \to \Gamma(\eta)$ there is a bundle homomorphism

$$\xi \oplus \xi' \xrightarrow{\begin{pmatrix} h_{11} & h_{12} \\ h_{21} & h_{22} \end{pmatrix}} \eta \oplus \eta'$$

with $\Gamma(h_{11}) = g, \Gamma(h_{12}) = \Gamma(h_{21}) = \Gamma(h_{22}) = 0$. We know already that this implies $h_{12} = h_{21} = h_{22} = 0$.

So the surjectivity of the map $\text{Hom}_X(\xi, \eta) \to \text{Hom}_{\mathcal{C}(X)}(\Gamma(\xi), \Gamma(\eta))$ is also clear.   □

5.4 Examples

We construct several vector bundles, whose nontriviality can be proven by topological means. In certain cases we can give here complete proofs of the nontriviality. And we use these topological examples to find nonfree projective modules over Noetherian rings.

5.4.1 The Möbius Bundle. Let $A := \mathbb{R}[X, Y]/(X^2 + Y^2 - 1)$ be the real coordinate ring of the circle and x, y the residue classes of X, Y respectively. Consider the maximal ideal I associated to the point $(1, 0) \in \mathbb{R}^2$. It is generated e.g. by $x + y - 1$, $x - y - 1$. It is an invertible non-principal ideal as we have seen in Example 2.4.15 b) iii). Here we give a topological proof of the

CLAIM: I is not free, i.e. not principal.

Consider A as a subring of $\mathcal{C}(S^1)$ and set $J := (x+y-1)\mathcal{C}(S^1)+(x-y-1)\mathcal{C}(S^1)$. Let $P_+ := (0, 1)$ and $P_- := (0, -1)$. Then on $S^1 - \{P_+\}$ the module J is generated by $x + y - 1$, hence corresponds there to a trivial line bundle. Analogously on $S^1 - \{P_-\}$ it is generated by $x - y - 1$. The glueing of these trivial line bundles is given by $S^1 - \{P_+, P_-\} \to GL_1(\mathbb{R}) = \mathbb{R}^\times$, $(x, y) \mapsto (x + y - 1)/(x - y - 1)$. This function is < 0 on $\{(x, y) \in S^1 \,|\, x > 0\}$, and > 0 on $\{(x, y) \in S^1 \,|\, x < 0\}$. So we obtain the Möbius bundle, which is not trivial.

If it were so, it would have a continuous section without zeros. This would lead to continuous functions $S^1 - P_+ \to \mathbb{R}^\times$ and $S^1 - P_- \to \mathbb{R}^\times$, both of which could not change their sign. But, according to the glueing, one of them must

5.4.2 A projective module over $A/\mathrm{Jac}(A)$, which is not induced from one over A. Let X be any subset of $\mathbb{R}^n$ and define $S_X \subset \mathbb{R}[t_1, \ldots, t_n]$ to be the set of all polynomials which have no zeros along X. Then the maximal ideals of $S_X^{-1}\mathbb{R}[t_1, \ldots, t_n]$ belong to the points of X. (i.e. they are generated by n-tuples of the form $t_1 - x_1, \ldots, t_n - x_n$ with $(x_1, \ldots, x_n) \in X$.)

Namely let the ideal $I = (f_1, \ldots, f_r)$ of $\mathbb{R}[t_1, \ldots, t_n]$ be not contained in any of the maximal ideals attached to the points of X. Then $f_1^2 + \cdots + f_r^2 \in S \cap I$. So $S^{-1}I = S^{-1}\mathbb{R}[t_1, \ldots, t_n]$.

Apply this to $X := S^1 = \{(x, y) \in \mathbb{R}^2 \,|\, x^2 + y^2 = 1\}$. Let $A := S_X^{-1}\mathbb{R}[t_1, t_2]$. Then $\mathrm{Jac}(A)$ is generated by $t_1^2 + t_2^2 - 1$, and $A/\mathrm{Jac}(A) \cong \mathbb{R}[t_1, t_2]/(t_1^2 + t_2^2 - 1)$. Since A – as a localization of a factorial ring – is factorial, every projective A-module of rank 1 is free. Hence the Möbius module over $A/\mathrm{Jac}(A)$ is not induced by any projective A-module.

5.4.3 Tangent bundles on spheres. Let $S^n := \{a \in \mathbb{R}^{n+1} \,|\, |a| = 1\}$ be the n-sphere – embedded in $\mathbb{R}^{n+1}$ in the usual way.

Consider the section σ of the trivial real vector bundle $\lambda := [\mathbb{R}^{n+1} \times S^n \to S^n]$ defined by $\sigma(x) := (x, x)$. This generates a trivial subbundle ν of rank 1 of λ, the normal bundle of S^n in $\mathbb{R}^{n+}$. (Recall that the vector x is normal to the sphere in the point x.) The tangent bundle of S^n is the orthogonal complement τ of ν in λ, i.e.

$$E(\tau) = \{(y, x) \in \mathbb{R}^{n+1} \times S^n \,|\, \langle y, x \rangle = 0\} \qquad (5.7)$$

So $\lambda = \nu \oplus \tau$ and $F_\tau(x)$ is orthogonal to $F_\nu(x)$ in $F_\lambda(x) = \mathbb{R}^{n+1}$.

In spite of the fact that λ and ν are always trivial, the tangent bundle τ is only trivial in the cases $n = 1, 3, 7$. This is a deep theorem in topology, whose proof is out of our scope. But at least – due to Milnor – we are able to prove the nontriviality of τ if $n > 0$ is even – using only higher dimensional calculus. (Also see [8] for another accessible argument.)

If τ is a trivial vector bundle, there are n elements $e_1, \ldots, e_n \in \Gamma(\tau)$, such that $e_1(x), \ldots, e_n(x)$ are linear independent for every $x \in S^n$.

Therefore the following proposition is stronger than the nontriviality of the tangent bundle of an even dimensional sphere.

Proposition 5.4.4 *Let τ be the tangent bundle of an even dimensional sphere S^{2n}. There does not exist any $v \in \Gamma(\tau)$ (a so called* **vector field***) such that $v(x) \neq 0$ for every $x \in S^{2n}$.*

The hypothesis that the dimension of the sphere is *even*, is essential, as otherwise

$$\varphi(x_1, \ldots, x_{2n}) = (x_2, -x_1, \ldots, x_{2n}, -x_{2n-1}) \tag{5.8}$$

gives a differentiable nowhere vanishing vector field on S^{2n-1}.

Proof. (Milnor) Assume there is such a vector field, Using Weierstrass' Approximation Theorem we see that there exists even a continuously differentiable vector field v on S^{2n} – see [57]. Replacing v by $|v|^{-1}v$, we may and will assume that $|v(x)| = 1$ for every $x \in S^{2n}$. We can extend v throughout the region $A := \{x \in \mathbb{R}^{2n+1} \mid a \leq |x| \leq b\}$, where $0 < a < 1 < b$, by setting $v(ru) = rv(u)$ for $a \leq r \leq b$ and $u \in S^{2n} = \{u \in \mathbb{R}^{2n+1} \mid |u| = 1\}$

For $t \in \mathbb{R}$ consider the function $f_t : A \to \mathbb{R}^{2n+1}$, $f_t(x) := x + tv(x)$.

We will show:

a) f_t is injective for small values of t,

b) f_t, for small values of t, transforms the annulus A onto a nearby region $f_t(A)$ whose volume can be expressed as a polynomial function of t.

c) f_t, for small values of t, transforms A onto the region

$$\{x \in \mathbb{R}^{2n+1} \mid a\sqrt{1+t^2} \leq |x| \leq b\sqrt{1+t^2}\},$$

whose volume equals $(\sqrt{1+t^2})^{2n+1}\mathrm{Vol}\,(A)$.

This implies the proposition. Namely $(\sqrt{1+t^2})^{2n+1}$ is not a polynomial function of t, and so c) contradicts b).

Proof the above items: a) Since v is continuously differentiable on the compact region A it satisfies a **Lipschitz condition** , i.e. there is a constant $c > 0$ so that

$$|v(x) - v(y)| \leq c|x - y| \quad \text{for all } x, y \in A.$$

In particular if $|t| < c^{-1}$ then f_t is injective: Namely let $x \neq y$, but $f_t(x) = f_t(y)$. Then $x - y = t(v(y) - v(x))$, hence

$$|x - y| = |t| \cdot |v(x) - v(y)| \leq c|t| \cdot |x - y| < |x - y|$$

a contradiction.

b) We have

$$\det(\frac{\partial_i f_t}{\partial x_j}(x)) = \det(I + t\frac{\partial v_i}{\partial x_j}(x)) = 1 + t\sigma_1(x) + \cdots + t^{2n+1}\sigma_{2n+1}(x),$$

where the σ_k are continuous functions of x. The determinant is > 0 for small t. So

$$\text{Vol } f_t(A) = a_0 + a_1 t + \cdots + a_{2n+1} t^{2n+1},$$

with $a_k = \int_A \sigma_k(x) dx_1 \cdots dx_{2n+1}$.

c) For small values of t, the sphere S of radius r around 0 is mapped by f_t onto the sphere T of radius $r\sqrt{1+t^2}$ around 0. Namely, the matrix of first derivatives of f_t is non-singular on A. By the Inverse Function Theorem f_t is an open mapping on A. Hence $f_t(S^{2n})$ is a relatively open subset of the sphere of radius $r(\sqrt{1+t^2})$. But the sphere S is compact and so $f_t(S)$ is compact and so closed. Since T is connected, any nonempty open and closed subset of T equals T. □

Applying the functor Γ we see that $\Gamma(\tau)$ is a stably free $\mathcal{C}(S^n)$-module for all $n > 0$. It is nonfree for $n \neq 1, 3, 7$. We have shown this only for even n.

Further we have seen that $\Gamma(\tau)$ has a free direct summand of rank 1 if n is odd by Formula (5.8) and that it is free if $n = 1, 3$ or 7 in Example 3.1.9.

5.4.5 One can apply this to rings, which are more of 'algebraic kind' than $\mathcal{C}(S^n)$. Consider e.g. $A := \mathbb{R}[X_1, \ldots, X_{n+1}]/(X_1^2 + \cdots + X_{n+1}^2 - 1)$, which is the ring of polynomial (real) functions on S^n. This is a Noetherian subring of $\mathcal{C}(S^n)$. (See Chapter 6, Section 1.) Let x_i be the residue classes of the X_i in A and consider the row $(x_1, \ldots, x_{n+1})$, which is unimodular, since $\sum x_i^2 = 1$. It defines a stably free module P, which, tensoring with $\mathcal{C}(S^n)$, becomes $\Gamma(\tau)$ by Equation (5.7). So P is nonfree, if τ is nontrivial, especially for even n.

5.4.6 Let $B = \mathbb{R}[X_1, \ldots, X_{n+1}]$ and $f = \sum X_i^2$. Then $(X_1, \ldots, X_{n+1})$ is unimodular over B_f and defines a projective A_f-module P, which is stably free but not free for even n. (More general for $n \neq 1, 3, 7$.)

Namely the elements of B_f define continuous maps on $\mathbb{R}^{n+1} \setminus \{0\}$ and so on S^n. Therefore we get an obvious ring homomorphism $B_f \to \mathcal{C}(S^n)$. Via this map $P \otimes_{B_f} \mathcal{C}(S^2)$ is the projective $\mathcal{C}(S^n)$-module that corresponds to the tangent bundle on S^n.

5.4.7 Let B, f be as in the last example with even n, and $\mathfrak{m} = BX_1 + \cdots + BX_{n+1}$. Then – again – the stably free $(B_\mathfrak{m})_f$-module P, defined by the unimodular row $(X_1, \ldots, X_{n+1})$, is not free.

Assume P is free. Then it is extended from $B_\mathfrak{m}$, Hence there is a $g \in B \setminus \mathfrak{m}$ such that P is extended from the subring B_{fg} of $(B_\mathfrak{m})_f$. (To describe P, only a finite set of denominators of $B \setminus \mathfrak{m}$ is required.) Let $\mathcal{Z}(g) \subset \mathbb{R}^{n+1}$ be the set of real zeros of g. It is closed and does not contain 0. So there is an n-sphere S round 0 of some (small) radius with $S \cap \mathcal{Z}(g) = \emptyset$. The elements of B_{fg} are continuous real functions on S (without poles). Therefore there is a ring embedding $B_{fg} \hookrightarrow \mathcal{C}(S)$. And in $\mathcal{C}(S)$ by $(X_1, \ldots, X_{n+1})$ our well known nonfree, stably free module is defined.

In these examples one may replace $\mathbb{R}$ by any subring of $\mathbb{R}$.

5.4.8 Extension Problem. In the last example f belongs to the square of the maximal ideal $\mathfrak{m}A_{\mathfrak{m}}$ of the regular local ring $A_{\mathfrak{m}}$. (For the definition of 'regular' see Chapter 8.) It is an open question, whether for a regular local ring $(R, \mathfrak{m})$ and an $f \in \mathfrak{m} \setminus \mathfrak{m}^2$ every projective R_f-module is free. This would imply that for every regular ring A every projective $A[X]$-module is extended from A - a conjecture of H. Bass, and now generally refered to as the Bass-Quillen conjecture, due to the above question raised by D. Quillen. Quillen's question was raised in the paper [71] where he solved Serre's conjecture, and is also known as Quillen's question! Some known cases of Quillen's question can be found in [7], [100]. The known cases of the Bass-Quillen conjecture can be found in [47], [74].

5.4.9 We will give here an example of a nontrivial *complex* vector bundle on each sphere of odd dimension ≥ 5. It would take us afar to prove here the topological facts we will use.

Let S^{2n+1} be interpreted as the subset of the $\mathbb{C}^{n+1}$ defined by

$$S^{2n+1} := \{(z_0, \ldots, z_n) \in \mathbb{C}^{n+1} \mid \sum_{j=0}^{n} z_j \overline{z_j} = 1\}$$

Consider the bundle with total space

$$E := \{(z, w) \in S^{2n+1} \times \mathbb{C}^{n+1} \mid \sum_{j=0}^{n} \overline{z_j} w_j = 0\}$$

and with the obvious projection to S^{2n+1}. Every fibre is an n-dimensional subspace of $\mathbb{C}^{n+1}$ and the fibre over z is orthogonal to z, also if z and w are interpreted as real vectors. Namely the real part of $\sum_j \overline{z_j} w_j$ is the usual real inner product if one interpretes $\mathbb{C}^{n+1}$ as $\mathbb{R}^{2n+2}$. So the vector bundle $\xi = [E \to S^{2n+1}]$ is a subbundle of the real tangent bundle over S^{2n+1} of real rank $2n$. For $z \in S^{2n+1}$ the vector iz also is orthogonal to z over $\mathbb{R}$, i.e. a tangent vector to S^{2n+1} in z. (The vector field $z \mapsto iz$ is the same as that in Formula 5.8 in Proposition 5.4.4!) The vector bundle $\{(z, riz) \in S^{2n+1} \times \mathbb{C}^{n+1} \mid r \in \mathbb{R}\}$ is a trivial real vector bundle of rank 1 over S^{2n+1}. Over every $z \in S^{2n+1}$ it is orthogonal to $F_z(\xi)$. The direct sum of this and the complex vector bundle

ξ is the full tangent bundle τ on S^{2n+1}. As we already have mentioned – but not proven – the latter is not trivial for $n = 2, n \geq 4$. So *a fortiori* ξ cannot be trivial as an $\mathbb{R}$-vector bundle since it has a trivial complement in τ. So ξ cannot be trivial as a $\mathbb{C}$-vector bundle. By other topological means on can even show that it is nontrivial for all $n \geq 2$.

5.4.10 We will exploit this in order to construct nontrivial projective modules. Define the ring $B_n := \mathbb{C}[z_0, \ldots, z_n, w_0, \ldots, w_n]/(\sum_j z_j w_j - 1)$ and a ring homomorphism $B_n \to \mathcal{C}(S^{2n+1})$ by $z_j \mapsto z_j$, $w_j \mapsto \overline{z_j}$. Then the stably free B_n-module defined by the unimodular row $(z_0, \ldots, z_n)$ extends to the $\mathcal{C}(S^{2n+1})$-module $\Gamma(\xi)$ and therefore is not free.

Clearly one can replace $\mathbb{C}$ by any subring of it.

5.5 Vector Bundles and Grassmannians

We are going to construct a correspondence between isomorphism classes of soft vector bundles over X and the homotopy classes of continuous maps from X to certain spaces.

But first we need some preparations.

5.5.1 The Direct Limit and Infinite Matrices

We define the so called direct limit in a special situation.

Definition 5.5.1 *Let*

$$G_0 \overset{g_0}{\to} G_1 \overset{g_1}{\to} G_2 \overset{g_2}{\to} \cdots \tag{5.9}$$

be a sequence of maps of sets (or group homomorphisms). Define $g_{ii} := \mathrm{id}_{G_i}$ *and* $g_{ik} := g_{k-1} \circ \cdots \circ g_i$ *for* $i < k$. *On the disjoint union* $\bigsqcup_{i \in \mathbb{N}} G_i$ *define the following equivalence relation for* $a \in G_i$, $b \in G_j$:

$$a \sim b : \iff \text{ there is a } k \text{ with } g_{ik}(a) = g_{jk}(b)$$

The **direct limit** *of the above sequence is then defined to be*

$$\varinjlim G_n := \bigsqcup_{n \in \mathbb{N}} G_n \Big/ \sim \tag{5.10}$$

Canonical maps $\kappa_j : G_j \to \varinjlim G_n$ *are defined in the obvious way.* $(G_j \to \bigsqcup_{n \in \mathbb{N}} G_n \to \bigsqcup_{n \in \mathbb{N}} G_n / \sim.)$

Remarks 5.5.2 a) The canonical maps commute with the g_n.

b) To get a good feeling of the meaning of the direct limit: An element of $\varinjlim G_n$ is one which appears in some G_n. Elements are equal in $\varinjlim G_n$ if and only if they 'become' equal in some G_n.

c) If the G_i are groups and the g_i homomorphisms, $\varinjlim G_i$ is also a group.

d) If the g_i are injective and one regards G_i as a subset of G_{i+1} due to the map g_i, then $\varinjlim G_i = \bigcup_i G_i$ and every G_n can be interpreted as a subset of $\varinjlim G_n$.

e) As an exercise the reader should try to show that the additive group of $\mathbb{Q}$ is isomorphic to the direct limit of the following sequence:

$$\mathbb{Z} \xrightarrow{\cdot 2} \mathbb{Z} \xrightarrow{\cdot 3} \mathbb{Z} \xrightarrow{\cdot 4} \cdots$$

where the maps are the homotheties of the integers ≥ 2.

Proposition 5.5.3 *In the situation of the definition let H be a set (or group) and $h_n : G_n \to H$ be maps (homomorphisms) which commute with the g_n. Then there is a unique map (homomorphism) $\varinjlim G_n \to H$, which commutes with the g_n.*

Since we do not need this proposition, we leave the proof to the reader.

5.5.4 Recall that $M_n(R)$ denotes the ring of $n \times n$-matrices over a ring R, and $GL_n(R)$ the group of invertible $n \times n$-matrices. We have injective maps

$$M_n(R) \to M_{n+1}(R), \ \alpha \mapsto \begin{pmatrix} \alpha & 0 \\ 0 & 0 \end{pmatrix} \tag{5.11}$$

$$GL_n(R) \to GL_{n+1}(R), \ \alpha \mapsto \begin{pmatrix} \alpha & 0 \\ 0 & 1 \end{pmatrix} \tag{5.12}$$

(Note the difference!) According to these maps we define
$M_\infty(R) := \varinjlim M_n(R)$ and $GL_\infty(R) := \varinjlim GL_n(R)$.

The map $M_n(R) \to M_{n+1}(R)$ is a homomorphism of additive groups and respects multiplication, but does not map I_n to I_{n+1}. Insofar it is not a ring homomorphism. So the direct limit $\varinjlim M_n(R)$ is nearly a ring, but lacks a 1. The maps $GL_n(R) \to GL_{n+1}(R)$ are group homomorphisms. So $GL_\infty(R)$ is a group.

The elements of $M_\infty(R)$ may be interpreteted as infinite matrices $(a_{ij})_{i,j>0}$ with the property that there is an n with $a_{ij} = 0$ if i or $j > n$.

Analogously the elements of $\mathrm{GL}_\infty(R)$ may be interpreteted as infinite matrices $(a_{ij})_{i,j}$ with the properties that $a_{ii} = 1$ for all i and that there is an n such that the submatrix $(a_{ij})_{i,j \le n}$ is invertible and that $a_{ij} = 0$, if $i \ne j$ and i or $j > n$.

One may perform the product $\alpha\beta$ if α, as well as β belongs to any of $\mathrm{M}_\infty(R)$ or $\mathrm{GL}_\infty(R)$. If both belong to $\mathrm{GL}_\infty(R)$ then $\alpha\beta \in \mathrm{GL}_\infty(R)$; in the three other cases $\alpha\beta \in \mathrm{M}_\infty(R)$.

For example for $\sigma \in \mathrm{GL}_\infty(R)$, $\alpha \in \mathrm{M}_\infty(R)$ we may define the **conjugated** or **similar element** $\sigma\alpha\sigma^{-1}$. This induces on $\mathrm{M}_\infty(R)$ the equivalence relation, to be similar, i.e. conjugated

$$\alpha \sim \beta \iff \text{ there is a } \sigma \in \mathrm{GL}_\infty(R) \text{ with } \beta = \sigma\alpha\sigma^{-1} \tag{5.13}$$

It is clear that every matrix, similar to an idempotent one, is again idempotent.

The equivalence classes of this relation are called the **conjugation classes** of $M_\infty(R)$.

5.5.2 Metrization of the Set of Continuous Maps

By $\mathcal{C}(C,D)$ we denote the set of continuous maps $f : C \to D$ if C, D are topological spaces. It is not always possible to equip $\mathcal{C}(C,D)$ with a useful topology. But if C is compact and D a metric space, we even get a good metric on $\mathcal{C}(C,D)$.

5.5.5 In the following let C be a compact topological space and D be a metric space with distance function d. The set $\mathcal{C}(C,D)$ of continuous maps $f : C \to D$ then inherits a metric according to the definition:

$$d(f,g) := \mathrm{Max}\{d(f(t),g(t)) \mid t \in C\}.$$

Namely the map $C \to \mathbb{R}$, $t \mapsto d(f(t),g(t))$ is continuous and so the maximum exists by the compactness of C. The axioms for a metric easily follow.

Let X be an arbitrary topological space. Every (not necessarily continuous) map $f : X \times C \to D$ induces a map $X \to \mathrm{Map}(C,D)$, $x \mapsto f(x,\cdot)$, where $\mathrm{Map}(C,D)$ denotes the set of all (not necessarily continuous) maps $C \to D$.

Lemma 5.5.6 *Under the above conditions on C and D a map $f : X \times C \to D$ is continuous if and only if every map $f(x,\cdot) : C \to D$ and the map $X \to \mathcal{C}(C,D), x \mapsto f(x,\cdot)$ are continuous.*

Proof. "$\Rightarrow$". The map $f(x,\cdot)$ is the composition of f with $i_x : C \to X \times C$, $c \mapsto (x,c)$, hence is continuous. Let further $x \in X$ and $\varepsilon > 0$. We have to show:

there is a neighbourhood U of x in X such that $d(f(x, \cdot)\,,\; f(x', \cdot)) < \varepsilon$ for all $x' \in U$. Since f is continuous, for every $c \in C$ there are open neighbourhoods V_c of x in X and W_c of c in C such that $f(V_c \times W_c)$ is contained in an ε-neighbourhood of $f(x, c)$. Since C is compact, there are finitely many of the W_c, say $W_1, \ldots, W_m$ with $\bigcup_{j=1}^{m} W_j = C$. Let $V_1, \ldots, V_m$ be the corresponding neighbourhoods of x. We claim that $U := \bigcap_{j=1}^{m} V_j$ has the desired property. We have to show: $d(f(x', c), f(x, c)) < \varepsilon$ for all $x' \in U$, $c \in C$. Now there is an i with $c \in W_i$. Then $f(U \times W_i) \subset f(V_i \times W_i)$ is contained in an ε-neighbourhood of $f(x, c)$.

"$\Leftarrow$". Let $(x, c) \in X \times C$ and $\varepsilon > 0$. Since the map $X \to \mathcal{C}(C, D), x' \mapsto f(x, \cdot)$ is continuous, there is a neighbourhood V of x in X such that $d(f(x', c'), f(x, c') < \varepsilon/2 \;\forall\; (x', c') \in V \times C$. Since $f(x, \cdot) : C \to D$ is continuous, there is a neighbourhood W of c in C such that $d(f(x, c'), f(x, c)) < \varepsilon/2$ for every $c' \in W$. Therefore $d(f(x', c'), f(x, c)) < \varepsilon \;\forall\; (x', c') \in V \times W$. $\qquad\square$

5.5.3 Correspondence of Vector Bundles and Classes of Maps

Lemma 5.5.7 *Let R be an arbitrary ring.*

a) *If M is an R-module and $M = P \oplus Q = P' \oplus Q'$ are splittings with $P \cong P'$, then: $Q \oplus M \cong Q' \oplus M$.*

b) *If $\varepsilon, \varepsilon' \in \mathrm{M}_n(R)$ are idempotents with $\mathrm{im}(\varepsilon) \cong \mathrm{im}(\varepsilon')$. Then there is a $\gamma \in \mathrm{GL}_{2n}(R)$ with $\gamma \varepsilon \gamma^{-1} = \varepsilon'$. Here $\varepsilon, \varepsilon'$ are regarded as elements of M_{2n} by filling up the matrices with zeros. (One may consider all matrices as elements of $\mathrm{M}_\infty(R)$.)*

Proof. a) If M is projective – which is the only case we need – this is a special case of Schanuel's Lemma 3.7.1. (Note that since there exist stably free nonfree modules one cannot expect $Q \cong Q'$.)

But the proof of the general case is easy: Since $P \oplus Q' \cong M \cong P \oplus Q$ we have $Q \oplus M \cong Q \oplus P \oplus Q' \cong M \oplus Q'$.

b) Let $P = \mathrm{im}(\varepsilon)$, $P' = \mathrm{im}(\varepsilon)$, $Q = \ker(\varepsilon)$, $Q' = \ker(\varepsilon')$. Then $P \cong P'$ by hypothesis and and $Q \oplus R^n \cong Q' \oplus R^n$ by a). So we have splittings

$$R^{2n} = P \oplus (Q \oplus R^n) = P' \oplus (Q' \oplus R^n).$$

Let $\tau_1 : P \to P'$, $\tau_2 : Q \oplus R^n \to Q' \oplus R^n$ be isomorphisms. Then $\sigma := \tau_1 \oplus \tau_2$ clearly fulfills the claim. Namely ε is zero on Q and the identity on P. So $\sigma \varepsilon \sigma^{-1}$ is $\tau_2 \circ 0 \circ \tau_2^{-1} = 0$ on Q' and $\tau_1 \mathrm{id}_P \tau_1^{-1} = \mathrm{id}_{P'}$ on P'. $\qquad\square$

We want to show the following:

Theorem 5.5.8 *The soft vector bundles on X correspond upto isomorphism bijectively to the elements of $\varinjlim \pi(X, G^n(\mathbb{F}))$. where $\pi(X, G^n(\mathbb{F}))$ denotes the set of homotopy classes of continuous maps $X \to G^n(\mathbb{F})$.*

Proof. Let $\alpha, \beta \in M_\infty(\mathcal{C}(X))$ be hermitian and idempotent. They can also be interpreted as elements of $\varinjlim \mathcal{C}(X, G^n(\mathbb{F}))$ We have to show that the following equivalence is true:

α, β are conjugated $\iff \alpha, \beta$ are homotopic.

Assume $\beta = \sigma\alpha\sigma^{-1}$. Then

$$\begin{pmatrix} \beta & 0 \\ 0 & 0 \end{pmatrix} = \begin{pmatrix} \sigma & 0 \\ 0 & \sigma^{-1} \end{pmatrix} \begin{pmatrix} \alpha & 0 \\ 0 & 0 \end{pmatrix} \begin{pmatrix} \sigma^{-1} & 0 \\ 0 & \sigma \end{pmatrix}$$

By Whitehead's Lemma the matrix $\begin{pmatrix} \sigma & 0 \\ 0 & \sigma^{-1} \end{pmatrix}$ is a product of elementary matrices, hence homotopic to the 1-matrix. Therefore also α and β are homotopic.

Conversely assume that $\alpha, \beta \in G^n(\mathbb{F})$ are homotopic. Then there is a continuous map

$$\gamma : X \times [0,1] \to G^n(\mathbb{F})$$

with $\gamma(\cdot, 0) = \alpha$, $\gamma(\cdot, 1) = \beta$. Regard γ as a map from X to the space $\pi G^n(\mathbb{F})$ of all continuous maps $[0,1] \to G^n(\mathbb{F})$. This space is metric by the distance function

$$d(f, g) := \mathrm{Max}_t\{d(f(t), g(t)) \mid 0 \le t \le 1\}$$

Therefore there is a locally finite partition of unity $(e'_i)_{i \in I}$ on $\pi G^n(\mathbb{F})$ with $d(u, v) < 1/9$ if $e'_i(u)e'_i(v) \ne 0$ for some $i \in I$. By the continuous map $X \to \pi G^n(\mathbb{F})$, $x \mapsto \gamma(x, \cdot)$ this can be lifted to a locally finite partition of unity $(e_i)_{i \in I}$ on X. So for $x, y \in X$ with $e_i(x)e_i(y) \ne 0$ we have $d(\gamma(x, t), \gamma(y, t)) < 1/9$ for all $t \in [0, 1]$. For every $i \in I$ we choose a point $x_i \in U_i := \{z \in X \mid e_i(z) \ne 0\}$ and a positive number ε_i such that $d(\gamma(x_i, t)\gamma(x_i, t')) < 1/9$ whenever $|t - t'| < \varepsilon_i$. Then $d(\gamma(z, t), \gamma(z, t')) \le d(\gamma(z, t), \gamma(x_i, t)) + d(\gamma(x_i, t), \gamma(x_i, t')) + d(\gamma(x_i, t'), \gamma(z, t')) < 1/9 + 1/9 + 1/9 = 1/3$ for every $z \in U_i$ whenever $|t - t'| < \varepsilon_i$.

Define $\delta(z) := \sum_{i \in I} \varepsilon_i e_i(z)$ for $z \in X$. Then clearly δ is a continuous positive function on X such that $d(\gamma(z, t), \gamma(z, t')) < 1/3$ whenever $|t - t'| < \delta(z)$.

We define continuous maps $t_k : X \to [0, 1]$ and $\gamma_k : X \to G^n(\mathbb{F})$ for every integer $k \ge 0$ by

$$t_k(z) := \mathrm{Min}\{1, k\delta(z)\}, \quad \gamma_k(z) := \gamma(z, t_k(z)).$$

Then $\gamma_0 = \alpha$ and $\gamma_k(z) = \beta(z)$ for $k \ge 1/\delta(z)$, further $d(\gamma_k(z), \gamma_{k+1}(z)) < 1/3$ for all z and k.

Interprete the α, β, γ_k as idempotent hermitian matrices in $M_n(\mathbb{F})$ for big enough n, let $I = I_n$ be the unit matrix and set

$$\sigma_k := \gamma_k \gamma_{k+1} - (I - \gamma_k)(I - \gamma_{k+1})$$

Then $\gamma_k \sigma_k = \gamma_k \gamma_{k+1} = \sigma_k \gamma_{k+1}$ Further $\sigma_k - I = (\gamma_k - \gamma_{k+1})(I - 2\gamma_{k+1})$, whence $|\sigma_k - 1| \le |\gamma_k - \gamma_{k+1}| \cdot (1 + 2|\gamma_{k+1}|) < (1/3) \cdot 3 = 1$. So σ_k is invertible. Define

$$\sigma := \sigma_0 \sigma_1 \sigma_2 \cdots , \text{ i.e. } \sigma(z) = \sigma_0(z)\sigma_1(z) \cdots \sigma_k(z) \text{ for } k > 1/\delta(z).$$

Then σ is an invertible matrix over $\mathcal{C}(X)$ and $\alpha(z)\sigma(z) = \gamma_0(z) \cdots \gamma_k(z) = \sigma(z)\beta(z)$ for $k > 1/\delta(z)$. This means $\alpha\sigma = \sigma\beta$, hence $\alpha = \sigma\beta\sigma^{-1}$. $\qquad\square$

5.5.9 A topological space X is called **contractible** if and only if there is a point $p \in X$ such that the constant map $X \to X, x \mapsto p$ is homotopic to id_X. For instance $\mathbb{R}^n$ is contractible by the homotopy: $h : \mathbb{R}^n \times [0,1] \to \mathbb{R}^n$, $(x, t) \mapsto tx$.

This was the reason why Serre's Conjecture was born! Namely $\mathcal{C}(\mathbb{R}^n)$ is the topological counterpart of $k[x_1, \ldots , x_n]$.

It is clear that every continuous map $X \to Y$ is homotopic to a constant map, if X is contractible. Therefore, by the theorem, on $\mathbb{R}^n$ there are only trivial soft vector bundles over $\mathbb{R}^n$. Indeed there are only trivial vector bundles on $\mathbb{R}^n$ at all.

5.6 Projective Spaces

5.6.1 The n-dimensional projective space $\mathbb{FP}^n$ over $\mathbb{F}$ is the Grassmannian of one dimensional linear subspaces of $\mathbb{F}^{n+1}$, i.e. $\mathbb{FP}^n = G_1^{n+1}(\mathbb{F})$.

Clearly we have a surjective map

$$p : \mathbb{F}^{n+1} \setminus \{0\} \longrightarrow \mathbb{FP}^n$$

which assigns to any nonzero $x \in \mathbb{F}^{n+1}$ the subspace, generated by x. Elements $x, y \in \mathbb{F}^{n+1} \setminus \{0\}$ belong to the same fibre of p, if and only if there is an $a \in \mathbb{F}^\times$ with $ax = y$. This means that $\mathbb{FP}^n$ is the orbit space of the obvious operation of $\mathbb{F}^\times$ on $\mathbb{F}^{n+1} \setminus \{0\}$.

The map p is continuous. Namely the topology on $\mathbb{FP}^n$ is given by assigning to any one dimensional subspace of $\mathbb{F}^{n+1}$ the orthogonal projection onto it, and then by using the operator norm on the space of endomorphisms of $\mathbb{F}^{n+1}$. But the orthogonal projection onto the subspace, generated by v is the map $w \mapsto \langle v, v \rangle^{-1} \langle v, w \rangle v$, which clearly depends continuously on v.

By $[x_0 : \dots : x_n]$ we denote the point $p(x_0, \dots, x_n)$. (Of course $[x_0 : \dots : x_n]$ is only defined for $(x_0, \dots, x_n) \neq (0, \dots, 0)$.)

The set $S^{nr+r-1} := \{(x_0, \dots, x_n) \in \mathbb{F}^{n+1} \mid \sum |x_i|^2 = 1\}$ is the (real) unit sphere in $\mathbb{R}^{nr+r}$ if this is identified with $\mathbb{F}^{n+1}$. Here $r = 1, 2, 4$, depending on whether $\mathbb{F} = \mathbb{R}, \mathbb{C}$ or $\mathbb{H}$. By p this sphere is mapped surjectively to $\mathbb{F}P$. Elements $x, y \in S^{nr+r-1}$ have the same image under p if and only if there is a $u \in \mathbb{F}$ of absolute value 1 with $ux = y$. The elements $u \in \mathbb{F}$ with $|u| = 1$ make up a subgroup of the multiplicative group of $\mathbb{F}$ and form the unit sphere S^{r-1} in $\mathbb{F}$.

Therefore $\mathbb{F}P^n$ is a quotient of the sphere S^{nr+r-1} by the operation of the group $S^{r-1} = \{x \in \mathbb{F} \mid |x| = 1\}$.

Note that $U_i := \{[x_0 : \dots : x_n] \mid x_i \neq 0\}$ for every $i = 0, \dots, n$ is an open subset of $\mathbb{F}P^n$. It is homeomorphic to $\mathbb{F}^n$. We describe the maps $\mathbb{F}^n \to U_i \to \mathbb{F}^n$ only for $i = 0$.

$$(y_1, \dots, y_n) \mapsto [1 : y_1 : \dots : y_n], \quad [x_0 : x_1 : \dots : x_n] \mapsto (\frac{x_1}{x_0}, \dots, \frac{x_n}{x_0})$$

(It is clear, how to do this for general i.)

Definition 5.6.2 *The* **universal line bundle** *on $\mathbb{F}P^n$ is the one dimensional subbundle of the trivial bundle $[\mathrm{pr}_2 : \mathbb{F}^{n+1} \times \mathbb{F}P^n \to \mathbb{F}P^n]$ whose total space is $\{(v, x) \in \mathbb{F}^{n+1} \times \mathbb{F}P^n \mid v \in x\}$. (Note that every $x \in \mathbb{F}P^n$ is a one dimensional subspace of $\mathbb{F}^{n+1}$ so that the relation $v \in x$ makes sense.) The universal bundle is also called the* **tautological bundle** *since the fibre over x is x itself, considered as vector space. Also the name* **canonical line bundle** *is in use.*

5.6.3 The universal line bundle γ is trivial on every one of the U_i, which we defined in Paragraph 5.6.1. Namely (for $i = 0$) we have the isomorphism

$$\mathbb{F} \times U_0 \longrightarrow E(\gamma|U_0), \quad (a, [x_0 : x_1 : \dots : x_n]) \mapsto (a, a\frac{x_1}{x_0}, \dots, a\frac{x_n}{x_0})$$

5.6.4 The complex projective line $\mathbb{C}P^1$ is the so called **Riemann sphere**. Indeed it is homeomorphic to the 2-sphere by the stereographic projection. We will explain this in detail.

$\mathbb{C}P^1$ is covered by the two open sets: $U_j := \{[z_0 : z_1] \mid z_j \neq 0\}$. And there are homeomorphisms $U_0 \to \mathbb{C}, [z_0 : z_1] \mapsto z_1/z_0$ and $U_1 \to \mathbb{C}, [z_0 : z_1] \mapsto z_0/z_1$.

Now let $S^2 := \{(z, t) \in \mathbb{C} \times \mathbb{R} \mid |z|^2 + t^2 = 1\}$ and let $n := (0, 1) \in S^2$, $s := (0, -1) \in S^2$ denote the 'north pole', resp. the 'south pole' of S^2 of the 2-sphere. We have two maps $\varphi_n, \varphi_s : \mathbb{C} \to S^2$ given by

$$\varphi_n(z) = \left(\frac{2z}{|z|^2 + 1}, \frac{|z|^2 - 1}{|z|^2 + 1} \right), \quad \varphi_s(z) = \left(\frac{2\bar{z}}{|z|^2 + 1}, \frac{-|z|^2 + 1}{|z|^2 + 1} \right).$$

(Here and some lines later $\bar{z}$ denotes the complex conjugate of z.) We will show that these are homeomorphisms from $\mathbb{C}$ with $S^2 \setminus \{n\}$, resp. with $S^2 \setminus \{s\}$. We do this by giving the inverse maps.

$$\psi_n : S^2 \setminus \{n\} \to \mathbb{C}, \; (z,t) \mapsto \frac{z}{1-t}, \quad \psi_s : S^2 \setminus \{s\} \to \mathbb{C}, (z,t) \mapsto \frac{\bar{z}}{1+t}.$$

We leave the computations to the reader. Also one computes $\psi_s \circ \varphi_n(z) = 1/z$ for $z \in \mathbb{C}^\times$. But this means that S^2 arises from two copies of $\mathbb{C}$ by the same glueing as $\mathbb{C}P^1$ does. Therefore they are homeomorphic to each other and can well be identified.

The above described map $p : S^3 \to \mathbb{C}P^1$ can now be viewed as a map $p : S^3 \to S^2$. This is the famous **Hopf map**.

Note 5.6.5 One may do the same in the cases $\mathbb{F} = \mathbb{R}$ and $\mathbb{F} = \mathbb{H}$. The first case is contained in the above computations. Only one has to restrict z to real numbers. One gets a map $S^1 \to S^1$ of degree 2. This means it can be described by $\zeta \mapsto \zeta^2$ when one identifies S^1 with the set of complex numbers of absolute value 1.

In the second case one proves as above that $\mathbb{H}P^1$ is homeomorphic to S^4 and one gets a Hopf map $S^7 \to S^4$.

5.6.6 Let us consider the universal $\mathbb{C}$-line bundle γ. On U_0 we have the trivialization

$$\mathbb{C} \times U_0 \to E(\gamma|U_0), \; (a, [z_0 : z_1]) \mapsto (a, a\frac{z_1}{z_0}),$$

and on U_1

$$\mathbb{C} \times U_1 \to E(\gamma|U_1), \; (b, [z_0 : z_1]) \mapsto (b\frac{z_0}{z_1}, b).$$

To compute the glueing map α^{01} we have to find the preimage of $(1, z_1/z_0)$ under the second map. It is $(z_1/z_0, [z_0 : z_1])$, since it clearly maps to $(1, z_1/z_0)$. Therefore $\alpha^{01}(z) = z$ if we identify $[z_0 : z_1]$ with $z = z_1/z_0$ on $U_0 \cap U_1$.

Fore later use, now we replace the U_j by the closed halfspheres $Y_0 := \{[z_0 : z_1] \mid |z_1/z_0| \le 1\}$, $Y_1 := \{[z_0 : z_1] \mid |z_0/z_1| \le 1\}$. Then $Y_0 \cap Y_1 = \{z \mid |z| = 1\}$ where $z = z_0/z_1$. Of course in this situation we also have $\alpha^{01}(z) = z$.

5.6.7 Now let us describe the tangent bundle of the S^2 in the same manner. It is the tangent $\mathbb{C}$-line bundle τ on $\mathbb{C}P^1$, considered as an $\mathbb{R}$-bundle of rank 2.

On each of the two halfspheres Y_j the tangent bundle is trivial and the glueing map β_{01} of the bundle is the derivative w.r.t. $z := z_1/z_0$ of the glueing map of the two charts $Y_0 \cap Y_1 \to Y_0 \cap Y_1$. The latter is $z \mapsto z^{-1}$. So the glueing map of the tangent bundle is $\beta^{01}(z) = -z^{-2}$. Since we may change the coordinate of Y_0 from z to $\overline{iz}$, we may take as glueing map of the bundle the map $\beta'^{01}(z) = z^2$. (The overbar here denotes the complex conjugation.) This means that $\gamma^{\otimes 2} := \gamma \otimes_{\mathbb{C}} \gamma = \tau$. Since τ as a real vector bundle is not trivial, *a fortiori* it is not trivial over $\mathbb{C}$. So the universal bundle γ on $\mathbb{CP}$ cannot be trivial.

5.6.8 Note, one has a similar situation in the case $\mathbb{F} = \mathbb{R}$. The one-dimensional real projective space is the 1-sphere S^1, the universal bundle is the Möbius bundle, its tensor square is the tangent bundle, which is trivial in this case.

5.6.9 Let $\mathbb{F} = \mathbb{R}$. On $\mathbb{RP}^n$ we consider the universal bundle γ. The map $p : S^n \to \mathbb{RP}^n$ induces a map $p^* : \mathcal{C}(\mathbb{RP}^n) \to \mathcal{C}(S^n)$, which is injective and identifies $\mathcal{C}(\mathbb{RP}^n)$ with the subring C_0 of $\mathcal{C}(S^2)$ consisting of those $f \in \mathcal{C}(S^n)$ which satisfy $f(-x) = f(x)$.

We want to interprete $\Gamma(\gamma)$ in a similar manner. Let $x \in S^n$, $p(x) \in \mathbb{RP}^n$. An element of the fibre of γ over $p(x)$ is of the form $(ax, p(x)) = ((-a)(-x), p(x))$ with $a \in \mathbb{R}$. So a section over $\mathbb{RP}^n$ is given by a continuous real function $g \in \mathcal{C}(S^n)$ with $g(-x) = -g(x)$. This clearly gives a good description of $\Gamma(\gamma)$ as a module over $C_0 = \mathcal{C}(\mathbb{RP}^n)$. Namely, If $f, g \in \mathcal{C}(S^n)$ satisfy $f(-x) = f(x)$ and $g(-x) = -g(x)$, then $(fg)(-x) = -(fg)(x)$.

We will now study the minimal number of generators of this projective C_0 module of rank 1. Unfortunately we only can do that by using some facts on **Stiefel-Whitney classes** we now cite here.

Stiefel-whitney classes $(w_0, w_1, w_2, \dots)$ assign to a real vector bundle ξ on a topological space X cohomology classes $w_i(\xi) \in H^i(X, \mathbb{Z}/2\mathbb{Z})$ of X with $w_0(\xi) = 1$. The total Stiefel-Whitney class of ξ is $W(\xi) = 1 + w_1(\xi) + w_2(\xi) + \cdots \in H^*(X, \mathbb{Z}/2\mathbb{Z})$. One has the following properties:

a) $W(\lambda) = 1$ if λ is trivial.

b) $W(\xi + \eta) = w(\xi)w(\eta)$.

c) $w_i(\xi) = 0$ for $i > \text{rk}(\xi)$.

The cohomology of the real projective space is of the form $H^*(\mathbb{RP}^n, \mathbb{Z}/2\mathbb{Z}) = (\mathbb{Z}/2\mathbb{Z})[T]/(T^{n+1})$ and, letting t denote the residue class of T, we have $t^i \in H^i(\mathbb{RP}^n)$ and $W(\gamma) = 1 + t$.

Proposition 5.6.10 *Let γ be as above and λ^r the trivial bundle of rank r on $\mathbb{RP}^n$. Then the $\mathcal{C}(\mathbb{RP}^n)$-module $\Gamma(\gamma \oplus \lambda^r)$ needs $n + r + 1$ generators.*

Proof. Assume that $\Gamma(\gamma \oplus \lambda^r)$ could be generated by $n + r$ generators. Then $\gamma \oplus \lambda^r$ is a direct summand of λ^{n+r} the trivial bundle of rank $n + r$, i.e. there is a bundle η with $\gamma \oplus \lambda^r \oplus \eta = \lambda^{n+r}$. Clearly $\mathrm{rk}(\eta) = n - 1$. So $W(\eta) = 1 + a_1 t + \cdots + a_{s-1} t^{s-1} + t^s$ for some $a_i \in \mathbb{Z}/2\mathbb{Z}$ and some $s < n$. Therefore $W(\gamma \oplus \lambda^r \oplus \eta) = 1 + \cdots + t^{s+1} \neq 1$, since $s + 1 \leq n$ and therefore $t^{s+1} \neq 0$. On the other hand $W(\lambda^{n+r}) = 1$. A contradiction. $\qquad\square$

5.6.11 Now we consider the ring $A := \mathbb{R}[X_0, \ldots, X_n]/(\sum X_i^2 - 1)$. It is the subring of $\mathcal{C}(S^n)$, consisting of the polynomial functions on S^n. By x_i we denote the residue class of X_i. Then $A_0 := C_0 \cap A$ consists of the residue classes of those polynomials, all of whose monomials have even (total) degree.

We have a splitting $A = A_0 \oplus A_1$ of the additive group of A, where A_1 is the subgroup of the residue classes of those polynomials, all of whose monomials have odd (total) degree. This splitting is a so called $\mathbb{Z}/2\mathbb{Z}$-grading of A, which means $A_0 A_0 \subset A_0$, $A_0 A_1 \subset A_1$, $A_1 A_1 \subset A_0$.

So A_1 is an A_0-module, and we will see now that it is a finitely generated projective one. One has a surjective linear map $f : A_0^{n+1} \to A_1$, given by $f(a_0, \ldots, a_n) = \sum_{i=0}^n x_i a_i$ and a section to this map $s : A_1 \to A_0^{n+1}$, $b \mapsto (x_0 b, x_1 b, \ldots, x_n b)$. Note that $f \circ s = \mathrm{id}$ follows directly from $\sum_{i=0}^n x_i^2 = 1$.

According to the above description of $\Gamma(\gamma)$ as a C_0-module we have parallely C_0-linear maps $f' : C_0^{n+1} \to \Gamma(\gamma)$, $(a_0, \ldots, a_n) \mapsto \sum_{i=0}^n x_i a_i$ and $s' : \Gamma(\gamma) \to C_0^{n+1}$, $b \mapsto (x_0 b, \ldots, x_n b)$ with $f' \circ s' = \mathrm{id}$.

So $\Gamma(\gamma) = C_0 \otimes_{A_0} A_1$. And therefore A_1 is a projective A_0-module of rank 1, such that $A_1 \oplus A_0^r$ cannot be generated by less than $n + 1 + r$ elements.

5.7 Algebraization of Vector Bundles

Let X be a compact subset of $\mathbb{R}^n$. We have seen how the continuous vector bundles on X may be interpreted as finitely generated projective $\mathcal{C}(X)$-modules. Algebraists might prefer to have examples of projective modules over rings which are more algebraic, i.e. 'smaller' in some good sense. At least they should be Noetherian. A ring is called (left) Noetherian if its (left) ideals are finitely generated. This property will be introduced and studied in the first section of Chapter 6. We will even show that the rings we construct, are essentially of finite type over $\mathbb{F}$.

Definition 5.7.1 *A ring A is called **essentially of finite type over a ring R**, if it is of the form $A = S^{-1} R[\alpha_1, \ldots, \alpha_n]$.*

If R is Noetherian, say a field, then every R-algebra, essentially of finite type, is Noetherian, too.

It is clear that for a single finitely generated projective $\mathcal{C}(X)$-module P (or finitely many ones) there are finitely many functions $f_1, \ldots, f_n \in \mathcal{C}(X)$ such that P is extended from $\mathbb{Z}[f_1, \ldots, f_n]$. Namely P is the image of an idempotent endomorphism of a finitely generated free $\mathcal{C}(X)$-module, i.e. P is given by a finite matrix. Let $f_1, \ldots, f_n$ be the entries of this matrix.

But clearly one prefers, to find Noetherian rings or algebras A, essentially of finite type, with interesting behaviour of the *totality* of their projective modules.

At first K. Lønsted in [48] gave a method to produce Noetherian such A. Later R.G. Swan in [99] constructed such A, essentially of finite type over $\mathbb{F}$. M. Carral in [16] then was able to replace a part of Swan's construction by a more efficient one. We reproduce here the method of Swan and Carral and Milnor's famous construction of projective modules, which is used for that.

5.7.1 Projective Modules over Topological Rings

Here, following Swan, we will study the behaviour of $\mathbb{P}$ under a ring homomorphism $A \to B$ with certain topological properties.

5.7.2 Let A be a topological ring and P, Q finitely generated projective A-modules. We define topologies on P and $\mathrm{Hom}(P, Q)$ as follows.

First identify $\mathrm{Hom}(A^r, A^s)$ with A^{rs} and give it the product topology. If σ, ρ are automorphisms of A^s and A^r respectively, then $\alpha \mapsto \sigma\alpha\rho$ is a homeomorphism of $\mathrm{Hom}(A^r, A^s)$ with itself, since it is continuous as well as $\beta \mapsto \sigma^{-1}\beta\rho^{-1}$. Therefore the topology on $\mathrm{Hom}(A^r, A^s)$ is independent of the chosen bases.

Now let P', Q' be so that $P \oplus P' \cong A^r$ and $Q \oplus Q' \cong A^s$. Then $\alpha \mapsto \alpha \oplus 0$ embeds $\mathrm{Hom}(P, Q)$ into $\mathrm{Hom}(A^r, A^s)$. The induced topology on $\mathrm{Hom}(P, Q)$ is then independent of the choices of P' and Q'. To see this note first that replacing P' by $P' \oplus A^n$ and Q' by $Q' \oplus A^m$ does not change the topology. Let $P \oplus P'' \cong A^n$ and $Q \oplus Q'' \cong A^m$. Then $P \oplus P' \oplus A^n \cong P \oplus P'' \oplus A^r$ and analogously $Q \oplus Q' \oplus A^m \cong Q \oplus Q'' \oplus A^s$. Hence the independence follows.

The topology on P can be defined via the identification $P = \mathrm{Hom}_A(A, P)$ or equivalently via the inclusion $P \hookrightarrow P \oplus P' = A^r$.

The following hypothesis is not always true, but in important cases it is:

(0) *The units of A form an open subset and $u \mapsto u^{-1}$ is continuous on it.*

This is equivalent to saying that some neighbourhood of 1 consists of units and u^{-1} is near 1 if u is near enough to 1. This is easy to show for $A = \mathcal{C}(X)$ with compact X. (Exercise.) More generally it holds for any Banach algebra.

The matrix ring $M_n(A)$ regarded as $\mathrm{Hom}(A^n, A^n)$ has the above defined topology, namely the product topology, if one regards $M_n(A)$ as A^{n^2}. With this topology $M_n(A)$ is a topological ring.

Lemma 5.7.3 *Let A satisfy (0). There are a neighbourhood U of I_n in $M_n(A)$ and continuous maps $e_i, f_i, d : U \to M_n(A)$, $i = 1, \ldots, r = n(n-1)/2$, such that*

a) all $e_i(x), f_i(x)$ are elementary and the $d(x)$ are invertible diagonal matrices for $x \in U$ which reduce to I_n for $x = I_n$ and

b) $x = e_1(x) \cdots e_r(x)d(x)f_1(x) \cdots f_r(x)$

Proof. Let $x = (a_{ij})$ be near enough I_n. Then a_{11} is near 1, hence a unit in A. So by $n-1$ elementary row and $n-1$ column operations we can transform x to a matrix of the form $\begin{pmatrix} a_{11} & 0 \\ 0 & N \end{pmatrix}$. Since a_{11}^{-1} is near 1 and the a_{i1}, a_{1j} for $i, j \neq 1$ were near 0, the matrix N is near I_{n-1} in $\mathrm{GL}_{n-1}(A)$. By induction on n we get through.

Proposition 5.7.4 *If A satisfies (0), so does $\mathrm{End}(P)$ for any finitely generated projective module P.*

Proof. At first we assume $P = A^n$ to be free. By Lemma 5.7.3 there is an open neighbourhood U of I_n in $M_n(A)$ which is contained in the general linear group and on which the map $x \to x^{-1}$ is continuous. Namely, for x near I_n we have $x = e_1(x) \cdots e_r(x)d(x)f_1(x) \cdots f_r(x)$, hence $x^{-1} = f_r(x)^{-1} \cdots f_1(x)^{-1}d(x)^{-1}e_r(x)^{-1} \cdots e_1(x)^{-1}$.

In the general case let $P \oplus Q = A^n$. Then by the above definition of the topology on $\mathrm{End}(P) = \mathrm{Hom}(P, P)$ an endomorphism g of P is near id_P if and only if $g \oplus \mathrm{id}_Q$ is near I_n in $M_n(A)$. $\qquad\square$

Proposition 5.7.5 *Let A satisfy (0) and $f, g : P \to Q$ be linear maps between finitely generated projective A-modules. If f is surjective and g sufficiently near f, then g is surjective too, and $\ker(g) \cong \ker(f)$.*

Proof. Let s be a section of f, i.e. $fs = \mathrm{id}_Q$. If g is sufficiently near f then gs is near $fs = \mathrm{id}_Q$ and so by Hypothesis (0), gs is an automorphism of Q. So g is surjective.

Let $t = s(gs)^{-1} : Q \to P$. Then $gt = \mathrm{id}_Q$ and therefore $\ker g \cong \mathrm{coker}\, t = \mathrm{coker}\, s \cong \ker f$. $\qquad\square$

Proposition 5.7.6 *Let A satisfy (0) and F be a finitely generated projective A-module, further $e : F \to F$ be an idempotent. If $f : F \to F$ is idempotent too and sufficiently near e, then $e(F) \cong f(F)$.*

Proof. Let $P = e(F)$, $Q = f(F)$ and $Q' = (\mathrm{id}_F - f)(F)$. If f is sufficiently near e, then $e + (\mathrm{id}_F - f)$ will be near enough $e + (\mathrm{id}_F - e) = \mathrm{id}_F$, hence an automorphism by Proposition 5.7.4. Therefore $F = P + Q'$ and so the composition α of the canonical maps $P \to F \to Q$ is surjective. We have $\alpha(x) = f(x)$ for $x \in P$. Define $\beta : Q \to P$ by $\beta(y) = e(y)$. If f is sufficiently near e, then $\beta\alpha : P \to P$ will be very near to the map $x \mapsto e^2(x) = x$. Hence again $\beta\alpha$ is an automorphism and so α will also be injective. $\square$

5.7.7 Now we consider a ring homomorphism $\varphi : A \to B$ and conditions

(1) *B is a topological ring which satisfies* (0) *above;*

(2) *$\varphi(A)$ is dense in B;*

(3) *there is an open neighbourhood U of 1 in B with $\varphi^{-1}(U) \subset A^\times$.*

Remarks 5.7.8 a) Let $I \subset \mathrm{Jac}(A)$ be an ideal and $B := A/I$ equipped with the discrete topology. Then (1) and (2) are obvious, and (3) means $1 + I \subset A^\times$, which follows easily from $I \subset \mathrm{Jac}(A)$.

b) If $\varphi : A \to B$ satisfies (1), (2) and
(3') $\varphi^{-1}(B^\times) \subset A^\times$, i.e. $\varphi^{-1}(B^\times) = A^\times$,

then φ satisfies (1) to (3).

Especially, if φ satisfies (1) and (2), and $S := \varphi^{-1}(B^\times)$ is contained in the center of A, then the induced homomorphism $S^{-1}A \to B$ satisfies (1) to (3).

(If A is a noncommutative ring and S is a multiplicative subset of the center of A, then $S^{-1}A$ can be defined in the same way as in the commutative case.)

c) Let $\varphi : A \to B$ satisfy (1). Define a topology on A by taking the neighbourhoods of 0 to be the sets $\varphi^{-1}(U)$ where U is a neighbourhood of 0 in B. Then we have the equivalence:

$$A^\times \text{ is open in } A \iff \varphi \text{ satisfies (3)}.$$

Since $\varphi(u^{-1}) = \varphi(u)^{-1}$ for units in A we also have the equivalence

$$A \text{ satisfies (0)} \iff \varphi \text{ satisfies (3)}.$$

d) Let X be a compact subset of some $\mathbb{R}^n$. The ring $\mathcal{R}(X)$ of regular functions on X is defined by

$$\mathcal{R}(X) := S_X^{-1}\mathbb{R}[t_1, \ldots, t_n]/I_X$$

where I_X is the zero ideal of X, i.e. consists of the polynomials vanishing along X, whereas S_X consists of those polynomials which are nowhere zero on X.

Then the inclusion $\varphi : \mathcal{R}(X) \to \mathcal{C}(X)$ satisfies (1) to (3). Namely (1) is trivial, whereas (2) follows by Weierstrass' Approximation Theorem, and (3) is clear, since we even have $\varphi^{-1}(\mathcal{C}(X)^\times) = \mathcal{R}(X)^\times$.

Lemma 5.7.9 *If* $\varphi : A \to B$ *satisfies* (1) *to* (3), *then so does* $\mathrm{M}_n(\varphi)$: $\mathrm{M}_n(A) \to \mathrm{M}_n(B)$.

Proof. (1) has already been proved in Proposition 5.7.4, and (2) is trivial.

For (3), define a topology on A resp. $\mathrm{M}_n(A)$ as in Remark c) above by taking the neighbourhoods of 0 to be the sets $\varphi^{-1}(U)$ where U is a neighbourhood of 0 in B, resp. of 0 in $\mathrm{M}_n(B)$. The resulting topology on $\mathrm{M}_n(A)$ is the same as the topology, induced by that on A. Now by Remark c), condition (3) for φ implies Hypothesis (0) for A, hence for $\mathrm{M}_n(A)$ by Proposition 5.7.4. Again by Remark c) we obtain (3) for $\mathrm{M}_n(\varphi)$. $\qquad\square$

Lemma 5.7.10 *Let* $\varphi : A \to B$ *satisfy* (2) *and* $f : B \otimes P \to B \otimes Q$ *be B-linear, where P, Q are finitely generated and projective over A. Then we can find $g : P \to Q$ such that $\mathrm{id}_B \otimes g$ is arbitrarily close to f.*

Proof. This is clear, if P, Q are free. Let $P \oplus P' = F$ and $Q \oplus Q' = G$ be free and find $h : F \to G$ such that h approximates $f \oplus 0$. Then the composition g of $P \xrightarrow{i} F \xrightarrow{h} G \xrightarrow{p} Q$ will be near f, where i, p are the canonical inclusion resp. projection. $\qquad\square$

Lemma 5.7.11 (Generalized Nakayama Lemma) *Let* $\varphi : A \to B$ *satisfy* (1) *to* (3) *and M be a finitely generated A-module with $B \otimes M = 0$. Then $M = 0$.*

Proof. Let $F' \xrightarrow{f} F \to M \to 0$ be an exact sequence of A-modules where F, F' are free and F is finitely generated By hypothesis $B \otimes f$ is surjective. Hence there is a section s of it: $(B \otimes f) \circ s = \mathrm{id}_F$. There is a finite subset of a base of F', generating a free submodule G, such that $s(B \otimes F)$ is contained in $B \otimes G$. By Lemma 5.7.10 find $t : F \to G$ with $B \otimes t$ near s. So $B \otimes (f \circ t)$ is near $(B \otimes f) \circ s = \mathrm{id}_F$. Since $\mathrm{M}_n(\varphi)$ satisfies (3), the endomorphism $f \circ t$ will be an automorphism, if $B \otimes t$ is sufficiently close to s. Therefore f is surjective. $\square$

Theorem 5.7.12 *Let* $\varphi : A \to B$ *satisfy* (1) *to* (3).

a) $\mathbb{P}(\varphi) : \mathbb{P}(A) \to \mathbb{P}(B)$ *is injective.*

b) *The isomorphism class of a finitely generated projective B-module belongs to the image of $\mathbb{P}(\varphi)$ if it is stably isomorphic to another finitely generated projective B-module whose isomorphism class is in the image. Especially the isomorphism class of any stably free module is in the image of $\mathbb{P}(\varphi)$.*

Proof. a) Let P, Q be finitely generated projective A-modules and let $f :$ $B \otimes P \to B \otimes Q$ be an isomorphism. Find a $g : P \to Q$ with $\mathrm{id}_B \otimes g$ so close to f, that $B \otimes g$ is an isomorphism, using Lemma 5.7.10 and Proposition 5.7.4. Then $B \otimes \mathrm{coker}(g) = 0$, whence $\mathrm{coker}(g) = 0$ by the Generalized Nakayama

Lemma. So the sequence $0 \to \ker(g) \to P \to Q \to 0$ is split exact, hence $\ker(g)$ as a factor of P is finitely generated. Therefore from $B \otimes \ker(g) = 0$ we obtain $\ker(g) = 0$. Hence $P \cong Q$.

b) Let P be a finitely generated projective B-module, Q a finitely generated projective A-module with $P \oplus B^r \cong (B \otimes_A Q) \oplus B^r$ for some r. With $Q' := Q \oplus A^r$ we get

$$P \oplus B^r \cong (B \otimes_A Q) \oplus B^r \cong B \otimes_A (Q \oplus A^r) = B \otimes_A Q'.$$

the projection $P \oplus B^r \to B^r$ composed with this isomorphism gives us a homomorphism $p : B \otimes_A Q' \to B^r$. Using Lemma 5.7.10, we find $g : Q' \to A^r$ with $\mathrm{id}_B \otimes g$ so close to p, that $\mathrm{id}_B \otimes g$ is surjective by Proposition 5.7.5. Then g is also surjective by the Generalized Nakayama Lemma. So $P' := \ker(g)$ is a direct summand of Q' and hence projective. Finally we get $B \otimes_A P' = \ker(\mathrm{id}_B \otimes g) \cong \ker(p) = P$ by Proposition 5.7.5 if only $B \otimes g$ was chosen close enough to p.   □

5.7.2 Projective Modules as Pull-Backs

We present here Milnor's construction of projective modules.

Definition 5.7.13 *A diagram of the form*

$$
\begin{array}{ccc}
A & \xrightarrow{\ i_1\ } & A_1 \\
{\scriptstyle i_2}\downarrow & & \downarrow{\scriptstyle j_1} \\
A_2 & \xrightarrow{\ j_1\ } & A'
\end{array}
$$

of ring or group homomorphisms is called a **pull-back diagram** *or* **Cartesian**, *if $A \cong \{(a_1, a_2) \in A_1 \times A_2 \mid j_1(a_1) = j_2(a_2)\}$ and i_1, i_2 are induced by the projections. (From this clearly the commutativity of the diagram follows.)*

In this case one calls A, or more exactly (A, i_1, i_2) the pull back of the pair of maps (j_1, j_2). (Compare Appendix B.)

Examples 5.7.14 a) Let I_1, I_2 be (two-sided) ideals of a ring R and let $A = R/(I_1 \cap I_2)$, $A_k = R/I_k$, $k = 1, 2$, $A' = R/(I_1 + I_2)$ and i_1, i_2, j_1, j_2 be canonical maps.

For our purpose this will be the most important case. Namely let V_1, V_2 be closed subsets of a normal topological space X and set $A := \mathcal{C}(V_1 \cup V_2)$, $A_k := \mathcal{C}(V_k)$ and $A' := \mathcal{C}(V_1 \cap V_2)$ Then the restriction maps $A \to A_k$, $A_k \to A'$ are surjective by Tietze's Extension Theorem. And if $I_k = \ker[A \to A_k]$, then $I_1 + I_2 = \ker[A \to A']$. Namely if $f : V_1 \cup V_2 \to \mathbb{F}$ is 0 on V_1 or on V_2, then clearly it is 0 on $V_1 \cap V_2$, whence $I_k \subset \ker[A \to A']$ for $k = 1, 2$. So we only have to show $I_1 + I_2 \supset \ker[A \to A']$. But if $f : V_1 \cup V_2 \to \mathbb{F}$ is 0 on $V_1 \cap V_2$, then define

$$f_k : V_1 \cup V_2 \to \mathbb{F} \text{ by } f_k(x) = \begin{cases} f(x) & \text{if } x \notin V_k \\ 0 & \text{if } x \in V_k. \end{cases}$$

Since f_k is continuous on the closed sets V_1 and V_2, it is continuous on $V_1 \cup V_2$. Further $f_k \in I_k$ and $f = f_1 + f_2$.

b) Let $A \subset B$ be a ring extension and $I \subset A$ be an ideal of B (so that I is a common ideal of A and B). Then the diagram with the obvious maps

$$
\begin{array}{ccc}
A & \longrightarrow & B \\
\downarrow & & \downarrow \\
A/I & \longrightarrow & B/I
\end{array}
$$

is Cartesian. We note that this situation for example arises in the case where A is a domain, B its integral closure and B is finite over A. Then the so called **conductor** $\mathfrak{C}(B/A) := \mathrm{Ann}_A(B/A)$, which is the maximal common ideal of A and B, is not zero. (We will use the concept of the conductor in Chapter 10, see Definition 10.4.6.)

c) Let A' be any ring, A_1, A_2 subrings of A' and A their intersection.

5.7.15 *Construction.* In the following let

$$
\begin{array}{ccc}
A & \xrightarrow{i_1} & A_1 \\
{\scriptstyle i_2}\downarrow & & \downarrow{\scriptstyle j_1} \\
A_2 & \xrightarrow{j_1} & A'
\end{array}
$$

be a pull back diagram and assume additionally that j_1 or j_2 is *surjective*.

(The last requirement is fulfilled in Examples a) and b) above, but not generally in Example c).)

If $f : A \to B$ is a ring homomorphism and P a projective A-module, we denote by $f_\# P$ the projective B-module $A \otimes_A P$ and by $f_* : P \to f_\# P$ the canonical A-linear map $m \mapsto 1 \otimes m$.

From a pair of projective A_k-modules P_k for $k = 1, 2$ and an isomorphism $h : j_{1\#} P_1 \to j_{2\#} P_2$ we construct $M = M(P_1, P_2, h)$ to be the pull back of the pair of group homomorphisms $(h \circ j_{1*}, j_{2*})$. So the diagram

$$
\begin{array}{ccc}
M & \longrightarrow & P_1 \\
\downarrow & & \downarrow{\scriptstyle h \circ j_{1*}} \\
P_2 & \xrightarrow{j_{2*}} & j_{2\#} P_2
\end{array}
$$

is Cartesian. i.e. M consists of the pairs $(p_1, p_2) \in P_1 \times P_2$ which fulfill $h \circ j_{1*}(p_1) = j_{2*}(p_2)$. It is a subgroup of $P_1 \times P_2$ and becomes an A-module by $a \cdot (p_1, p_2) = (i_1(a)p_1, i_2(a)p_2)$.

Theorem 5.7.16 *In the above situation the following holds:*

a) $M = M(P_1, P_2, h)$ *is a projective A-module, and it is finitely generated if P_1 and P_2 are so.*

b) *Every (finitely generated) projective A-module is isomorphic to such an $M(P_1, P_2, h)$ for suitably chosen (finitely generated) P_1, P_2, h.*

c) $P_k \cong i_{k\#} M(P_1, P_2, h)$ *for $k = 1, 2$.*

In the following proof we will restrict ourselves to the case where all rings have Invariant Basis Property (see Definition 3.1.3), and all modules are finitely generated. We leave the general case to the reader, who may consult Milnor's book [58].

Proof. The first and biggest part of the proof will be to show a). We do this in several steps:

1) Let P_1, P_2 be free (of the same rank). Thereover we assume that P_1 has a basis $x_1, \ldots, x_n$ and P_2 has a basis $y_1, \ldots, y_n$ such that $h \circ j_{1*}(x_k) = j_{2*}(y_k)$ for $k = 1, \ldots, n$. Then clearly M is free with basis $(x_1, y_1), \ldots, (x_n, y_n)$.

2) From 1) we immediately derive

Lemma 5.7.17 *Let P_1, P_2 be free with bases $x_1, \ldots, x_n$, resp. $y_1, \ldots, y_n$ and let the matrix $\alpha = (a_{rs})$ over A' describe h with respect to these bases. We assume that α is the image under j_2 of an invertible matrix γ over A_2. Then M is also free of rank n.*

Namely, using the matrix α, one may change the basis $y_1, \ldots, y_n$ in such a way that one arrives at the situation of 1).

3) Let still P_1, P_2 be free and j_2 be surjective. Let again α describe h. Then let Q_1, Q_2 be free over A_1, resp. A_2 of rank n. Then one has an isomorphism $P_1 \oplus Q_1 \longrightarrow P_2 \oplus Q_2$ given by the matrix $\beta = \begin{pmatrix} \alpha & 0 \\ 0 & \alpha^{-1} \end{pmatrix}$.

By Whitehead's Lemma 3.4.4 we know that β is the product of elementary matrices, which are images of invertible matrices over A_2, since j_2 is surjective. Therefore $M(P_1 \oplus Q_1, \ P_2 \oplus Q_2, \ \beta)$ is free over A, according to Lemma 5.7.17.

On the other hand let $g : j_{1\#}Q_1 \longrightarrow j_{2\#}Q_2$ be described by α^{-1}. Then

$$M(P_1 \oplus Q_1, \ P_2 \oplus Q_2, \ \beta) = M(P_1, P_2, h) \oplus M(Q_1, Q_2, g).$$

Therefore $M(P_1, P_2, h)$ is a finitely generated projective A-module.

4) Now consider the general case (only with the restrictions indicated before the proof).

CLAIM: There are finitely generated projective modules Q_1, Q_2 over A_1, resp. A_2, such that $P_k \oplus Q_k$ are free for $k = 1, 2$ and $f_{1\#}Q_1 \cong f_{2\#}Q_2$.

Namely there are N_1, N_2 with $P_1 \oplus N_1 \cong A_1^r$ and $P_2 \oplus N_2 \cong A_2^s$. Writing $P' = j_{1\#}P_1 \cong j_{2\#}P_2$, we get

$$j_{1\#}N_1 \oplus (A')^s \cong j_{1\#}N_1 \oplus P' \oplus j_{2\#}N_2 \cong j_{2\#}N_2 \oplus (A')^r \ .$$

Now clearly $Q_1 := N_1 \oplus A_1^s$, $Q_2 := N_2 \oplus A_2^r$ fulfil the claim.

5) To show that $M(P_1, P_2, h)$ is projective, let Q_k be as under 4) and choose some isomorphism $h_1 : Q_1 \to Q_2$. Then

$$M(P_1, P_2, h) \oplus M(Q_1, Q_2, h_1) \cong M(P_1 \oplus Q_1, \ P_2 \oplus Q_2, \ h \oplus h_1) \ .$$

Now the module on the right hand side is finitely generated and projective by 3) and therefore $M(P_1, P_2, h)$ is so. So we have proven a).

We prove b). Let P be a finitely generated projective A-module. Then set $P_k := i_{k\#}P$ and let $h : j_{1\#}\circ i_{1\#}P \to j_{2\#}\circ i_{2\#}P$ be the canonical isomorphism. Then clearly $P = M(P_1, P_2, h)$.

We prove c). There is a natural A-linear map $M = M(P_1, P_2, h) \to P_1$. This induces an A_1-module homomorphism $f : i_{1\#}M \to P_1$. We have to show that f is an isomorphism.

If the conditions of Lemma 5.7.17 are fulfilled, the statement clearly is true. In the general case during the proof of a) we constructed projective modules Q_k such that $P_k \oplus Q_k$ are free and the conditions of Lemma 5.7.17 are satisfied for some isomorphism $h \oplus h' : j_{1\#}P_1 \oplus j_{1\#}Q_1 \to j_{2\#}P_2 \oplus j_{2\#}Q_2$. Therefore, setting $M' := M(Q_1, Q_2, h')$, the canonical map $f' : j_{1\#}M' \to Q_1$ has the property that $f \oplus f' : j_{1\#}M \oplus j_{1\#}M' \longrightarrow P_1 \oplus Q_1$ is an isomorphism. This implies that f is one.   □

Remark 5.7.18 Note for later use that is f is an automorphism of P_2 and $f' : A' \otimes P_2 \to A' \otimes P_2$ the induced automorphism, then $M(P_1, P_2, h) \cong M(P_1, P_2, f' \circ h)$.

5.7.3 Construction of a Noetherian Subalgebra

5.7.19 We will now consider the following situation. Let $\mathbb{F} = \mathbb{R}$ or $\mathbb{C}$. (We exclude the case $\mathbb{F} = \mathbb{H}$.) Let $A \subset B$ be an extension of $\mathbb{F}$-algebras with the following properties:

(1) $B = \mathcal{C}(X)$ where $X \subset \mathbb{R}^n$ is a compact set. B is a topological ring via the maximum norm.

(2) Every polynomial function on X belongs to A.

(3) $B^\times \cap A = A^\times$.

(Later we will consider A of the form $A = S^{-1}\mathbb{F}[x_1, \ldots, x_n, g_1, \ldots, g_m]$, where $x_1, \ldots, x_n$ are the coordinate functions, restricted to X and $g_1, \ldots, g_m$ are finitely many continuous functions on X, and finally S consists of all $s \in \mathbb{F}[x_1, \ldots, x_n, g_1, \ldots, g_m]$ which have no zero on X. For certain X by a skilful choice of the g_j we will achieve that the finitely generated B-modules – i.e. vector bundles on X – are extended from A.)

5.7.20 We will draw some conclusions from the assumptions (1), (2), (3).

The extension of rings $A \subset B$ satisfies (1), (2), (3') and (3) of Paragraph 5.7.7 and Remark 5.7.8 b).

Note that $A = B$ is not excluded. Therefore true statements for general A are true for B.

To every point $x \in X$ we associate a maximal ideal $\mathfrak{m}_x$ of A by $\mathfrak{m}_x := \{f \in A \mid f(x) = 0\}$, which is the kernel of the ring homomorphism $A \to \mathbb{F}$, $f \mapsto f(x)$. This homomorphism is surjective, since $\mathbb{F} \subset A$.

We define for every $x \in X$ a maximal ideal $\mathfrak{n}_x$ of B analogously. We have $\mathfrak{m}_x = A \cap \mathfrak{n}_x$.

If $x, y \in X$ are different, then $\mathfrak{m}_x \neq \mathfrak{m}_y$. Namely there is a polynomial function $f \in A$ with $f(x) = 0$, $f(y) \neq 0$.

CLAIM: Every maximal ideal of A is of the form $\mathfrak{m}_x$. The same holds for B, since $A = B$ is not excluded ...

PROOF: Let $\mathfrak{m}$ be any maximal ideal of A. It is enough to show that $\mathfrak{m} \subset \mathfrak{m}_x$ for some $x \in X$. Assume this were not the case. Then for every $x \in X$ there were an $f \in \mathfrak{m}$ with $f(x) \neq 0$. Since, for a fixed f, the set of the $x \in X$ with $f(x) \neq 0$ is open, and since X is compact there were finitely many $f_1, \ldots, f_r \in \mathfrak{m}$ without common zero. But then $f = f_1 \overline{f_1} + \cdots + f_r \overline{f_r}$ would have no zero on X, i.e. it would belong to $B^\times$, hence to $A^\times$. But on the other hand f belongs to the proper ideal $\mathfrak{m}$, a contradiction. (Here $\overline{f_k}$ denotes the complex conjugate of f_k. If $\mathbb{F} = \mathbb{R}$, then $\overline{f_k} = f_k$.)

In all, there are canonical bijective maps $X \longrightarrow \mathrm{Spmax}(B) \longrightarrow \mathrm{Spmax}(A)$.

If one equips $\mathrm{Spmax}(B)$ and $\mathrm{Spmax}(A)$ with their Zariski topology, these maps become continuous.

The first one, $X \to \mathrm{Spmax}(B)$ then even is a homeomorphism. Namely if $Y \subset X$ is a closed subset, the distance function $f(x) = \mathrm{dist}(x, Y)$ is continuous and has the property that $f(x) = 0$, if and only if $x \in Y$, since Y is a compact subset of $\mathbb{R}^n$. So $V(f) \cap \mathrm{Spmax}(B)$ is the set of the maximal ideals $\mathfrak{m}_x$ of B with $x \in Y$.

5.7.21 For an ideal I of a commutative ring, by $\mathrm{rad}(I)$ we will denote the intersection of all maximal ideals which contain I. (This means that $\mathrm{rad}(I)/I = \mathrm{Jac}(A/I)$.) Clearly $\sqrt{\mathrm{rad}(I)} = \mathrm{rad}(I)$ and $\sqrt{I} \subset \mathrm{rad}(I)$.

If I is an ideal of A, then $\mathrm{rad}(IB) \cap A = \mathrm{rad}(I)$. Namely for $x \in X$ we have the equivalences

$$I \subset \mathfrak{m}_x \iff I \subset \mathfrak{n}_x \iff IB \subset \mathfrak{n}_x$$

Further we know already $\mathfrak{n}_x \cap A = \mathfrak{m}_x$.

Proposition 5.7.22 *In the above situation let I be an ideal of A with $\mathrm{rad}(I) = I$. Set $J := \mathrm{rad}(IB)$. Then the extension of $\mathbb{F}$-algebras $A/I \subset B/J$ has the properties* (i), (ii) *and* (iii).

Proof. Let $Y := \{y \in X \mid f(y) = 0 \text{ for all } f \in I\} = \{y \in X \mid \mathfrak{m}_y \supset I\} = \{y \in X \mid \mathfrak{n}_y \supset IB\}$. Then Y is a closed subset of X. By Tietze's extension theorem, every continuous function on Y can be extended to X. This means that the restriction homomorphism $\mathcal{C}(X) \to \mathcal{C}(Y)$ is surjective. Its kernel clearly is $\mathrm{rad}(I\mathcal{C}(X)) = J$. Therefore $B/J = \mathcal{C}(Y)$ in a canonical way. This is (i).

The polynomial functions on Y extend to polynomial functions on X. Polynomial functions on X restrict to equal functions on Y, if and only if they are congruent modulo I. Therefore A/I contains the polynomial functions on Y. This is (ii).

We have seen at the beginning of the proof that the maximal ideals of A/I are the $\mathfrak{m}_y/I$ with $y \in Y$. Let $a \in A$ such that its residue class $\bar{a}$ belongs to $(B/J)^{\times}$. This means that $a \notin \mathfrak{n}_y$ for all $y \in Y$. But then $a \notin \mathfrak{m}_y$ for all $y \in Y$. So $\bar{a}$ is a unit in A/I. This is (iii). $\qquad \square$

Lemma 5.7.23 *Let $B = \mathcal{C}(X)$ be as above ($X \subset \mathbb{R}^n$ compact) and J an ideal of B. Let $\alpha \in \mathrm{GL}_m(B/J)$. There is an $\varepsilon > 0$ such that for any $m \times m$-matrix $\delta = (d_{ij})$ over B with $|d_{ij}| < \varepsilon$ for all i, j there is a $\beta \in \mathrm{GL}_m(B)$ with $\bar{\beta}\alpha = \alpha + \delta$. (Here $\bar{\beta}$ denotes the residue class modulo J.)*

Proof. Let α' be an $m \times m$-matrix over B such that its residue class modulo J is α^{-1}, and set $\beta = I_m + \delta\alpha'$, which is invertible if ε is small enough. $\qquad \square$

Theorem 5.7.24 (Carral) *Let $A \subset B$ be a ring extension which fulfills (i), (ii) and (iii) above. Let further Q be a finitely generated projective B-module for which there are finitely many ideals $I_1, \ldots, I_d$ of A with $\mathrm{rad}(I_k) = I_k$ and $\bigcap_{k=1}^{d} I_k = (0)$ such that $Q/I_k Q$ is free of constant rank over $B/I_k B$ for $k = 1, \ldots, d$. Then there is a projective A-module P with $B \otimes P \cong Q$.*

Proof. For $k = 1, \ldots, d$ set $J_k := \mathrm{rad}(I_k B)$. There are free A/I_k-modules P_k with $B \otimes P_k \cong Q/J_k Q \ (= Q/I_k Q)$. We will glue these by Milnor's method step by step.

So let $1 < r \leq d$, set $I = \bigcap_{k=1}^{r-1} I_k$, $J := \mathrm{rad}(IB)$ and assume, there is a projective A/I-module P_1 with an isomorphism $\varphi : Q/JQ \xrightarrow{\sim} B \otimes P_1$.

We have a Cartesian diagram of rings with surjective homomorphisms

$$
\begin{array}{ccc}
B/(J \cap J_r) & \longrightarrow & B/J \\
\downarrow & & \downarrow \\
B/J_r & \longrightarrow & B/(J + J_r)
\end{array}
$$

Write $Q' := Q/JQ$, $Q_r := Q/J_r Q$. Then there is an isomorphism $h : Q'/J_r Q' \to Q_r/JQ_r$ with $Q/(J \cap J_r)Q = M(Q', Q_r, h)$. Note that $Q_r \cong (B/J_r)^m$.

Hence $Q/(J \cap J_r)Q \cong M\big((B/J) \otimes P_1, \ (B/J_r)^m, \ h \circ \overline{\varphi}\big)$, where here and later on the overbar means 'modulo $J + J_r$.'

Since $Q/(J + J_r)Q$ is free, by Theorem 5.7.12 also $P_1/(J + J_r)P_1$ is free over $B/(J + J_r)$. Hence $h \circ \overline{\varphi}$ is given by a matrix $\alpha \in \mathrm{GL}_m(\overline{B})$. There is a matrix $\alpha' \in \mathrm{GL}_m(A/(I + I_r))$, arbitrarily near to α. By Lemma 5.7.23 there is a matrix $\beta \in \mathrm{GL}_m(B/J_r)$ such that $\overline{\beta}\alpha = \alpha'$. By Remark 5.7.18 we see that

$$
Q/(J \cap J_r)Q \cong M\big((B/J) \otimes P_1, \ (B/J)^m, \ \overline{\beta} \circ h \circ \overline{\varphi}\big) =
$$

$$
M\big((B/J) \otimes P_1, \ (B/J)^m, \ \alpha'\big) \cong B/(J \cap J_r) \otimes M\big(P_1, (A/I)^m, \alpha'\big).
$$

$\square$

Corollary 5.7.25 *Let $X \subset \mathbb{R}^n$ be a compact subset which is the union of finitely many contractible closed subsets. Then there is an $\mathbb{F}$-subalgebra A of $B = \mathcal{C}(X)$, which is essentially of finite type over $\mathbb{F}$ such that the induced map $\mathbb{P}(A) \to \mathbb{P}(B)$ is bijective.*

Proof. Let $X = \bigcup_{k=1}^{d} Y_k$ with closed contractible Y_k. Further for $k = 1, \ldots, d$ let $f_k \in \mathcal{C}(X)$ be the distance function $f_k(x) := \mathrm{dist}(x, Y_k)$. Then set $A := S^{-1}\mathbb{F}[x_1, \ldots, x_n, f_1, \ldots, f_d]$, where the x_j are the coordinate functions restricted to X. For $k = 1, \ldots, d$ set $I_k := \mathrm{rad}(f_k A)$. Then the hypotheses of Theorem 5.7.24 are fulfilled for every finitely generated projective B-module Q. So the corollary follows. $\square$

Remarks 5.7.26 a) This corollary clearly applies to the spheres, but also to the infinite union X of the following n-spheres: the k-th of them has the radius $1/k$ and the midpoint $(1/k, 0, 0, \dots, 0)$. Then $X = Y_1 \cup Y_2$ with $Y_1 = \{(x_0, \dots, x_n) \in X \mid x_n \geq 0\}$, $Y_2 = \{(x_0, \dots, x_n) \in X \mid x \leq 0\}$ and Y_1, Y_2 both are contractible. For instance if $n = 2$, one gets a vast variety of projective modules.

b) One may embed two n-spheres into $\mathbb{R}^{n+}$ such that their intersection is a common equator. Namely let one sphere be defined by the polynomial $f_1 := X_0^1 + \dots + X_n^2 - 1$ the other by $f_2 := 2X_0^2 + X_1^2 + \dots + X_n^2 - 1$. Let $X \subset \mathbb{R}^{n+1}$ be the union of the upper half sphere Y_1 of $V_{\mathbb{R}}(f_1)$ (where $x_0 \geq 0$) and the lower half sphere Y_2 of $V_{\mathbb{R}}(f_2)$ (where $x_0 \leq 0$). Then X is homeomorphic to an n-sphere. Now define $A := S^{-1}\mathbb{R}[X_0, \dots, X_n]/(f_1 f_2)$ where S consists of those functions which have no zero along X. Then $V(\overline{f_k}) = Y_k$ in the maximal spectrum of A, and we have used only rational functions to build A.

This can be generalized to finite 'CW complexes', see [16].

Remark 5.7.27 R.G. Swan in his paper [99] constructs rings A, essentially of finite type over $\mathbb{F}$ with an interesting behaviour of their projective modules. For instance for every $m \equiv 2 \mod 4$ he constructs a Noetherian ring A of Krull dimension m (see Chapter 6) such that there is a nonfree projective A-module of rank m, but of no other rank.

To show this, he uses deep results on homotopy theory we cannot present here. Therefore we recommend the interested reader to read his nicely written original paper.

6

Basic Commutative Algebra II

In the rest of the book our main interest will be, to study the minimal numbers of generators of ideals. A natural beginning is to turn the attention to those rings whose ideals are all finitely generated.

Apart from the very beginning in this chapter all rings are supposed to be *commutative*, unless stated otherwise.

6.1 Noetherian Rings and Modules

Proposition 6.1.1 *Let R be a (not necessarily commutative) ring and M an R-module. The following are equivalent statements.*

(1) **Finite generation:** *Every submodule N of M (including M itself) is finitely generated.*

(2) **Ascending chain condition:** *Every ascending chain of submodules*

$$M_1 \subseteq M_2 \subseteq \cdots \subseteq M_n \subseteq \cdots$$

becomes stationary, i.e. there is a k such that $M_n = M_k$ for all $n \geq k$.

(3) **Maximal condition:** *Every non-empty subset of the set of all submodules of M has a maximal element.*

The proof is a nice exercise.

Definition 6.1.2 *A module satisfying the above conditions is called a* **Noetherian R-module.** *A ring R is called a* **(left) Noetherian ring** *if it is a Noetherian as a (left) R-module.*

Remarks 6.1.3 a) Every sub-module and factor module of a Noetherian module clearly is Noetherian. And every residue class ring of a Noetherian ring is so, too. But, as we will see under c), a subring A of a Noetherian ring B need not be Noetherian.

Conversely, if M is a module and U a submodule of M such that U and M/U are Noetherian, then M is so.

Namely, let E be any submodule of M. If $x_1, \ldots, x_n$ generate $E \cap U$ and residue classes $\overline{y_1}, \ldots, \overline{y_m}$ of $y_1, \ldots, y_m \in E$ generate $E/(E \cap U) \cong (E + U)/U$, then $x_1, \ldots, x_n, y_1, \ldots, y_m$ generate E, as one easily shows.

b) Consequently, if R is a Noetherian ring, then any finitely generated free module R^n is a Noetherian module. (Induction on n.) And therefore any finitely generated module over a Noetherian ring R – being a factor module of some R^n – is Noetherian.

By this we see that every finitely generated module M over a Noetherian ring R has a resolution (see Remark 1.2.19) by finitely generated free modules –

$$\cdots \longrightarrow R^{n_1} \longrightarrow R^{n_0} \longrightarrow M \longrightarrow 0.$$

c) Consider ring extensions $A \subset B$. Since an ideal of A need not be one of B, there are cases, where B is Noetherian but A not. One may take A to be any non-Noetherian domain – e.g. a polynomial ring in infinitely many indeterminates over a field, or the ring of holomorphic functions on a connected open non-empty subset of $\mathbb{C}$ – and B to be the quotient field of A.

(We indicate why the ring A of holomorphic functions on a non-empty connected open subset U of $\mathbb{C}$ is a domain and not Noetherian. First let $fg = 0$ on U, then at least one of the functions f, g is zero on a non-discrete subset of U and must therefore be identically zero by the Identity Theorem. Secondly let (z_n) be a sequence of points which is discrete in U. We define an ascending chain of ideals $I_m := \{f \in A \mid f(z_n) = 0 \text{ for all } n \geq m\}$. By Weierstrass' Product Theorem for every m there is an $f \in A$ whose zeros are exactly the z_n with $n \geq m$. Therefore the I_m form an infinite strictly ascending chain of ideals.)

Theorem 6.1.4 (Hilbert's Basissatz) *Let R be a Noetherian ring. Then $R[x]$ is Noetherian ring.*

Proof. If not, then we can find an ideal I of $R[x]$ which is not finitely generated Choose $f_1 \in I \setminus \{0\}$ of *least* degree, say n_1. Inductively choose $f_k \in I \setminus (f_1, \ldots, f_{k-1})$ of least degree n_k, for all $k \geq 1$. This is possible since I is not finitely generated Clearly $n_i \leq n_{i+1}$ for all i. Let a_i be the leading coefficient of f_i, and $J_i := Ra_1 + \cdots + Ra_i$ for all i. Since R is Noetherian we have $J_k = J_{k+1}$ for some k. But then $a_{k+1} = \sum_{i=1}^{k} \lambda_i a_i$ for some $\lambda_i \in R$, and

$$g_{k+1} := f_{k+1} - \sum_{i=1}^{k} \lambda_i f_i x^{n_{k+1}-n_i} \in I \backslash (f_1, \ldots, f_k), \text{ whereas } \deg(g_{k+1}) < n_{k+1} \,.$$

This contradicts the choice of f_{k+1} . □

Corollary 6.1.5 *An algebra of finite type* $k[x_1, \ldots , x_n]/I$ *over a field or principal domain* k *is a Noetherian ring.* □

We now give I.N. Herstein's proof ([37]) of Krull's Intersection Theorem and begin with the following

Lemma 6.1.6 *Let* R *be a Noetherian ring,* I *an ideal of* R, *further* M *a finitely generated* R-module *and* N *a submodule of* M. *Let* N' *be a submodule of* M *which is maximal with respect to the property* $N' \cap N = IN$. *Then* $I^n M \subset N'$ *for some* n.

Proof. Since I is finitely generated it suffices to show that for any $a \in I$ there is an m with $a^m M \subset N'$. (If $I = (a_1, \ldots , a_r)$ and $a_i^{m_i} M \subset N'$, then, with $m = \text{Max}_i\{m_i\}$, we have $I^{mr} M \subset N'$.) Define an ascending chain of submodules D_r of M by:

$$D_r := \{x \in M \mid a^r x \in N'\}.$$

By the Noetherian property the chain D_r becomes stationary, say at $r = m$.

CLAIM: $(a^m M + N') \cap N = IN.$

Clearly, $IN = N' \cap N \subset (a^m M + N') \cap N$. Conversely, if $t = a^m x + y \in$ l.h.s., for some $x \in M$, $y \in N'$, then $at = a^{m+1}x + ay \in aN \subset IN \subset N'$ and so $a^{m+1}x \in N'$. Hence $a^m x \in N'$, because $D_{m+1} = D_m$. And so $t \in N' \cap N = IN$.

By the maximal property of N' we see $a^m M \subset N'$!   □

Theorem 6.1.7 (Krull's Intersection Theorem) *Let* R *be a Noetherian ring,* I *an ideal of* R, M *a finitely generated* R-module, *and* $M_0 = \bigcap_{n>0} I^n M$. *Then* $IM_0 = M_0$. *In particular, if* R *is a domain or a local ring and* $I \neq R$, *then* $M_0 = 0$.

Proof. Let $\mathcal{S} := \{N \text{ submodule of } M \mid N \supset IM_0, \text{ and } N \cap M_0 = IM_0\}$. By the Noetherian property $\mathcal{S}$ has a maximal element N'. By Lemma 6.1.6 $I^n M \subset N'$, for some n; and so $M_0 \subset I^n M \subset N' \subset IM_0$ and so $M_0 = N' \cap M_0 = IM_0$. The last assertion follows via Nakayama's Lemma.   □

Noetherianity behaves well under localization.

Proposition 6.1.8 *Let R be a commutative ring and $S \subset R$ a multiplicative subset. If M is a Noetherian R-module, then $S^{-1}M$ is a Noetherian $S^{-1}R$-module. Especially, if R is Noetherian, then so is $S^{-1}R$.*

Proof. Let $E_1 \subset E_2$ be $S^{-1}R$-submodules of $S^{-1}M$. Let $E_j' := i_{M,S}^{-1}(E_j)$. Then $S^{-1}E_j' = E_j$ and $E_1' \subset E_2'$. So we have an injective order preserving map from the set of all $S^{-1}R$-submodules of $S^{-1}M$ to the set of all R-submodules of M.

From this the statement follows.   □

Note that as an R-module, $S^{-1}M$ need not be Noetherian. For example $\mathbb{Q}$ is finitely generated over $\mathbb{Q}$; but it is not so over $\mathbb{Z}$.

6.2 Irreducible Sets

6.2.1 A topological space X is called **irreducible** if it is nonempty and either of the following equivalent statements holds

(1) every non-empty open set is dense in it,

(2) any two non-empty open sets intersect non-trivially,

(3) $X = F_1 \cup F_2$, F_1, F_2 closed, implies $X = F_1$ or $X = F_2$.

Proposition 6.2.2 *Let A be a ring. The irreducible closed subsets of $\operatorname{Spec}(A)$ are those of the form $\mathrm{V}(\mathfrak{p})$ with $\mathfrak{p} \in \operatorname{Spec}(A)$.*

Namely, let $\mathrm{V}(\mathfrak{p}) = \mathrm{V}(I) \cup \mathrm{V}(J)$. Then $\mathfrak{p} = \sqrt{\mathfrak{p}} = \sqrt{IJ}$. So $I \subset \mathfrak{p}$ or $J \subset \mathfrak{p}$, i.e. $\mathrm{V}(I) \supset \mathrm{V}(\mathfrak{p})$ or $\mathrm{V}(J) \supset \mathrm{V}\mathfrak{p})$.

Conversely, assume that $\mathrm{V}(I)$ is irreducible for a radical ideal I. If I were not prime, there would be $a_1, a_2 \notin I$ with $a_1 a_2 \in I$. Then $\mathrm{V}(I, a_i) \subsetneq \mathrm{V}(I)$ for $i = 1, 2$, since $I = \sqrt{I}$, and also $\mathrm{V}(I) = \mathrm{V}(I, a_1) \cup \mathrm{V}(I, a_2)$, contradicting the irreducibility.   □

Thus, the map V gives an inclusion reversing bijection between prime ideals and irreducible closed subsets of $\operatorname{Spec}(R)$.

6.2.3 An **irreducible component** of a topological space X is a maximal irreducible subset of X. If Y is an irreducible subset of X then so is its closure $\overline{Y}$. Therefore, any irreducible component of X is a closed set.

A topological space X is called **Noetherian** if it satisfies the **descending chain condition** on closed subsets, i.e. any sequence

$$Y_1 \supseteq Y_2 \supseteq \cdots \supseteq Y_r \supseteq \cdots$$

of closed subsets Y_i, $i \geq 1$, is eventually stationary.

Example 6.2.4 If R is a Noetherian ring then $\mathrm{Spec}(R)$ is a Noetherian space. The converse is not true. See Exercise 22.

Proposition 6.2.5 *In a Noetherian topological space* X *every non-empty closed subset* Y *can be expressed as a finite union* $Y = Y_1 \cup Y_2 \cup \cdots \cup Y_r$ *of irreducible closed subsets* Y_i. *If in addition we assume that* $Y_i \not\subset Y_j$, *for all* $i \neq j$, *then the* Y_i *are uniquely determined. These are the irreducible components of* Y.

Proof. Let S be the set of non-empty closed subsets of X which cannot be written as a finite union of irreducible closed subsets. If S is non-empty, then, since X is Noetherian, it must contain a minimal element, say Y. Since Y, belonging to S, is not irreducible we may write $Y = Y_1 \cup Y_2$ where Y_1 and Y_2 are proper closed subsets of Y. By the minimality of Y both Y_1, Y_2 can be expressed as finite union of closed irreducible subsets, hence also Y, a contradiction. Hence $S = \emptyset$ and so every closed set Y has an expression as a finite union of irreducible closed subsets Y_i. By throwing away a few we may assume that $Y_i \not\subset Y_j$ if $i \neq j$.

Now let $Y = \bigcup_{i=1}^{r} Y_i$ such that $Y_i \not\subset Y_j$ for $i \neq j$. If Z is any irreducible subset of Y, by

$$Z = \bigcup_{i=1}^{r} (Z \cap Y_i)$$

we get $Z \subset Y_i$ for some i, since Z is irreducible. So $Y_1, \ldots, Y_r$ are **the** maximal irreducible subsets of Y, i.e. the irreducible components of Y. From this the uniqueness follows directly. $\qquad\square$

Corollary 6.2.6 (Finiteness of minimal prime over-ideals) *Let* R *be a ring with a Noetherian spectrum (e.g.* R *Noetherian) and* $\mathfrak{a}$ *an ideal of* R. *Then there are finitely many prime ideals* $\mathfrak{p}_1, \ldots, \mathfrak{p}_r$ *such that* $\sqrt{I} = \bigcap_{i=1}^{r} \mathfrak{p}_i$.

Proof. $\mathrm{Spec}(R)$ is a Noetherian topological space and $V(I) = \bigcup_{i=1}^{r} Y_i$ with Y_i irreducible closed sets. Therefore $Y_i = V(\mathfrak{p}_i)$ for some prime ideals $\mathfrak{p}_i$ of R. Hence, $\mathrm{I}(V(I)) = \sqrt{I} = \mathrm{I}(\bigcup_{i=1}^{r}) V(\mathfrak{p}_i) = \bigcap_{i=1}^{r} \mathrm{I}(V(\mathfrak{p}_i)) = \bigcap_{i=1}^{r} \mathfrak{p}_i$. $\qquad\square$

6.3 Dimension of Topological Spaces and Rings

6.3.1 If X is a topological space we define the **dimension** of X (denoted by dim X) to be the supremum of all integers n such that there is a chain

$$Z_0 \subsetneq Z_1 \subsetneq \cdots \subsetneq Z_n$$

of distinct irreducible closed subsets of X.

Note that this definition of dimension is only suited to spaces with a 'Zariski'-like topology. Consider e.g. $\mathbb{C}^n$ with its usual (Euclidean) topology. Then its irreducible sets are the one-point sets. So its dimension, as defined above, would be 0. The **'Zariski topology'** on $\mathbb{C}^n$ is defined as follows: The closed subsets are defined to be the zero sets of systems of polynomials (in n variables). $\mathbb{C}^n$ equipped with this topology is homeomorphic to the maximal spectrum of $\mathbb{C}[X_1, \ldots, X_n]$ and has dimension n. (See Section 8. on Hilbert's Nullstellensatz . This section also will explain the geometric meaning of 'dimension'.)

We define the **Krull dimension** of a ring R to be $\dim(\mathrm{Spec}(R))$. Thus, the Krull dimension of R is the length n of a chain of maximal length, if any, of prime ideals in R:

$$\mathfrak{p}_0 \subsetneq \mathfrak{p}_1 \subsetneq \cdots \subsetneq \mathfrak{p}_n.$$

If no such chain exists (and $R \not\cong 0$) then we say that R has infinite Krull dimension.

The dimension of a vector space V over a field K is denoted by $\dim_K V$.

Remark 6.3.2 If a ring B is integral over a subring A, then $\dim A = \dim B$. This follows from Proposition 1.4.18 and Theorem 1.4.19.

The polynomial ring in infinitely many variables is an example of a ring with infinite dimension. It is, however, not Noetherian. The following example of M. Nagata shows that there are Noetherian rings of infinite dimension.

Example 6.3.3 Let $R = k[x_1, x_2, \ldots]$ be the polynomial ring over a field k in infinitely, but countably many indeterminates and $\mathfrak{p}_1 = (x_1)$, $\mathfrak{p}_2 = (x_2, x_3)$, $\mathfrak{p}_3 = (x_4, x_5, x_6)$, $\ldots$. Let $S = R \backslash \bigcup_{i \geq 1} \mathfrak{p}_i$, which is a multiplicatively closed subset of R. Set $A := S^{-1}R$ and $\mathfrak{m}_i := S^{-1}\mathfrak{p}_i$.

First we show that the $\mathfrak{m}_i$ are the maximal ideals of A. Let $\mathfrak{a} \neq (0)$ be an ideal of R which is contained in $\bigcup_{i \geq 1} \mathfrak{p}_i$. It suffices to show that $\mathfrak{a}$ is already contained in one of the $\mathfrak{p}_i$. Assume it were not so. For any n we have $\mathfrak{a} \subset \mathfrak{p}_1 \cup \cdots \cup \mathfrak{p}_n \cup (\sum_{i > n} \mathfrak{p}_i)$. So by Proposition 1.1.8 we get $\mathfrak{a} \subset \sum_{i > n} \mathfrak{p}_i$, since we assumed $\mathfrak{a} \not\subset \mathfrak{p}_i$ for all i. But clearly $\bigcap_{n \geq 1}(\sum_{i > n} \mathfrak{p}_i) = (0)$, which leads to a contradiction. (We owe this argument to H.J. Fendrich (Mainz).)

Note that any non-zero element in A belongs to only finitely many $\mathfrak{m}_i$ as $f(x_1, \ldots, x_r) \in R \setminus (0)$ is not in $\mathfrak{p}_i$ for large i. So every non-zero ideal I of A is contained in only finitely many $\mathfrak{m}_i$. Now every $A_{\mathfrak{m}_i} = R_{\mathfrak{p}_i}$ is a localization of a polynomial ring over some field (of infinite transcendence degree over k), hence Noetherian. Let $E_i := \{a_{i1}, \ldots, a_{ir_i}\}$ with $a_{ij} \in I$ generate $I_{\mathfrak{m}_i}$ in $A_{\mathfrak{m}_i}$, and let J be the ideal of A which is generated by the union of those E_i with $I \subset \mathfrak{m}_i$. Then J is finitely generated $J \subset I$ and $I_\mathfrak{m} = J_\mathfrak{m}$ for all maximal ideals of A, whence $I = J$ is finitely generated.

An important theorem in Commutative Algebra is that a Noetherian semilocal ring has finite dimension. This will be established in the next two sections.

Definition 6.3.4 *We say that a prime ideal $\mathfrak{p}$ has* **height** *r if there is a chain of prime ideals*

$$\mathfrak{p} = \mathfrak{p}_r \supsetneq \mathfrak{p}_{r-1} \supsetneq \cdots \supsetneq \mathfrak{p}_0$$

and there is no such chain of longer length. Notation: $\mathrm{ht}(\mathfrak{p})$ or $\mathrm{ht}_R(\mathfrak{p})$ if the ring R is to be specified.

It may happen that there are maximal (i.e. non refinable and non elongable) such chains of different length!

In view of the bijective correspondence between prime ideals of R contained in $\mathfrak{p}$ and prime ideals of $R_\mathfrak{p}$ we have $\mathrm{ht}(\mathfrak{p}) = \dim(R_\mathfrak{p}) = \dim(\mathrm{Spec}(R_\mathfrak{p}))$.

For example if R is a domain then (0) has height 0. If R is principal ideal domain and not a field then $\dim R = 1$.

In general a principal prime ideal of Noetherian ring has height ≤ 1.

PROOF: Let $\mathfrak{p} := xR$ and $\mathfrak{q}$ be different prime ideals with $xR \supset \mathfrak{q}$ then $\mathfrak{q} = x\mathfrak{q}$. Namely every element of $\mathfrak{q}$ is of the form xr with $r \in \mathfrak{q}$, since $x \notin \mathfrak{q}$ and $\mathfrak{q}$ is prime. In the local ring $R_\mathfrak{p}$ Nakayama's Lemma implies then $\mathfrak{q}_\mathfrak{p} = (0)$, whence $\mathfrak{q}$ is minimal.

(We will soon generalize this to prime ideals which are not necessarily principal themselves, but only minimal prime over-ideals of a principal ideal.)

The chain of prime ideals

$$(0) \subsetneq (x_1) \subsetneq (x_1, x_2) \subsetneq \cdots \subsetneq (x_1, \ldots, x_n)$$

in $k[x_1, \ldots, x_n]$ shows that $\dim(k[x_1, \ldots, x_n]) \geq n$. (Later we show equality.)

Definition 6.3.5 *If I is an ideal of R we define its height to be $\mathrm{ht}(I) = \inf\limits_{\mathfrak{p} \in V(I)} \mathrm{ht}(\mathfrak{p})$. Especially if $I = R$ then $\mathrm{ht}(I) = \infty$.*

If S is a multiplicatively closed subset of R and I is an ideal of R then $\mathrm{ht}(I_S) \geq \mathrm{ht}(I)$ and $\mathrm{ht}(I_S) = \mathrm{ht}(I)$ if and only if some minimal prime over-ideal of I which has the same height as I does not intersect S.

Later on we will show how the concept of dimension suffices to establish a valuable bound on the number $\mu(M)$ of generators of a finitely generated R-module M in terms of the spectrum of R. Further we present a proof of Serre's Conjecture due to H. Lindel, which only needs the concept of dimension of a Noetherian ring and its main properties!

6.4 Artinian Rings: 0-Dimensional Noetherian Rings

We first recall the **Jordan - Hölder theorem** (cf. [110], [45]).

A **Jordan-Hölder series** for an R-module M is a chain

$$M = M_0 \supseteq M_1 \supseteq \cdots \supseteq M_n = 0 \qquad (*)$$

of submodules M_i of M, $0 \leq i \leq n$, such that for every i the factor M_i/M_{i+1} is simple, i.e. has no proper submodule and is not isomorphic to 0. We say that $(*)$ is a Jordan-Hölder series of **length** n.

Theorem 6.4.1 (Jordan - Hölder) *If M has a Jordan-Hölder series then any chain of submodules of M can be refined to a Jordan-Hölder series, and any two Jordan-Hölder series have the same length.*

Definition 6.4.2 *We say, an R-module M has* **finite length** l *if it has a Jordan-Hölder series of length l.*

Examples 6.4.3 A finite dimensional vector space has a finite length; any finite abelian group G has finite length; for any ring R and finitely generated maximal ideal $\mathfrak{m}$ of R the ring $R/\mathfrak{m}^n$ has finite length as

$$R/\mathfrak{m}^n \supset \mathfrak{m}^1/\mathfrak{m}^n \supset \mathfrak{m}^2/\mathfrak{m}^n \supset \cdots \supset (0)$$

is a chain of ideals where the i-th factor is isomorphic to $\mathfrak{m}^i/\mathfrak{m}^{i+1}$, which is a finite dimensional vector space over $R/\mathfrak{m}$ as $\mathfrak{m}$ is finitely generated The series can therefore be refined to a Jordan-Hölder series for $R/\mathfrak{m}^n$.

The next proposition shows that this is a typical prototype of rings of finite length:

Proposition 6.4.4 *Let R be a ring. The following statements are equivalent:*

(1) *R is a Noetherian ring with every prime ideal a maximal ideal.*

(2) *Every finitely generated R-module has finite length.*

(3) *R has finite length as a R-module.*

(4) **Descending chain condition on ideals:** *Every descending chain*

$$I_1 \supseteq I_2 \supseteq \cdots \supseteq I_r \supseteq \cdots$$

of ideals I_i of R is stationary after some stage, i.e. there is a $k > 0$ such that $I_k = I_{k+r}$ for all $r \geq 0$.

(5) **Minimal condition:** *Any non-empty subset of the set of all ideals of R has a minimal element.*

(6) *$(0) = \mathfrak{m}_1^{n_1} \cdots \mathfrak{m}_r^{n_r}$ for finitely many maximal ideals $\mathfrak{m}_i$ of R and $n_i \in \mathbb{N}$.*

To show this, we suggest $(1) \Rightarrow (2) \Rightarrow (3) \Rightarrow (4) \Leftrightarrow (5) \Rightarrow (6) \Rightarrow (1)$. For details we refer to [110, §2, §3 see pages 203 - 208]. It is much easier, and totally sufficient for our purposes, to prove the equivalence of (1) to (6) under the general hypothesis, that R is Noetherian. The proof that (4) or (5) imply the Noetherian property of R is not at all automatic. (Note that *modules*, satisfying the analogue of (4) or (5) need not be Noetherian! For example let p be a prime number and $A \subset \mathbb{Q}$ the additive group consisting of the fractions of the form m/p^n. Then $A/\mathbb{Z}$ is an Artinian not Noetherian $\mathbb{Z}$-module.)

Definition 6.4.5 *A ring satisfying any, hence all, of the above six condition is called an* **Artinian ring**.

Thus an Artinian ring is a zero-dimensional Noetherian ring and conversely. If $\mathrm{Jac}(R) = \bigcap_{i=1}^{r} \mathfrak{m}_i = \mathfrak{m}_1 \cdots \mathfrak{m}_r$ is the Jacobson radical of an Artinian ring R then due to (vi) $\mathrm{Jac}(R)$ is a nilpotent ideal, i.e. $\mathrm{Jac}(R)^l = 0$ for some $l \geq 0$.

Note that by the Chinese Remainder Theorem for an Artinian ring R we have $R \cong \prod_{i=1}^{r} R/\mathfrak{m}^{n_i}$ with the notation of (vi) above.

Corollary 6.4.6 *A reduced Artinian ring is a finite direct product of fields.*

6.5 Small Dimension Theorem

A central theorem of Commutative Algebra says that the dimension of a local and hence of a semilocal Noetherian ring is finite. We recall a proof of Kaplansky (cf. [37]) here. A preliminary lemma:

Lemma 6.5.1 *Let R be a domain and let $x, y \in R \setminus \{0\}$. Assume that the R-module $(x, y)/(x^2)$ has finite length and that $((y) : x) = ((y) : x^2)$. Then $(x, y) = (x^2, y)$.*

Proof. Since R is a domain, multiplication by x induces an isomorphism

$$\frac{(x, y)}{(x)} \xrightarrow{\cdot x}_{\sim} \frac{(x^2, xy)}{(x^2)}.$$

Since $((y) : x) = ((y) : x^2)$ it follows that the kernel of the map $R \to (x^2, y)/(x^2, xy)$, $1 \mapsto \overline{y}$ is (x). Hence,

$$\frac{R}{(x)} \xrightarrow{\sim} \frac{(x^2, y)}{(x^2, xy)}.$$

Since R is a domain, $R/(x) \xrightarrow{\sim} (x)/(x^2)$. Therefore, $l((x, y)/(x^2)) = l((x, y)/(x)) + l((x)/(x^2)) = l((x^2, xy)/(x^2)) + l((x^2, y)/(x^2, xy)) = l((x^2, y)/(x^2))$. Hence, $(x^2, y) = (x, y)$. $\qquad\square$

Proposition 6.5.2 *Let R be a Noetherian local domain with maximal ideal $\mathfrak{m}$. Assume that there is a $z \in \mathfrak{m}$ such that $\mathfrak{m}$ is a minimal prime over-ideal of (z). Then $\mathrm{ht}(\mathfrak{m}) \leq 1$. In fact $\mathrm{Spec}(R) = \{(0), \mathfrak{m}\}$.*

Proof. Note that by Proposition 6.4.4 (i) the ring $R/(z)$ is Artinian since $\mathrm{Spec}(R/(x)) = \{\mathfrak{m}\}$ and $R/(z)$ is Noetherian. Let $y \in \mathfrak{m} \setminus \{0\}$.

The ascending chain $\{((y) : z^k)\}$ becomes stationary say

$$((y) : z^n) = ((y) : z^{n+1}) = \cdots = ((y) : z^{2n}) = \cdots$$

Let $x = z^n$. Consider the R/x^2-module $(x, y)/(x^2)$. Since $R/(x^2) = R/(z^{2n})$ is clearly Artinian, $(x, y)/(x^2)$ has finite length. Now $(x, y) = (x^2, y)$. By Lemma 6.5.1. Therefore, $x = \lambda x^2 + \mu y$, for some $\lambda, \mu \in R$. Hence, $x(1 - \lambda x) \in (y) \Rightarrow x \in (y)$ as $1 - \lambda x$ is a unit in R.

If there were a prime ideal $\mathfrak{p}$ different from (0) and $\mathfrak{m}$, we may choose $y \in \mathfrak{p} \setminus \{0\}$. From $z^n = x \in (y) \in \mathfrak{p}$ we get $z \in \mathfrak{p}$, contradicting the hypothesis that $\mathfrak{m}$ should be a minimal prime over-ideal of (z). $\qquad\square$

Definition 6.5.3 a) *If M is a finitely generated R-module, for example a finitely generated ideal, we denote by $\mu_R(M)$ or $\mu(M)$ the 'minimal number of generators' i.e. the minimal cardinality of all generating sets of M.*

b) *If further $\mathfrak{p}$ is a prime ideal we write $\mu_{\mathfrak{p}}(M) := \mu_{R_{\mathfrak{p}}}(M_{\mathfrak{p}})$.*

To find bounds of this number will be one of our aims!

Remark 6.5.4 It may happen that there is a minimal set of generators of M with more then $\mu(M)$ elements. For example 2, 3 generate the $\mathbb{Z}$-module $\mathbb{Z}$, but not 2 or 3 alone. On the other hand, if $(R, \mathfrak{m})$ is local and $x_1, \dots, x_n$ form a minimal generating set (with different x_i), then their residue classes $\overline{x_1}, \dots, \overline{x_n}$ modulo $\mathfrak{m}M$ form a basis over $R/\mathfrak{m}$, whence $\mu(M) = n$ in this case.

Theorem 6.5.5 (Small Dimension Theorem) *Let R be a Noetherian ring and let $\mathfrak{p}$ be a minimal prime over-ideal of an ideal I generated by r elements. Then $\operatorname{ht}(\mathfrak{p}) \leq r$. In particular, $\operatorname{ht}(\mathfrak{p}) \leq \mu(\mathfrak{p})$.*

The case $r = 1$ is also called the Principal Ideal Theorem.

Proof. We prove the result by induction on r.

Since $\operatorname{ht}(\mathfrak{p}) = \operatorname{ht}(\mathfrak{p}R_{\mathfrak{p}})$ and $\mathfrak{p}R_{\mathfrak{p}}$ is minimal prime over-ideal over $IR_{\mathfrak{p}}$ we may assume that R is a local ring with maximal ideal $\mathfrak{p}$ minimal over $I = (a_1, \dots, a_r)$. If $r = 1$ apply Proposition 6.5.2. Assume the result for $r - 1$. Suppose now that you have a strictly ascending chain of prime ideals

$$\mathfrak{p}_0 \subsetneq \mathfrak{p}_1 \subsetneq \cdots \subsetneq \mathfrak{p}_r \subsetneq \mathfrak{p}_{r+1} = \mathfrak{p}.$$

We may further ensure (since R is Noetherian) that there is no prime ideal properly between $\mathfrak{p}_r$ and $\mathfrak{p}_{r+1}$. We may also assume w.l.g. that $a_1 \notin \mathfrak{p}_r$.

Note that $\operatorname{Spec}(R/(a_1, \mathfrak{p}_r)) = \{\mathfrak{p}/(a_1, \mathfrak{p}_r)\}$, whence by $R/(a_1, \mathfrak{p}_r)$ is an Artinian local ring by Proposition 6.4.4 *(i)*. Hence its maximal ideal is nilpotent, due to which we can find a $t > 0$ such that

$$a_i^t = \lambda_i a_1 + b_i,$$

$i = 2, \dots, r$, for some $\lambda_i \in R$, $b_i \in \mathfrak{p}_r$. Let $J = (b_2, \dots, b_r) \subset \mathfrak{p}_r$ and let $\mathfrak{q}$ be a minimal prime over-ideal of J contained in $\mathfrak{p}_r$. By induction hypothesis $\operatorname{ht}(\mathfrak{q}) \leq r - 1$. Since $\operatorname{ht}(\mathfrak{p}_r) \geq r$, $\mathfrak{q} \subsetneq \mathfrak{p}_r$.

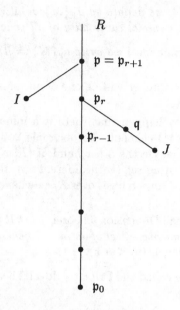

Note that $I^l \subset (a_1, J)$ for some l. Therefore any prime over-ideal of (a_1, J) contains I and therefore $\mathfrak{p}$ is the unique minimal prime over-ideal of (a_1, J).

Now in the ring R/J the prime ideal $\mathfrak{p}/J$ is of height at least 2 and also minimal over the principal ideal generated by $(a_1 \bmod J)$. A contradiction to Proposition 6.5.2. $\qquad\square$

The above theorem has an important converse.

Proposition 6.5.6 *Let $\mathfrak{p}$ be a prime ideal of height r in a Noetherian ring R. Then there are $a_1, \dots, a_r \in R$ such that $\mathfrak{p}$ is a minimal prime over-ideal of $(a_1, \dots, a_r)$.*

Proof. Induction on r, the case $r = 0$ being obvious. Let $\mathfrak{p}_1, \dots, \mathfrak{p}_s$ be the finitely many minimal prime ideals of R. If $r = \mathrm{ht}(\mathfrak{p}) \geq 1$ then $\mathfrak{p} \not\subset \bigcup_{i=1}^s \mathfrak{p}_i$ as by Lemma 1.1.8 it would follow that $\mathfrak{p} \subset \mathfrak{p}_i$ for some i. Let $a_1 \in \mathfrak{p} \setminus \bigcup_{i=1}^r \mathfrak{p}_i$. In the ring $R/(a_1)$ the prime ideal $\mathfrak{p}$ is of height less than r, since the minimal prime ideals $\mathfrak{p}_1, \dots, \mathfrak{p}_s$ have disappeared. Now use induction. $\qquad\square$

Corollary 6.5.7 *Let R be a Noetherian local ring with maximal ideal $\mathfrak{m}$. Then $\dim R = \mathrm{ht}(\mathfrak{m})$ equals the minimal number r of elements $a_1, \dots, a_r$, such that $\mathfrak{m}$ is a minimal prime over-ideal of $(a_1, \dots, a_r)$. In particular $\dim R$ is finite.*

Also if $\mathfrak{p}$ is any prime ideal in any Noetherian ring then $\mathrm{ht}(\mathfrak{p})$ is finite.

Proposition 6.5.8 *Let $R = k[x_1, \ldots, x_n]$ be a polynomial ring in n variables over a field k. Then $\dim(R) = n$.*

Proof. We prove the result by induction on n, it being well known for $n \leq 1$. The chain of prime ideals

$$(0) \subsetneqq (x_1) \subsetneqq \cdots \subsetneqq (x_1, \ldots, x_n)$$

shows that $\dim(R) \geq n$. Let $\mathfrak{m}$ be a maximal ideal of R of height $\geq n$, and let $S = k[x_1] \setminus \{0\}$. If $S \cap \mathfrak{m} = \emptyset$, then $\mathrm{ht}(\mathfrak{m}) = \mathrm{ht}(S^{-1}\mathfrak{m}) \leq n - 1$ by the induction hypothesis; hence assume $S \cap \mathfrak{m} \neq \emptyset$. Let $f(x_1) \in \mathfrak{m}$ be an irreducible polynomial. Then $R/(f(x_1)) \cong (k[x_1]/(f(x_1)))[x_2, \ldots, x_n]$. Since $k[x_1]/(f(x_1))$ is a field we get by induction that $\dim(R/(f(x_1))) = n - 1$. By Proposition 6.5.6 the ideal $\mathfrak{m}/(f(x_1))$ is minimal over an ideal generated by $n - 1$ elements, hence $\mathfrak{m}$ is minimal over an ideal generated by n elements. Hence $\mathrm{ht}(\mathfrak{m}) \leq n$ by 6.5.5. $\qquad\square$

Another consequence of Theorem 6.5.5 is

Corollary 6.5.9 *A Noetherian domain R is factorial if and only if every prime ideal of height 1 of R is principal.*

Proof. Let R be factorial and $\mathfrak{p}$ a prime ideal of height 1. There is an $a \in \mathfrak{p} \setminus (0)$. Let $a = p_1 \cdots p_n$ with prime elements p_i. Then $p_i \in \mathfrak{p}$ for at least one i. But $p_1 R$ is a non zero prime ideal, contained in $\mathfrak{p}$. Since the latter is of height 1, we see $p_i R = \mathfrak{p}$.

Conversely, let $a \in R$ be neither 0 nor a unit. Then there is a prime ideal $\mathfrak{p}$ of height 1, containing a. By assumption $\mathfrak{p} = p_1 R$ for some element $p_1 \in R$. This, generating a prime ideal, is a so called **prime element**. (We define it by this property.) $a \in p_1 R$ means $a = p_1 a_1$ for some $a_1 \in R$. If a_1 is not yet a unit, then by the same argument $a_1 = p_2 a_2$ for some prime element p_2 and some $a_2 \in R$. One gets a strictly ascending chain of ideals

$$(a) \subsetneqq (a_1) \subsetneqq (a_2) \subsetneqq \cdots$$

By Noetherianity this must stop with $(a_n) = R$. So $a = p_1 \cdots p_n a_n$ with prime elements p_i and a unit a_n. A decomposition of an element into prime elements always is essentially unique. This the reader should either know or prove as an exercise. $\qquad\square$

6.6 Noether Normalization

Definitions 6.6.1 *Let $K \supseteq k$ be a field extension.*

a) *An n-tuple $x_1, \ldots, x_n$ of elements in K is called* **algebraically indepen-dent over** k, *if there is no non-trivial polynomial relation between the x_i, i.e. if the k-algebra homomorphism*

$$k[X_1, \ldots, X_n] \to K, \qquad X_i \mapsto x_i,$$

where the $X_1, \ldots, X_n$ are indeterminates, is injective. In this case we also use the phrase: the $x_1, \ldots, x_n$ are algebraically independent. (This is unprecise insofar, as algebraic independence is indeed a property of the n-tuple, and not of the individual x_i.)

b) *We say that the* **transcendence degree** *of the field extension $K \supseteq k$ is n and write $\mathrm{trdg}_k K = n$, if the following holds: There are algebraically independent $x_1, \ldots, x_n \in K$ such that K is algebraic over $k(x_1, \ldots, x_n)$.*

Note that one can easily replace the above n by any – also transfinite – cardinal number. It is shown in ([45], VIII, 1) that $\mathrm{trdg}_k K$ is uniquely defined.

Remark 6.6.2 If $K' \supseteq K \supseteq k$ are field extensions and K' is algebraic over K, then clearly $\mathrm{trdg}_k K = \mathrm{trdg}_k K'$.

Definition 6.6.3 *An* **affine algebra** *is an algebra of finite type over a field. If we want to specify this field, we say* **affine k-algebra**. *It is clear what we mean by an* **affine domain**.

The name is explained by the fact that such algebras are function algebras on so called affine varieties – at least if k is algebraically closed and the algebra reduced – as we will see later.

Theorem 6.6.4 (Noether Normalization Theorem) *Let A be a k-algebra of finite type over a field k. Let I be an ideal of A. Then there are natural numbers $\delta \leq d$ and elements $y_1, \ldots, y_d \in A$ such that*

(1) *the $y_1, \ldots, y_d$ are algebraically independent over k;*

(2) *A is finite over $k[y_1, \ldots, y_d]$ (i.e. it is finitely generated as a $k[y_1, \ldots, y_d]$-module).*

(3) *$I \cap k[y_1, \ldots, y_d] = (y_{\delta+1}, \ldots, y_d)$.*

Proof. We consider three cases :

CASE 1: Let A be the polynomial ring $k[x_1, \ldots, x_n]$ and $I = (f)$ a principal ideal, where f is a non-constant polynomial.

Set $y_n = f$, and $y_i = x_i - x_n^{r_i}$ $(1 \leq i < n)$. As in Lemma 4.3.13 we may transform f, for suitable r_i, to an element of the form

$$f = a x_n^m + a_1 x_n^{m-1} + \cdots + a_m$$

where $a \in k^*$ and $a_i \in k[y_1, \ldots, y_{n-1}]$ for $1 \leq i \leq n - 1$. Clearly $A = k[y_1, \ldots, y_n][x_n]$. Since

$$0 = f - y_n = a x_n^m + a_1 x_n^{m-1} + \cdots + a_m - y_n,$$

x_n is integral over $k[y_1, \ldots, y_n]$, and therefore $A = k[y_1, \ldots, y_n][x_n]$ is finite over $k[y_1, \ldots, y_n]$ by Proposition 1.4.4.

If the $y_1, \ldots, y_n$ were not algebraic independent over k, then $n > \mathrm{trdg}_k k(y_1, \ldots, y_n) = \mathrm{trdg}_k k(x_1, \ldots, x_n) = n$, a contradiction.

We now prove that $I \cap k[y_1, \ldots, y_n] = (y_n)$. Clearly $(y_n) \subset I \cap k[y_1, \ldots, y_n]$. Let $g \in I \cap k[y_1, \ldots, y_n]$, then $g = h y_n$ for some $h \in A$. Since A is finite, hence integral over $k[y_1, \ldots, y_n]$, the polynomial h satisfies an equation of the form

$$h^s + b_1 h^{s-1} + \cdots + b_s = 0 \text{ with } b_i \in k[y_1, \ldots, y_n],$$

from which we get

$$g^s + b_1 y_n g^{s-1} + \cdots + b_s y_n^s = 0.$$

This implies that y_n divides g^s in the factorial ring $k[y_1, \ldots, y_n]$, whence y_n divides g, i.e. $g = h' y_n$ for some $h' \in k[y_1, \ldots, y_n]$. So $g \in (y_n)$ as required.

CASE 2: Now let I be an arbitrary ideal in $A = k[x_1, \ldots, x_n]$. The result is clear for $I = (0)$. So we may assume that there is a non-constant polynomial $f \in I$. We now proceed by induction on n, the case $n = 1$ being contained in Case 1. So assume $n > 1$. Let $k[y_1, \ldots, y_n]$ with $y_n = f$ be constructed as in Case 1.

By induction we may assume that there are elements $t_1, \ldots, t_{d-1} \in k[y_1, \ldots, y_{n-1}]$ algebraically independent over k such that $k[y_1, \ldots, y_{n-1}]$ is finite over $k[t_1, \ldots, t_{d-1}]$ and $I \cap k[t_1, \ldots, t_{d-1}] = (t_{\delta+1}, \ldots, t_{d-1})$ with some $\delta < d$. Then $k[y_1, \ldots, y_n]$ is finite over $k[t_1, \ldots, t_{d-1}, y_n]$, whence also A, being finite over $k[y_1, \ldots, y_n]$, is finite over $k[t_1, \ldots, t_{d-1}, y_n]$.

Further $n - 1 = \dim \ k[y_1, \ldots, y_{n-1}] = \dim \ k[t_1, \ldots, t_{d-1}] = d - 1$, and so $n = d$. Therefore and since $k(x_1, \ldots, x_n)$ is algebraic over $k(t_1, \ldots, t_{n-1}, y_n)$, the $t_1, \ldots, t_{n-1}, y_n$ are algebraically independent over k.

Since $y_n = f \in I$, clearly $I \cap k[t_1, \ldots, t_{n-1}, y_n] = (t_{\delta+1}, \ldots, t_{d-1}, y_n)$. Conversely let $g \in I \cap k[t_1, \ldots, t_{n-1}, y_n]$ then $g = g^* + hy_n$ for some $g^* \in I \cap k[t_1, \ldots, t_{n-1}] = (t_{\delta+1}, \ldots, t_{d-1})$, and $h \in k[t_1, \ldots, t_{n-1}, y_n]$. So $I \cap k[t_1, \ldots, t_{d-1}, y_n]$ is generated by $t_{\delta+1}, \ldots, t_{d-1}, y_n$.

CASE 3: For the general case let $A = k[x_1, \ldots, x_n]/J$. As in Case 2 determine a subalgebra $k[y_1, \ldots, y_n]$ of $k[x_1, \ldots, x_n]$ with $J \cap k[y_1, \ldots, y_n] = (y_{d+1}, \ldots, y_n)$. The image of $k[y_1, \ldots, y_n]$ in A can be identified with the polynomial algebra $k[y_1, \ldots, y_d]$. And A is then finite over this ring. Again apply Case 2 to $I' = I \cap k[y_1, \ldots, y_d]$ we get a polynomial subalgebra $k[t_1, \ldots, t_d] \subset k[y_1, \ldots, y_d]$ over which $k[y_1, \ldots, y_d]$ is finite and $I \cap k[t_1, \ldots, t_d] = I' \cap k[t_1, \ldots, t_d] = (t_{\delta+1}, \ldots, t_d)$ with some $\delta \leq d$.

Since A is finite over $k[t_1, \ldots, t_d]$ the elements $t_1, \ldots, t_d$ satisfies all the requirements of the theorem. $\qquad \square$

Corollary 6.6.5 *Let A be a k-algebra of finite type of dimension d. Assume that A is a domain. Then the transcendence degree of $Q(A)$, the quotient field of A, over k is d.*

Proof. By the Noether Normalization Theorem there is an integral extension

$$k[y_1, \ldots, y_n] \hookrightarrow A$$

with algebraically independent $y_1, \ldots, y_n$. By Remark 6.3.2 we see dim $A =$ dim $k[y_1, \ldots, y_n]$. By Remark6.5.8 we have dim $k[y_1, \ldots, y_n] = n$. Hence $n = d$. Also $Q(A)$ is an algebraic extension of $k[y_1, \ldots, y_d]$ and so $\mathrm{trdg}_k Q(A) = \mathrm{trdg}_k k(y_1, \ldots, y_d) = d$ by Remark 6.6.2. $\qquad \square$

6.7 Affine Algebras

Theorem 6.7.1 (Algebraic form of Hilbert's Nullstellensatz) *Let k be a field. A an affine k-algebra and $\mathfrak{m}$ a maximal ideal of A. Then the canonical inclusion $k \hookrightarrow A/\mathfrak{m}$ is a finite field extension.*

Proof. The field $A/\mathfrak{m}$ is a k-algebra of finite type. By Noether's Normalization Lemma 6.6.5 we have $\mathrm{trdg}_k(A/\mathfrak{m}) = \dim(A/\mathfrak{m}) = 0$. So $A/\mathfrak{m}$ is algebraic and of finite type, hence finite over k. $\qquad \square$

Corollary 6.7.2 *Let k be a field, $f : A \to B$ a homomorphism of affine k-algebras and $\mathfrak{m}$ a maximal ideal of B. Then $f^{-1}(\mathfrak{m})$ is a maximal ideal of A.*

Proof. We have inclusions $k \subset A/f^{-1}(\mathfrak{m}) \subset B/\mathfrak{m}$. By the Theorem $B/\mathfrak{m}$ is finite over k, hence finite over $A/f^{-1}(\mathfrak{m})$. Therefore the latter must be a field by Proposition 1.4.13. $\square$

This is a remarkable fact for affine k-algebras, and certainly not true for general rings.

Corollary 6.7.3 *Every maximal ideal* $\mathfrak{m}$ *of the polynomial ring* $k[X_1, \ldots, X_n]$ *over a field* k *is of height* n *and generated by* n *and not by less elements. (Symbolically* $\mathrm{ht}\,\mathfrak{m} = \mu(\mathfrak{m}) = n$.)

Proof. Induction on n, the case $n \leq 1$ being clear. So let $n > 1$, $B' := k[X_1, \ldots, X_{n-1}] \subset B := k[X_1, \ldots, X_n]$, $\mathfrak{m} \in \mathrm{Spmax}\,B$ and $\mathfrak{m}' := B' \cap \mathfrak{m}$. By Corollary 6.7.2 $\mathfrak{m}'$ is a maximal ideal of B'. So by induction $\mathfrak{m}'$ is generated by $n-1$ elements and of height $n-1$. Then $\mathfrak{p} := \mathfrak{m}'B = \mathfrak{m}'[X_n]$ is also generated by $n - 1$ elements and of height $\geq n - 1$. Namely a chain

$$(0) = \mathfrak{p}_0 \subsetneq \mathfrak{p}_1 \subsetneq \cdots \subsetneq \mathfrak{p}_{n-1} = \mathfrak{m}'$$

of prime ideals in B' induces the chain

$$\mathfrak{p}_0[X_n] \subsetneq \cdots \subsetneq \mathfrak{p}_{n-1}[X_n] = \mathfrak{p}$$

in B. Now $B/\mathfrak{p} = B'[X_n]/\mathfrak{m}'[X_n] \cong (B'/\mathfrak{m}')[X_n]$. So $\mathfrak{m}/\mathfrak{p}$ is a maximal ideal in a principal domain and therefore generated by 1 element. Therefore $\mathfrak{m}$ can be generated by $n - 1 + 1 = n$ elements. Further $\mathrm{ht}(\mathfrak{m}) = \mathrm{ht}(\mathfrak{m}'[X_n]) + 1 \geq n - 1 + 1 = n$. So we have $n \leq \mathrm{ht}(\mathfrak{m})$, $\mu(\mathfrak{m}) \leq n$ and, by dimension theory, $\mathrm{ht}(\mathfrak{m}) \leq \mu(\mathfrak{m})$, which finishes the proof. $\square$

This shows once more, but – using Noether's Normalization – with more effort, that $\dim(k[X_1, \ldots, X_n]) = n$. The reader will realize from the above proof that $\mathfrak{m}$ is of the form $(f_1(X_1), f_2(X_1, X_2), \ldots, f_n(X_1, \ldots, X_n))$.

Corollary 6.7.4 *Let* $1 \leq \delta \leq n$. *Then the prime ideal* $(X_{\delta+1}, \ldots, X_n)$ *of* $k[X_1, \ldots, X_n]$ *is of height* $n - \delta$.

Proof. Let $S := k[X_1, \ldots, X_\delta] \setminus \{0\}$. Then $S^{-1}k[X_1, \ldots, X_n] = k(X_1, \ldots, X_\delta)[X_{\delta+1}, \ldots, X_n]$ is a polynomial ring over a field in $n - \delta$ indeterminates $X_{\delta+1}, \ldots, X_n$, where the ideal generated by these is of height $n - \delta$. The analogous prime ideal $(X_{\delta+1}, \ldots, X_n)$ in the non localized ring $k[X_1, \ldots, X_n]$ then must be of the same height. $\square$

Proposition 6.7.5 *Let* A *be an affine domain and* $\mathfrak{p} \in \mathrm{Spec}(A)$. *Then*

$$\mathrm{ht}\,\mathfrak{p} + \dim(A/\mathfrak{p}) = \dim(A).$$

Proof. There is a Noether Normalization $k[y_1, \ldots, y_n] \subset A$ with $n = \dim A$, such that for some $\delta \leq n$ we have $\mathfrak{p}' := \mathfrak{p} \cap k[y_1, \ldots, y_n] = (y_{\delta+1}, \ldots, y_n)$. Since A is finite over the integrally closed $k[y_1, \ldots, y_n]$ we can apply "going down" and see ht $\mathfrak{p} = $ ht $\mathfrak{p}' = n - \delta$.

On the other hand $A/\mathfrak{p}$ is finite over $k[y_1, \ldots, y_n]/\mathfrak{p}' \cong k[y_1, \ldots, y_\delta]$ and hence of dimension δ.

Since $(n - \delta) + \delta = n$ we are through. $\qquad\qquad\qquad\qquad\qquad\qquad\square$

Corollary 6.7.6 *Let A be as above. Then all maximal chains of prime ideals in A (i.e. those which are neither refinable nor extendable) have the length $\dim A$.*

Proof. Let A be a counterexample of minimal dimension and

$$(0) = \mathfrak{p}_0 \subsetneq \mathfrak{p}_1 \subsetneq \cdots \subsetneq \mathfrak{p}_m$$

be a maximal chain of prime ideals in A with $m < n := \dim A$. Clearly ht $\mathfrak{p}_1 = 1$. Since, by the minimality of $\dim A$, the ring $A/\mathfrak{p}_1$ fulfills the corollary, every maximal chain of prime ideals in $A/\mathfrak{p}_1$ has length $m - 1$. Therefore $\dim(A/\mathfrak{p}_1) = m - 1$. But by Proposition 6.7.5 we get $\dim(A/\mathfrak{p}_1) = n - 1 > m - 1$. Contradiction. $\qquad\qquad\qquad\qquad\qquad\qquad\qquad\qquad\square$

Examples 6.7.7 a) If A is not a domain, Proposition 6.7.5 and its corollary may fail to hold. For example let $A := k[X, Y, Z]/(XZ, YZ)$. Denote the residue classes of X, Y, Z by x, y, z. Then A has the minimal prime ideals (x, y) and (z), and $\dim(A/(x, y)) = 1$, whereas $\dim(A/(z)) = 2$. One may also replace A by its localization in the maximal ideal (x, y, z) to get a local ring B with a minimal prime ideal $\mathfrak{p}$ such that $\dim(B/\mathfrak{p}) < \dim(B)$. (Geometrically speaking, A is the ring of the union of a plane and a line, intersecting in one point. After having read the section on Hilbert's Nullstellensatz, you will understand this better.)

b) Let A be a discrete valuation ring, i.e. a principal domain with exactly one non-zero maximal ideal (p). (Up to associates then p is the only prime element of A.) Consider the ideal $(pX - 1)A[X]$. As a principal ideal it is of height 1. On the other hand it is maximal, since the residue class ring $A[X]/(pX - 1)$ is the field $Q(A)$. So the 2-dimensional domain $A[X]$ has maximal ideals of height 1. (Other maximal ideals of height 1 are $(pX^n - 1)$. The maximal ideals of height 2 are those of the form (p, f), where f is irreducible modulo p.)

Definition 6.7.8 *A (Noetherian) ring A is called **catenary** if for every pair of prime ideals $\mathfrak{q} \subset \mathfrak{p}$ in A any two maximal (i.e. non refinable) chains of prime ideals starting with $\mathfrak{q}$ and ending with $\mathfrak{p}$:*

$$\mathfrak{q} = \mathfrak{p}_0 \subsetneq \cdots \subsetneq \mathfrak{p}_n = \mathfrak{p}$$

have the same length.

Note that the examples above are catenary. Remarkably enough there do exist non catenary Noetherian rings. But:

Corollary 6.7.9 *Every affine algebra A is catenary.*

Proof. Let $\mathfrak{q} \subset \mathfrak{p}$ be as in the definition above. We may assume that $\mathfrak{q} = (0)$ and so that A is a domain. If there were two maximal chains of prime ideals

$$(0) = \mathfrak{p}_0 \subsetneq \cdots \subsetneq \mathfrak{p}_r = \mathfrak{p}, \tag{6.1}$$

$$(0) = \mathfrak{p}_0' \subsetneq \cdots \subsetneq \mathfrak{p}_s' = \mathfrak{p} \tag{6.2}$$

with $r \neq s$, they could be prolonged by one maximal chain beginning in $\mathfrak{p}$ and ending in any maximal ideal $\mathfrak{m} \supseteq \mathfrak{p}$. In this way we get two maximal chains of prime ideals in A of different length, contradicting Corollary 6.7.6.   □

The following proposition shows that the spectrum and the maximal spectrum of an affine algebra are highly related. We prepare for this by two lemmas.

Lemma 6.7.10 *Let A be a Noetherian domain of dimension at least 2. Then A possesses infinitely many prime ideals of height 1, and (0) is their intersection.*

Proof. By the Small Dimension Theorem in the case $r = 1$, every nonunit lies in some prime ideal of height 1. So any maximal ideal $\mathfrak{m}$ is contained in the union of the prime ideals of height 1. If there were only finitely many of them, by the Prime Avoidance Lemma 1.1.8, at least one of them would contain $\mathfrak{m}$, whence $\operatorname{ht} \mathfrak{m} \leq 1$. But by hypothesis there are maximal ideals of height ≥ 2.

Now assume there were an $x \neq 0$ contained in every prime ideal of height 1. Then the Noetherian ring $A/(x)$ would possess infinitely many minimal prime ideals, contradicting Corollary 6.2.6.   □

Lemma 6.7.11 *Let A be an affine domain of dimension 1. It possesses infinitely many maximal ideals and their intersection is (0).*

Proof. There is a finite ring extension $k[X] \subset A$ with a field k. First we show that $k[X]$ has infinitely many maximal ideals. These are in bijective correspondence to the irreducible monic polynomials. If there were only finitely many, say $f_1, \ldots, f_r$ of them, following Euclid form $g := 1 + f_1 \cdots f_r$. Since g is not a unit, it is divisible by a monic irreducible f, which clearly is different from all f_i.

Over every maximal ideal of $k[X]$ lies one of A. And these are all of height 1. By the same argument as in the proof above, their intersection is (0).   □

Proposition 6.7.12 *Every radical ideal I of an affine algebra A is the intersection of the maximal ideals containing I.*

Proof. Since I is the intersection of prime ideals, it is enough to prove the proposition for prime ideals I. Let $\mathfrak{p}$ be a maximal counterexample in A. Then $\mathfrak{p}$ is no maximal ideal.

If $\dim(A/\mathfrak{p}) = 1$, then $\mathfrak{p}$ is the intersection of the maximal ideals containing it by Lemma 6.7.11.

If $\dim(A/\mathfrak{p}) > 1$, then $\mathfrak{p}$ is an intersection of strictly bigger prime ideals by Lemma 6.7.10, the latter being intersections of maximal ideals, since $\mathfrak{p}$ was a maximal counterexample. So $\mathfrak{p}$ is an intersection of maximal ideals. $\qquad\Box$

Corollary 6.7.13 *Let A be an affine algebra. The relation $Z \mapsto Z \cap \mathrm{Spmax}(A)$ gives a bijection of the set of closed subsets of $\mathrm{Spec}(A)$ to the set of those of $\mathrm{Spmax}(A)$. This map is compatible with the forming of unions and intersections and with the relation of inclusion. Therefore a closed subset Z of $\mathrm{Spec}(A)$ is irreducible if and only if $Z \cap \mathrm{Spmax}(A)$ is irreducible in $\mathrm{Spmax}(A)$. Especially $\dim(A) = \dim(\mathrm{Spmax}\, A)$.*

Note that, if A is semilocal of dimension > 0, things are totally different.

Proof. By definition a subset $Z' \subset \mathrm{Spmax}(A)$ is closed if and only if it is of the form $Z \cap \mathrm{Spmax}(A)$ with Z closed in $\mathrm{Spec}(A)$. So we just have to show that different closed subsets of $\mathrm{Spec}(A)$ have different intersections with $\mathrm{Spmax}(A)$. But if $V(\mathfrak{a})$ is closed in $\mathrm{Spec}(A)$ we know by Proposition 6.7.12 that $\sqrt{\mathfrak{a}}$ is the intersection of the maximal ideals $\mathfrak{m} \in V(\mathfrak{a}) \cap \mathrm{Spmax}(A)$. So we recover $V(\mathfrak{a}) = V(\sqrt{\mathfrak{a}})$ from $V(\mathfrak{a}) \cap \mathrm{Spmax}(A)$.

The rest is then clear. $\qquad\Box$

6.8 Hilbert's Nullstellensatz

It is well-known that the maximal ideals in the ring $\mathcal{C}[0,1]$ of continuous functions $[0,1] \to \mathbb{R}$ are the 'points' $\mathfrak{m}_x = \{f \mid f(x) = 0\}$ for $x \in [0,1]$. The algebraic analogue of this is the '**Zero-point Theorem**' or '**Nullstellensatz**' of D. Hilbert, which classifies the maximal ideals in a polynomial ring $k[X_1, \dots, X_n]$ over an algebraically closed field k (for example $\mathbb{C}$, the field of complex numbers) as 'points' $(X_1 - a_1, \dots, X_n - a_n)$. Note that the maximal ideal $(X^2 + 1)$ in $\mathbb{R}[X]$ is not a 'point' – so the condition that the field is algebraically closed is essential.

Proposition 6.8.1 (Weak Hilbert's Nullstellensatz) *Let k be an algebraically closed field. Then the maximal ideals of $k[X_1, \dots, X_n]$ are those of the form $(X_1 - a_1, \dots, X_n - a_n)$ where $(a_1, \dots, a_n) \in k^n$.*

This means that the injective map

$$k^n \longrightarrow \mathrm{Spmax}(k[X_1, \dots, X_n],$$

which we know from Paragraph 1.1.7, is bijective if k is algebraically closed.

(The reader should not be confused by the notational ambiguity: As usual by $(X_1 - a_1, \dots, X_n - a_n)$ we mean an ideal of $k[X_1, \dots, X_n]$, and by $(a_1, \dots, a_n)$ an n-tuple in k^n.)

Proof. First note that the ideal $(X_1, \dots, X_n)$ is a maximal ideal of $A :=$ $k[X_1, \dots, X_n]$, since the residue class ring is the field k. So the ideal $(X_1 - a_1, \dots, X_n - a_n)$ for $(a_1, \dots, a_n) \in k^n$ is also maximal, since it is the image of $(X_1, \dots, X_n)$ under the k-algebra automorphism $A \to A$ defined by $X_i \mapsto X_i - a_i$.

For the converse we need the hypothesis that k is algebraically closed. Let $\mathfrak{m}$ be any maximal ideal of A. Then the inclusion $k \hookrightarrow A/\mathfrak{m}$ is a finite field extension by Theorem 6.7.1, hence an isomorphism, since k is algebraically closed.

So, for every $i = 1, \dots, n$ there is an $a_i \in k$ such that $X_i \equiv a_i \pmod{\mathfrak{m}}$. Therefore $(X_1 - a_1, \dots, X_n - a_n) \subset \mathfrak{m}$. Equality holds as $(X_1 - a_1, \dots, X_n - a_n)$ is maximal. $\qquad\square$

Note 6.8.2 Let $f \in A := k[X_1, \dots, X_n]$. Recall from Paragraph 1.1.7 that then

$$f \in (X_1 - a_1, \dots, X_n - a_n) \iff f(a_1, \dots, a_n) = 0,$$

(also if k is not necessarily algebraically closed).

We will repeat the argument. Here and in the sequel let us write $\underline{a} :=$ $(a_1, \dots, a_n) \in k^n$ and $\mathfrak{m}_{\underline{a}} := (X_1 - a_1, \dots, X_n - a_n) \in \mathrm{Spmax}(A)$, The composition of canonical maps

$$k \longrightarrow A \xrightarrow{\kappa} A/\mathfrak{m}_{\underline{a}}$$

is an isomorphism, by which we identify k with $A/\mathfrak{m}_{\underline{a}}$. Now $\kappa(X_i) = \kappa(a_i)$ the latter being equal to a_i under the above identification. So, if $f \in k[X_1, \dots, X_n]$ is arbitrary, $\kappa(f) = f(\underline{a})$. This implies the above statement.

This note enables us to derive from the proposition the following corollary, which deserves as well to be called 'Nullstellensatz'.

Corollary 6.8.3 *Let k be algebraically closed and $E \subset A$ generate a proper ideal $(E) \subsetneq A$. Then there is an $\underline{a} \in k^n$ with $f(\underline{a}) = 0$ for all $f \in E$.*

Proof. By hypothesis there is a maximal ideal m of A containing the ideal (E). By Proposition 6.8.1 there is an $\underline{a} \in k^n$ with $m = m_{\underline{a}}$. So, by the above note, $f(\underline{a}) = 0$ for all $f \in (E)$. $\qquad\square$

Definition 6.8.4 *Let E be an arbitrary subset of $A = k[X_1, \dots, X_n]$. Then define*
$$V(E) := \{\underline{a} \in k^n \mid f(\underline{a}) = 0 \text{ for all } f \in E\}.$$
*It is called the **algebraic (sub)set** (of k^n) defined by E. (Sometimes one calls it the set of zeros of E.)*

The letter 'V' stands for 'variety'. And indeed the algebraic sets – equipped with suitable rings of functions – are the so called 'affine varieties' of Algebraic Geometry.

On the other hand the notion '$V(E)$' is also used to denote the set of prime ideals containing E. This makes sense, since under the correspondence of points in k^n and maximal ideals of A the set $V(E)$ corresponds to the set of maximal ideals containing E. This follows directly from the above note.

Definition 6.8.5 *Let M be an arbitrary subset of k^n. Then define*

$$I(M) := \{f \in A \mid f(\underline{a}) = 0 \text{ for all } \underline{a} \in M\}.$$

*It is called the **zero ideal** of M.*

Remarks 6.8.6 a) Clearly $I(M)$ is an ideal, even a radical ideal of A.

b) If $\mathfrak{a} := (E)$ is the ideal generated by E, then clearly

$$V(E) = V(\mathfrak{a}) = V(\sqrt{\mathfrak{a}}).$$

$V(E) \neq \emptyset$, if $\mathfrak{a} \neq A$ by the above corollary.

c) $E \subset I(V(E))$ and $M \subset V(I(M))$.

d) $E_1 \subset E_2 \Longrightarrow V(E_1) \supseteq V(E_2)$ and $M_1 \subset M_2 \Longrightarrow I(M_1) \supseteq I(M_2)$.

Theorem 6.8.7 (Strong Hilbert's Nullstellensatz) *Let E be any subset of $A = k[X_1, \dots, X_n]$ and $\mathfrak{a} = (E)$ the ideal generated by E. Then*

$$I(V(E)) = \sqrt{\mathfrak{a}}.$$

Proof. By Remark b) we see that $V(E)$ consists of those points $\underline{a} \in k^n$ for which $f(\underline{a}) = 0$ for all $f \in \sqrt{\mathfrak{a}}$. The maximal ideals $\mathfrak{m}_{\underline{a}}$ of A corresponding to the points $\underline{a} \in V(E)$ are therefore exactly those which contain $\sqrt{\mathfrak{a}}$. And by Proposition 6.7.12 their intersection is $\sqrt{\mathfrak{a}}$. Therefore the set of $f \in A$ with $f(\underline{a}) = 0$, i.e. with $f \in \mathfrak{m}_{\underline{a}}$ for all $\underline{a} \in V(E)$ equals $\sqrt{\mathfrak{a}}$.   □

Corollary 6.8.8 *By the maps* V *and* I *we get a bijective correspondence between the radical ideals of A and the algebraic subsets of k^n, provided k is algebraically closed.*

Proof. If $\mathfrak{a} = \sqrt{\mathfrak{a}}$ is a radical ideal then $I(V(\mathfrak{a})) = \mathfrak{a}$ by Theorem 6.8.7. Conversely let $M = V(E)$ be an algebraic subset of k^n. Then $E \subset I(M)$ and so $M = V(E) \supseteq V(I(M))$, hence $M = V(I(M))$, since trivially $M \subset V(I(M))$ as we already know.   □

Definition 6.8.9 *A subset of k^n is called* **(Zariski) closed** *if it is an algebraic subset.*

Clearly a subset of k^n, with an algebraically closed field k, is closed if and only if it corresponds to a closed subset of $\mathrm{Spmax}(A)$ where $\mathrm{Spmax}(A) \subset \mathrm{Spec}(A)$ is equipped with the induced topology. So k^n by the definition of closed subsets gets a topology, also called **Zariski topology**, with respect to which it is homeomorphic to $\mathrm{Spmax}(A)$.

Note that a subset of $k = k^1$ is Zariski closed if and only if it is either finite or equal to k. Therefore (for $n > 1$) on k^n the product topology is different from the Zariski topology. The latter is finer. (Exercise!) Further, in the case $k = \mathbb{C}$, the Euclidean topology on $\mathbb{C}^n$ is strictly finer than the Zariski topology. (Exercise!) The Euclidean topology is given by the Euclidean metric on $\mathbb{C}^n = \mathbb{R}^{2n}$.) In the following we only consider Zariski topology.

6.8.10 Recall the definition of irreducible subsets and the dimension of a topological space in Section 6.3.1.

The closed irreducible subsets of k^n are in bijective correspondence to the prime ideals of A. This is easily shown with the help of Corollary 6.8.8. So the dimension of $V(\mathfrak{a})$ equals $\dim(A/\mathfrak{a})$ for every ideal $\mathfrak{a}$ of A.

6.9 Dimension of a Polynomial Ring

It is fundamental to expect that the dimension of a polynomial ring $R[X]$ is one more than that of the base ring R, when R is a Noetherian ring!

The contents of this section will be needed in Lindel's proof of Serre's Conjecture in Chapter 7, and in the proof of the Theorem of Cowsik and Nori in Chapter 10.

We will study the height of prime ideals in a polynomial ring $R[x]$ over a Noetherian ring R and, as a consequence, show that $\dim(R[x]) = \dim(R) + 1$. This generalizes the fact that $\dim(k[X_1, \ldots, X_n]) = n$ for a field k, which we already know. See Proposition 6.5.8.

Let $\mathfrak{p}$ be a (prime) ideal of R . Then $\mathfrak{p}[x] = \mathfrak{p}R[x]$ is a (prime) ideal of $R[x]$. Here $\mathfrak{p}[x]$ is defined to be the set of polynomials whose coefficients belong to $\mathfrak{p}$. Clearly $\mathfrak{p}[x] \cap R = \mathfrak{p}$ and $(\mathfrak{p}' \cap R)[x] \subset \mathfrak{p}'$ for $\mathfrak{p}' \in \mathrm{Spec}(R[x])$.

In the following lemmas R need not necessarily be Noetherian.

Lemma 6.9.1 *Let* $\mathfrak{p}' \subsetneq \mathfrak{p}''$ *be prime ideals of* $R[x]$ *with* $\mathfrak{p}' \cap R = \mathfrak{p}'' \cap R$. *Then* $\mathfrak{p}' = (\mathfrak{p}' \cap R)[x]$.

Proof. Set $\mathfrak{p} := \mathfrak{p}' \cap R$. The inclusion $\mathfrak{p}[x] \subset \mathfrak{p}'$ is clear. Were $\mathfrak{p}[x] \neq \mathfrak{p}'$, the chain $\mathfrak{p}[x] \subset \mathfrak{p}' \subset \mathfrak{p}''$ would strictly increase.

We may assume $\mathfrak{p} = (0)$, since we have the same situation for $R/\mathfrak{p} \subset (R/\mathfrak{p})[x] = R[x]/\mathfrak{p}[x]$. Now set $S := R \setminus (0)$. Since by hypothesis $\mathfrak{p}' \cap R = \mathfrak{p}'' \cap R = (0)$, we would have $\mathfrak{p}' \cap S = \mathfrak{p}'' \cap S = \emptyset$ and so get a strictly increasing chain of prime ideals

$$(0) \subsetneq S^{-1}\mathfrak{p}' \subsetneq S^{-1}\mathfrak{p}''$$

in $S^{-1}R[x]$, whence $\dim(S^{-1}R[x]) \geq 2$. But the latter ring is a principal domain, since $S^{-1}R$ is a field. A contradiction. $\qquad\square$

Lemma 6.9.2 *Let* $\mathfrak{p}$ *be a minimal prime over-ideal of an ideal* I *in* R. *Then* $\mathfrak{p}[x]$ *is a minimal prime over-ideal of* $I[x]$ *in* $R[x]$.

Proof. Assume there were a $\mathfrak{p}' \in \mathrm{Spec}(R[x])$ with $I[x] \subset \mathfrak{p}' \subsetneq \mathfrak{p}[x]$. So $\mathfrak{p} = \mathfrak{p}[x] \cap R \supseteq \mathfrak{p}' \cap R$. Since $\mathfrak{p}$ is minimal over I, we would get $\mathfrak{p} = \mathfrak{p}' \cap R$, and by Lemma 6.9.1 this would imply $\mathfrak{p}' = \mathfrak{p}[x]$, a contradiction. $\qquad\square$

Theorem 6.9.3 *Let* R *be Noetherian,* $\mathfrak{p}' \in \mathrm{Spec}(R[x])$ *and* $\mathfrak{p} := \mathfrak{p}' \cap R$. *Then*

a) $\mathrm{ht}_{R[x]}\mathfrak{p}' = \mathrm{ht}_R\mathfrak{p}$ *in the case* $\mathfrak{p}' = \mathfrak{p}[x]$;

b) $\mathrm{ht}_{R[x]}\mathfrak{p}' = \mathrm{ht}_R\mathfrak{p} + 1$ *in the case* $\mathfrak{p}' \neq \mathfrak{p}[x]$.

c) $\dim(R[x]) = \dim(R) + 1$.

Proof. Let $\mathrm{ht}_R \mathfrak{p} = n$. There is a chain

$$\mathfrak{p}_0 \subsetneq \cdots \subsetneq \mathfrak{p}_n = \mathfrak{p} \tag{6.3}$$

of prime ideals in R, from which we derive the following chain of prime ideals in $R[x]$

$$\mathfrak{p}_0[x] \subsetneq \cdots \subsetneq \mathfrak{p}_n[x] = \mathfrak{p}[x].$$

Therefore $\mathrm{ht}_{R[x]} \mathfrak{p}' \geq \mathrm{ht}_R \mathfrak{p}$ if $\mathfrak{p}[x] = \mathfrak{p}'$ and $\mathrm{ht}_{R[x]} \mathfrak{p}' \geq \mathrm{ht}_R \mathfrak{p} + 1$ if not.

Since $\mathrm{ht}_R \mathfrak{p} = n$, there are n elements $a_1, \ldots, a_n \in R$ such that $\mathfrak{p}$ is a minimal prime over-ideal of $Ra_1 + \cdots + Ra_n$ in R. By Lemma 6.9.2, $\mathfrak{p}[x]$ is a minimal prime over-ideal of $a_1 R[x] + \cdots + a_n R[x]$. Hence $\mathrm{ht}_{R[x]} \mathfrak{p}' \leq n$ if $\mathfrak{p}' = \mathfrak{p}[x]$. This proves a).

Now assume $\mathfrak{p}' \neq \mathfrak{p}[x]$ and $\mathrm{ht}_{R[x]} \mathfrak{p}' = r$. In $R[x]$ there is a chain of prime ideals

$$\mathfrak{p}_0' \subsetneq \mathfrak{p}_1' \subsetneq \cdots \subsetneq \mathfrak{p}_r' = \mathfrak{p}'.$$

Set $\mathfrak{q}_i = \mathfrak{p}_i' \cap R$. Then $\mathfrak{q}_r = \mathfrak{p}' \cap R = \mathfrak{p}_n = \mathfrak{p}$. And we have to show $r \leq n+1$. If

$$\mathfrak{q}_0 \subsetneq \mathfrak{q}_1 \subsetneq \cdots \subsetneq \mathfrak{q}_r,$$

then even $r \leq n$. So let j be maximal such that $\mathfrak{q}_j = \mathfrak{q}_{j+1}$. By Lemma 6.9.1 this implies $\mathfrak{p}_j' = \mathfrak{q}_j[x]$, hence $\mathrm{ht}_R \mathfrak{q}_j = \mathrm{ht}_{R[x]} \mathfrak{p}_j \geq j$. By the choice of j we have $\mathfrak{q}_j = \mathfrak{q}_{j+1} \subsetneq \mathfrak{q}_{j+2} \subsetneq \cdots \subsetneq \mathfrak{q}_r$. Therefore $n = \mathrm{ht}_R \mathfrak{q}_r \geq r - (j+1) + \mathrm{ht}_R \mathfrak{q}_j \geq r - j - 1 + j = r - 1$. (One might think, these arguments are unnecessarily complicated. But it is not clear from the beginning that there is a chain of prime ideals of length r in $R[x]$ ending with $\mathfrak{p}'$ and passing through $\mathfrak{p}[x]$.)

We are done with a) and b). And these clearly imply $\dim(R[x] \leq \dim(R) + 1$. To show the converse inequality assume $\dim(R) = n < \infty$. Then a chain of prime ideals of length n in R like Equation (6.3) produces in $R[x]$ the following chain of prime ideals of length $n+1$

$$\mathfrak{p}_0[x] \subsetneq \cdots \subsetneq \mathfrak{p}_n[x] \subsetneq \mathfrak{p}_n[x] + x R[x].$$

$\square$

7

Serre's Splitting Theorem and Lindel's Proof of Serre's Conjecture

In this chapter all rings are supposed to be *commutative*. We shall present a proof of the Quillen-Suslin theorem due to H. Lindel which only uses the concept of the Krull dimension of a noetherian ring, and its property that it increases by one for a polynomial extension in one variable, and that it decreases by one if we go modulo a non-zero divisor.

7.1 Serre's Splitting Theorem

In Section 1.1 we had proved the Prime Avoidance Lemma: *Let R be a ring, $\mathfrak{p}_1, \mathfrak{p}_2, \ldots, \mathfrak{p}_r \in \operatorname{Spec}(R)$, I an ideal of R and $x \in R$. If $x + I \subset \bigcup_{i=1}^r \mathfrak{p}_i$, then $(x, I) \subset \mathfrak{p}_{i_0}$ for some i_0.*

The above statement implies

Lemma 7.1.1 *Let $(a_1, \ldots, a_n) \not\subset \mathfrak{p}_i$, for prime ideals $\mathfrak{p}_1, \ldots, \mathfrak{p}_r$. Then there exist $b_2, \ldots, b_n \in R$ such that $c = a_1 + b_2 a_2 + \cdots + b_n a_n \notin \bigcup_{i=1}^r \mathfrak{p}_i$.*

Indeed, set $x = a_1$ and $I = (a_2, \ldots, a_n)$. □

Proposition 7.1.2 *Let A be a Noetherian ring and $I \subset A$ an ideal of height $\geq n$ generated by n elements $a_1, \ldots, a_n \in A$. Then there is a matrix $\varepsilon \in \mathrm{E}_n(A)$ such that the n-tuple $(b_1, \ldots, b_n) := (a_1, \ldots, a_n)\varepsilon$ generates I, and satisfies*

(1) $b_i = a_i + \sum_{j>i} c_{ij} a_j$, *for all* $1 \leq i \leq n$, *for some* $c_{ij} \in A$,

(2) $\operatorname{ht}(b_1, \ldots, b_i) = i$, *for* $1 \leq i \leq n$.

In particular, if $n \geq \dim(A) + 2$, then $(a_1, \ldots, a_n)$ is completable to a matrix of $\mathrm{E}_n(A)$.

Proof. By the above lemma we can find $c_{12} \dots , c_{1n}$ in A, such that the element $b_1 = a_1 + c_{12}a_2 + \cdots + c_{1n}a_n$ does not belong to any of the minimal prime ideals of A. Applying the above lemma again we can find $c_{21}, c_{23}, \dots , c_{2n}$ in A, such that the element $b_2' = a_2 + c_{21}b_1 + c_{23}a_3 + \cdots + c_{2n}a_n$ does not belong to any of the minimal prime overideals of (b_1). But then $b_2 := b_2' - c_{21}b_1 = a_2 + c_{23}a_3 + \cdots + c_{2n}a_n$ also does not belong to any of the minimal prime overideals of (b_1). In particular $\mathrm{ht}(b_1, b_2) = 2$. Proceeding as above we can obtain a set of generators as required.

Now to the last sentence. Clearly $(a_1, \dots , a_n) \sim_{\mathrm{E}_n(A)} (b_1, \dots , b_n)$. Let $d := \dim(A)$. Then $d + 1 < n$. Since $\mathrm{ht}(b_1, \dots , b_{d+1}) > \dim(A)$, already $(b_1, \dots , b_{d+1})$ is unimodular. So $(b_1, \dots , b_{d+1}, \dots , b_n)$ is 'elementarily' completable, since a proper subrow is unimodular. $\qquad\square$

Remark 7.1.3 From Proposition 7.1.2 we can conclude that if $(a_1, \dots , a_r, s)$ is in $\mathrm{Um}_{r+1}(A)$ then there are elements $c_1, \dots , c_r \in A$ such that $\mathrm{ht}(a_1 + sc_1, \dots , a_r + sc_r) \geq r$.

Let us illustate this in the special case when $r = 2$; the general case is similarly done. By Proposition 7.1.2 there are $\lambda, \mu_1, \mu_2 \in A$ such that the ideal $(a_1 + \lambda a_2 + \mu_1 s, a_2 + \mu_2 s)$ has height ≥ 2. Note that we can write $a_1 + \lambda a_2 + \mu_1 s = a_1 + \lambda(a_2 + \mu_2 s) + \mu_1' s$. But then $\mathrm{ht}(a_1 + \mu_1' s, a_2 + \mu_2 s) \geq 2$.

Lemma 7.1.4 *Let* $a_1, \dots , a_n, s \in A$. *Then there are elements* $c_i \in A$, $1 \leq i \leq n$ *such that* $\mathrm{ht}(a_1 + sc_1, \dots , a_i + sc_i)A_s \geq i$ *for* $1 \leq i \leq n$ *in the ring* A_s.

(If I is an ideal of A, by IA_s we clearly denote the ideal of A_s which is generated by the image of I.)

Proof. Let $\mathfrak{p}_1, \dots , \mathfrak{p}_m$ be those minimal prime ideals of A which do not contain s. Since $s \notin \mathfrak{p}_i$ for all i, we conclude that $a_1 + sA \not\subset \bigcup_{i=1}^m \mathfrak{p}_i$, i.e. there is a $c_1 \in A$ such that $(a_1 + sc_1)A_s$ has height ≥ 1.

Let $\mathfrak{p}_1', \dots , \mathfrak{p}_{m'}'$ be those minimal prime overideals of $(a_1 + sc_1)$ which do not contain s. Again, as above, $a_2 + sA \not\subset \bigcup \mathfrak{p}_i'$, i.e. there is a $c_2 \in A$ such that $a_2 + sc_2 \notin \bigcup \mathfrak{p}_i'$ whence $\mathrm{ht}(a_1 + sc_1, a_2 + sc_2)A_s \geq 2$.

Continuing thus we get $\mathrm{ht}(a_1 + sc_1, \dots , a_i + sc_i)A_s \geq i$ for all $1 \leq i \leq n$. $\qquad\square$

We shall 'globalize' the above lemma next.

Definitions 7.1.5 *Let* M *be an* A-module *and* $z \in M$.

a) z *is called* **unimodular** *if* Az *is a direct summand of* M *and* $\mathrm{Ann}(z) = (0)$. *The set of all unimodular elements of* M *is denoted by* $\mathrm{Um}(M)$.

b) *The* **order ideal** $\mathcal{O}_M(z)$ *of* z *is defined by*

$$\mathcal{O}_M(z) := \{\varphi(z) \,|\, \varphi \in \mathrm{Hom}_A(M, A)\}.$$

Clearly $\mathcal{O}_M(z)$ is an ideal. And z is unimodular, if and only if $\mathcal{O}_M(z) = A$.

Remarks 7.1.6 a) If $S \subset A$ is multiplicative and M finitely presented, then $S^{-1}\mathcal{O}_M(z) = \mathcal{O}_{S^{-1}M}(z_S)$.

This follows from the fact that $S^{-1}\mathrm{Hom}_A(M, A) \cong \mathrm{Hom}_{A_S}(M_S, A_S)$ canonically, if M is finitely presented.

b) Let $f = (f_1, \dots, f_n) \in A^n =: F$.   Then $\mathcal{O}_F(f) = \sum_{i=1}^n Af_i$.

Lemma 7.1.7 *Let P be a projective A-module of rank r and $p \in P$. Let further $s \in A$ such that P_s is free. Then there exists a $q \in P$ such that $\mathrm{ht}\,\mathcal{O}_P(p + sq)A_s \geq r$. In particular, if $\bar{p} \in \mathrm{Um}(P/sP)$ over $A/(s)$, then $\mathrm{ht}\,\mathcal{O}_P(p + sq) \geq r$.*

Proof. Let $p_1, \dots, p_r \in P$ such that their images in P_s make up a basis of P_s over A_s, and write

$$p_s = \sum_{i=1}^r \frac{a_i}{s^n} \cdot \frac{p_i}{1} \quad \text{with } a_i \in A, \ n \in \mathbb{N}.$$

We may assume $n > 0$. By Lemma 7.1.4 there are elements $c_i \in A$ such that $\mathrm{ht}(a_1 + c_1 s^n, \dots, a_r + c_r s^n)A_s \geq r$. Let $q := s^{n-1}\sum c_i p_i$.

By the above remarks

$$\mathcal{O}_P(p + sq)A_s = \mathcal{O}_{P_s}((p + sq)_s) \supset (a_1 + c_1 s^n, \dots, a_r + c_r s^n)A_s,$$

which implies $\mathrm{ht}\,\mathcal{O}_P(p + sq)A_s \geq r$.

If $\bar{p} \in \mathrm{Um}(P/sP)$ then $1 + sx' \in \mathcal{O}_P(p)$ for some $x' \in A$, whence $1 + sx \in \mathcal{O}_P(p + sq)$ for some $x \in A$. Therefore, if a prime ideal $\mathfrak{p} \supset \mathcal{O}_P(p + sq)$ then $s \notin \mathfrak{p}$. Hence $\mathrm{ht}\,\mathcal{O}_P(p + sq) = \mathrm{ht}\,\mathcal{O}_P(p + sq)A_s \geq r$. □

Theorem 7.1.8 (Serre's Splitting Theorem) *Let A be a commutative Noetherian ring of finite Krull dimension. Let P be a finitely generated projective A-module of rank $> \dim A$. Then P has a unimodular element. Moreover, if $s \in A$ such that P_s is free and $(p, s) \in \mathrm{Um}(P \oplus A)$ then there is a $q \in P$ such that $p + sq \in \mathrm{Um}(P)$.*

Proof 1. (H. Lindel): We may assume that A is a reduced ring with connected spectrum, whence $\mathrm{rk}\,P$ is constant. We prove the result by induction on $\dim A$. If $\dim A = 0$, then A is a finite direct product of fields and P is free. The second assertion is then easily verified.

If S is the multiplicatively closed subset of all non-zero-divisors of A then by Corollary 6.4.6 the ring $S^{-1}A$ is a finite direct product of fields. Hence $S^{-1}P$ is free; so P_s is free for some $s \in S$. Since $\dim(A/(s)) < \dim A$, by the induction hypothesis $\mathrm{Um}(P/sP) \neq \emptyset$. Let $\bar{p} \in \mathrm{Um}(P/sP)$ where $\bar{p}$ is the residue class of $p \in P$. By Lemma 7.1.7 there is a $q \in P$ such that ht $\mathcal{O}_P(p + sq) \geq \mathrm{rk}\ P$. By $\mathrm{rk}\ P > \dim\ A$ we see $p + sq \in \mathrm{Um}(P)$. $\qquad\square$

Proof 2. (R. A. Rao): We sketch the idea of the proof. As before we may assume that the ring A is reduced. View the ring A as a fibre product

$$A = A_s \times_{A_{sT}} A_T = \{(x, y) \in A_s \times A_T \mid x_T = y_s\},$$

where $T = 1 + sA$. (It is an easy check that the fibre product is equal to A.) Similarly, the projective module P is the fibre product

$$P_s \times_{A_{sT}} P_T = \{(m, n) \in P_s \times P_T \mid m_T = n_s\}.$$

We use induction on the dimension of A.

We choose a non-zero-divisor s so that P_s is free. Let $p_1 \in \mathrm{Um}(P_s)$, and let $p_2 \in \mathrm{Um}(P_T)$ which will exist by induction. Suppose that there is an $\alpha \in \mathrm{Aut}(P_{sT})$ such that

1. $(p_1)_T \alpha = (p_2)_s$,
2. $\alpha = (\alpha_1)_T (\alpha_2)_s$, $\alpha_1 \in \mathrm{Aut}(P_s)$, $\alpha_2 \in \mathrm{Aut}(P_T)$.

Then $(p_1 \alpha_1)_T = (p_2 \alpha_1)_s$, and so $p = (p_1 \alpha_1, p_2 \alpha_2^{-1}) \in P_s \times_{I_{sT}} P_T = P$. Since p is "locally unimodular" it is unimodular.

To check the existence of an α with the desired properties one observes that by the general position arguments there is an elementary matrix $\varepsilon \in \mathrm{E}_r(A_{sT})$ such that $(p_1)_T \varepsilon = (p_2)_s$. So we only have to verify that the second property holds.

It is well known that the elementary matrices have the desired 'splitting property'. More generally, one can show that

Proposition 7.1.9 *Let $s, t \in R$ with $(s, t) = 1$. Let P be a projective R-module such that P_{st} is free of rank n, and let $\varepsilon \in \mathrm{E}_n(R_{st})$ (regarded as a subset of $\mathrm{Aut}(P_{st}) \cong \mathrm{Aut}(R_{st}^n) = \mathrm{GL}_n(R_{st})$.) Then $\varepsilon = (\varepsilon_1)_t (\varepsilon_2)_s$, for some $\varepsilon_1 \in \mathrm{Aut}(P_s)$, $\varepsilon_2 \in \mathrm{Aut}(P_t)$.* $\qquad\square$

This follows from the fact that elementary matrices are homotopic to the identity. For then one can apply Quillen's Splitting Lemma 4.3.8 to a homotopy.

Corollary 7.1.10 *Let R be a one-dimensional Noetherian ring and P a finitely generated projective R-module of rank r. Then $P \cong \bigwedge^r P \oplus R^{r-1}$.*

Proof. By repeatedly using Serre's Splitting Theorem we get $P \cong Q \oplus R^{r-1}$ with Q projective of rank 1. Taking the r-th exterior power will now give us $\bigwedge^r P \cong Q$.   □

Let P be a projective R-module. We first describe some automorphisms of $P \oplus R$ called **Flips** which correspond to the elementary transformations of a free module. Let $p, q \in P$, $\varphi \in P^* = \mathrm{Hom}_R(P, R)$, $a \in R$. The following automorphisms of $P \oplus R$ are called Flips:

$$(p, a) \mapsto (p + aq, a),$$

$$(p, a) \mapsto (p, a + \varphi(p)).$$

These have the following nice property: If I is an ideal of A, then every Flip of $(P/IP) \oplus (R/I)$ over the ring R/I can be lifted to one of $P \oplus R$.

As a consequence we now derive the famous Cancellation Theorem of Hyman Bass, which was proved in the early sixties:

Theorem 7.1.11 (Bass' Cancellation Theorem) *Let R be a Noetherian ring of dimension d and P a finitely generated projective R-module of rank $> d$. Then P is "cancellative", i.e. $P \oplus Q \cong P' \oplus Q$, for some finitely generated projective module Q implies that $P \cong P'$. In fact, if $(p, a) \in \mathrm{Um}(P \oplus R)$ then there is a product τ of Flips such that $(p, a)\tau = (0, 1)$.*

Proof. We may assume that R is a reduced ring.

Let $Q \oplus Q' = F$ be free; then $P \oplus F \cong P' \oplus F$. So we may assume Q is free above. Therefore, it will suffice to show that $P \oplus R \cong P' \oplus R$ implies that $P \cong P'$. Let $(p, a) \in P \oplus R$ denote the image of $(0, 1) \in P' \oplus R$. Then $(p, a) \in \mathrm{Um}(P \oplus R)$; i.e. $\mathcal{O}_P(p) + Ra = R$. To show that P is cancellative, it suffices to show that there is an automorphism $\tau \in \mathrm{Aut}(P \oplus R)$ such that $(p, a)\tau = (0, 1)$. For then $P' \cong \mathrm{coker}(p, a) \cong \mathrm{coker}(0, 1) = P$. We show that there is a flip τ satisfying this, by induction on d. If $d = 0$ then R is a finite direct product of fields, and the assertion is easy.

To perform the induction step, let us first recall that by the Splitting Theorem there is a $p_0 \in \mathrm{Um}(P)$. Further there is a non-zero-divisor $s \in R$ such that P_s is free. Since $\dim(R/(s)) < \dim R$, by induction hypothesis by applying a product of flips we can map $(\overline{p}, \overline{a})$ to $(\overline{0}, \overline{1})$. By further flips we map $(\overline{0}, \overline{1})$ first to $(\overline{p_0}, \overline{1})$ and finally to $(\overline{p_0}, \overline{0})$. (The 'overline' denotes 'modulo (s)'.) Then we lift the composition of these flips to get an automorphism τ' such that, if $(p, a)\tau' = (p', a')$ then $\overline{p'} \in \mathrm{Um}(\overline{P})$, and $a' = sr \in sR$.

By Lemma 7.1.7 there are $q \in P$, $n \in \mathbb{N}$ such that $\mathrm{ht}(\mathcal{O}_{P_{sr}}(p' + (sr)^n q)) \geq d+1$. Note that if a prime ideal $\mathfrak{p} \supset \mathcal{O}_P(p' + (sr)^n q)$ then $sr \notin \mathfrak{p}$. This is due to the unimodularity of $(p' + (sr)^n q, sr)$! Therefore, any minimal prime overideal of

$\mathcal{O}_P(p'+(sr)^n q)$ does not contain sr. Consequently, $\mathrm{ht}(\mathcal{O}_P(p'+(sr)^n q) \geq d+1$; whence $p' + (sr)^n q \in \mathrm{Um}(P)$.

Now we can perform the following sequence of flips

$$(p', sr) \mapsto (p' + (sr)^n q, sr) \mapsto (p' + (sr)^n q, 1) \mapsto (0, 1).$$

$\square$

7.2 Lindel's Proof of Serre's Conjecture

Theorem 7.2.1 *Let R be a Noetherian ring and P a finitely genrated projective module over the polynomial ring $A := R[X_1, \ldots, X_n]$ of rank $>$ dim R. Then P possesses a unimodular element. (cf. Definition 7.1.5.)*

This means that P admits a free direct summand of rank 1. By induction on rk P this implies that P is free, if R is a field. If dim $R=1$ and $\mathrm{Spec}(R)$ is connected, then P splits into a direct sum of a free module and a rank-1-projective one. So it is free, if R is a principal domain. (It is extended from R, if R is a Dedekind ring, more generally, if R is a so called seminormal Noetherian ring of dimension 1. We will not go into this.)

It was asked in the early seventies by H. Bass, whether Theorem 7.2.1 might be true. It was established by S. M. Bhatwadekar and A. Roy in [6], where they used the Quillen-Suslin Theorem to start the induction process on dim(R). Later H. Lindel could do without this, and so gave a new proof of Serre's Conjecture. Here we outline a variant of his argument.

Remarks 7.2.2 a) Let P be a projective A-module and I an ideal of A. Consider $\overline{P} := P/IP$ as an A/I-module and let $a \in \mathcal{O}_{\overline{P}}(\overline{z})$, where $\overline{z}$ denotes the residue class of z. Then there is a $b \in \mathcal{O}_P(z)$ whose residue class in A/I is a.

Namely, since P is projective, one can lift every homomorphism $P/IP \longrightarrow A/I$ to a homomorphism $P \to A$.

b) Recall that, if $f = (f_1, \ldots, f_n) \in A^n =: F$, then $\mathcal{O}_F(f) = \sum_{i=1}^n Af_i$. Especially if $\eta : A \to A$ is a ring endomorphism, f as above and $f_\eta := (\eta(f_1), \ldots, \eta(f_n))$. Then $\mathcal{O}_F(f_\eta) = \eta(\mathcal{O}_F(f))A$, where we denote by the latter term the ideal of A, generated by $\eta(\mathcal{O}_F(f))$.

Let us apply Remark b) to prove

Lemma 7.2.3 *Let $f(X) = (f_1(X), \ldots, f_n(X)) \in R[X]^n$, $a \in R \cap \mathcal{O}_F(f)$. If there is a $g(X) \in R[X]$ with $1 + Xg(X) \in \mathcal{O}_F(f)$, then $f(abX) := (f_1(abX), \ldots, f_n(abX))$ is unimodular in $F := R[X]^n$ for every $b \in R$.*

Proof. Clearly the map $\eta : R[X] \to R[X]$, $h(X) \mapsto h(abX)$ is an endomorphism of the ring $R[X]$. So by Remark b) we get

$$1 + abXg(abX) = \eta(1 + Xg(X)) \in \mathcal{O}_F(f(abX)) \text{ and also } a = \eta(a) \in \mathcal{O}_F(f(abX)).$$

But a and $1 + abXg(abX)$ generate the unit ideal of $R[X]$.   □

Proposition 7.2.4 *Let M be a finitely presented $A[X]$-module, $m \in M$, $s \in A$ such that*

(1) M_s *is free over $A_s[X]$,*

(2) $(1 + XA[X]) \cap \mathcal{O}_M(m) \neq \emptyset$,

(3) $(1 + sA) \cap \mathcal{O}_M(m) \neq \emptyset$.

Then there is a unimodular $m' \in M$ with $m' \equiv m \,(\mathrm{mod}\ sXM)$.

Proof. By (3) there is an $r \in A$ such that $1 - sr \in \mathcal{O}_M(m)$. We may rename sr by s. Namely, if (1) holds for s it holds for rs; and if the assertion holds for rs it holds for s. So we may assume

$$1 - s \in \mathcal{O}_M(m) \tag{7.1}$$

Fix an identification $M_s = A_s[X]^n$ and let $m_s = (f_1, \dots, f_n)$ with $f_i \in A_s[X]$. For $N \in \mathbb{N}$, consider the endomorphism

$$g(X) \mapsto g((1 - s^N)X) = g(X - s^N X)$$

of $A_s[X]$. We can write

$$\left(f_1((1 - s^N)X), \dots, f_n((1 - s^N)X) \right) = (f_1, \dots, f_n) + s^N X v$$

with some $v \in M_s$. We choose N big enough such that there is a $w \in M$ with $s^N Xv = sXw_s$. We set $m' := m + sXw$. Then

$$m \equiv m' \quad (\mathrm{mod}\ sXM) \quad \text{and} \quad m'_s = \left(f_1((1 - s^N)X), \dots, f_n((1 - s^N)X) \right)$$

Since M is finitely presented, we see $(1-s)_s \in \mathcal{O}_{M_s}(m_s)$ by Remark 7.1.6 and Relation (7.1). So, using (ii), the lemma, applied to the ring A_s, to $a = (1-s)_s$ and $ab = (1 - s^N)_s$, gives that m'_s is unimodular in M_s over $A_s[X]$. But this means that $s^r \in \mathcal{O}_M(m')$ for some $r \in \mathbb{N}$.

Finally by (7.1) there is a linear map $\alpha : M \to A$ with $\alpha(m) = 1 - s$. Therefore $\alpha(m') = \alpha(m - sXw) \in 1 + sA$. Together with $s^r \in \mathcal{O}_M(m')$ this implies $1 \in \mathcal{O}_M(m')$, i.e. that m' is unimodular.   □

Definition 7.2.5 *Let I be an ideal of a polynomial ring $A[X]$ (in one indeterminate). By $l(I)$ we denote the set consisting of 0 and all leading coefficients of $f \in I \setminus \{0\}$. Obviously $l(I)$ is an ideal of A.*

Lemma 7.2.6 *(H. Bass, A. Suslin)*
Let A be a Noetherian ring and I an ideal of $A[X]$. Then $\operatorname{ht}_A l(I) \geq \operatorname{ht}_{A[X]} I$.

Proof. Let first I be a prime ideal and $\mathfrak{p} = I \cap A$. If $I = \mathfrak{p}[X]$, then clearly $l(I) = \mathfrak{p}$. With $\operatorname{ht}(\mathfrak{p}) = \operatorname{ht}(\mathfrak{p}[X])$ the lemma follows in this case. If on the other hand $I \neq \mathfrak{p}[X]$, i.e. $I \supsetneq \mathfrak{p}[X]$, let $g \in I \setminus \mathfrak{p}[X]$. Then there is an $h \in \mathfrak{p}[X]$, such that the leading coefficient of $f = g - h$ does not belong to $\mathfrak{p}$. Since $f \in I$ and $\mathfrak{p} \subset l(I)$, we have $\operatorname{ht} l(I) \geq \operatorname{ht}(\mathfrak{p}) + 1 = \operatorname{ht} I$. (The latter equation is Theorem 6.9.3 b)).

Now let I be arbitrary and $\mathfrak{q}_1, \ldots, \mathfrak{q}_r$ the prime ideals of $A[X]$ which are minimal over I. (Their number is finite, since $A[X]$ is Noetherian.) The case $I = A[X]$ being obvious, we may assume $r > 0$. Then $(\prod_i \mathfrak{q}_i)^N \subset I$ for some $N \in \mathbb{N}$. Since apparently $l(I) \cdot l(J) \subset l(IJ)$, one derives $\prod l(\mathfrak{q}_i)^N \subset l(I)$.

Let $\mathfrak{p} \supset l(I)$ be a prime ideal of A with $\operatorname{ht} \mathfrak{p} = \operatorname{ht} l(I)$. Then $l(\mathfrak{q}_i) \subset \mathfrak{p}$ for some I, and so

$$\operatorname{ht} I \leq \operatorname{ht} \mathfrak{q}_i \leq \operatorname{ht} \mathfrak{p} = \operatorname{ht} l(I).$$

$\square$

Proposition 7.2.7 *Let I be an ideal of a polynomial ring $R[X_1, \ldots, X_n]$ with $\operatorname{ht} I > \dim R$. Then there exists a so called Nagata transformation of variables, fixing X_n, and sending $X_i \mapsto X_i' = X_i + X_n^{r_i}$, for suitable $r_i \in \mathbb{N}$, $1 \leq i \leq n-1$, such that I contains a polynomial that is monic in X_n with coefficients in $R[X_1', \ldots, X_{n-1}']$.*

Proof. Induction on n. The case $n = 1$ follows directly from the Lemma 7.2.6. So assume $n > 1$.

Set $B := R[X_1, \ldots, X_{n-1}]$ and view A as a polynomial ring in one indeterminate X_n over B. Then $l(I) \subset B$ is of height $> \dim R$. By induction hypothesis we may assume, that there is a $g \in l(I)$ which is monic in X_1, i.e. of the form

$$g = X_1^T + g_{T-1} X_1^{T-1} + \cdots + g_0 \quad \text{with} \quad g_i \in R[X_2, \ldots, X_{n-1}].$$

By the definition of $l(I)$, in I there is a polynomial $f \in A$ of the form

$$f = g \cdot X_n^N + b_{N-1} X_n^{N-1} + \cdots + b_0 \quad \text{with} \quad b_i \in B.$$

Let M be the highest power of X_1 occurring in the b_i and let $K \in \mathbb{N}$ be specified later. Then set

$$Y_i := X_i \quad \text{for} \quad i < n \quad \text{and} \quad Y_n := X_n - X_1^K.$$

Now $g \cdot X_n^N = (Y_1^T + g_{T-1} Y_1^{T-1} + \cdots + g_0)(Y_n + Y_1^K)^N$ is monic of degree $T + KN$ in Y_1, whereas Y_1 occurs in $b_{N-1} X_n^{N-1} + \cdots + b_0$ with exponent $\leq M + K(N-1)$. If we choose K sufficiently large, $T + KN > M + K(N-1)$, and so f is monic in Y_1. $\square$

Proof of the theorem: Induction on on n (the number of variables). If $n = 0$, this is Serre's Splitting Theorem 7.1.8. We may assume that R hence $R[X_1, \ldots, X_n]$ is a reduced ring with connected spectrum.

Assume first that $R = k$ is a field. Then $R[X_1, \ldots, X_n] = k[X_1][X_2, \ldots, X_n]$. Then by the induction hypothesis $P \cong L \oplus R[X_1, \ldots, X_n]^{r-1}$ where L is a projective $k[X_1, \ldots, X_n]$-module of rank 1. Since $k[X_1, \ldots, X_n]$ is a factorial, L, and hence P is free.

For general R let S be the set of all non-zero-divisors in R. Since R is reduced and Noetherian, $S^{-1}R$ is a finite direct product of fields. By the case, handled above, $S^{-1}P$ is free. Since P is finitely generated. there is a non-zero-divisor $s \in R$ such that P_s is free.

Consider P/sX_nP over $\left(R[X_n]/(sX_n)\right)[X_1, \ldots, X_{n-1}]$. And note that P_{sX_n} is free over $R[X_1, \ldots, X_n]_{sX_n}$. Since the rank of P/sX_nP over $R[X_1, \ldots, X_n]/(sX_n)$ equals that of P and so is bigger than $\dim R \geq \dim R[X]/(sX_n)$ (*cf.* Theorem 6.9.3 c)), by induction hypothesis there is a $p \in P$, such that its residue class $\bar{p}$ is unimodular in P/sX_nP over $R[X_1, \ldots, X_n]/(sX_n)$. Therefore for every $q \in P$ there is an $h \in R[X_1, \ldots, X_n]$ with

$$1 + sX_nh \in \mathcal{O}_P(p + sX_nq). \tag{7.2}$$

Now we write $t := sX_n$. By Lemma 7.1.7 there is a $q \in P$ such that $\text{ht}(\mathcal{O}_{P_t}(p + tq)_t) \geq \text{rk}(P)$.

But $\overline{p + tq} = \bar{p}$, and so again by Lemma 7.1.7 we have $\text{ht}(\mathcal{O}_P(p + tq)) \geq \text{rk}(P) > \dim(R)$. Applying Proposition 7.2.7, we see that $\mathcal{O}_P(p+tq)$ contains a monic polynomial $f(X_n) \in R[X_1', \ldots, X_{n-1}'][X_n]$ with coefficients in $A := R[X_1', \ldots, X_{n-1}']$, for some suitable variables $X_1', \ldots, X_{n-1}'$.

This implies that $A[X_n]/\mathcal{O}_P(p+tq)$ is integral over $A/A \cap \mathcal{O}_P(p+tq)$. So, since s by the relation (7.2) is invertible in the first ring, it is invertible in the second one. This means $(1 + sA) \cap \mathcal{O}_P(p + tq) \neq \emptyset$. So the hypotheses of Proposition 7.2.4 are fulfilled for $M = P$ and $m = p + tq$. Therefore $\text{Um}(P) \neq \emptyset$. $\square$

8

Regular Rings

8.1 Definition and Jacobian Criterion

Let $(R, \mathfrak{m})$ be a Noetherian local ring. By the Small Dimension Theorem $\dim(R) = \operatorname{ht}\mathfrak{m} \leq \mu(\mathfrak{m}) = \dim_{R/\mathfrak{m}}(\mathfrak{m}/\mathfrak{m}^2)$ by Nakayama's Lemma. The case when there is equality is an event of much algebraic geometric significance and one isolates it as

Definition 8.1.1 *A local ring $(R, \mathfrak{m})$, which is Noetherian and for which $\dim R = dim_{R/\mathfrak{m}}(\mathfrak{m}/\mathfrak{m}^2)$, is called a* **regular local ring**. *A ring R is said to be be* **regular** *if it is Noetherian and $R_\mathfrak{m}$ is regular for every maximal ideal $\mathfrak{m}$ of R.*

Examples 8.1.2 a) A regular local ring of dimension 0 is a field and vice versa.

b) $k[x]_{(x)}$, $k[[x]]$ are regular local rings of dimension 1 if k is a field. More generally, a local principal domain is a regular local ring of dimension ≤ 1. The converse is also true, but will be proved later as Corollary 8.2.3 One-dimensional regular local rings are called **discrete valuation rings**. Namely one can define a discrete valuation

$$v : R \setminus \{0\} \to \mathbb{Z} \text{ by } v(a) = n \text{ where } a \in \mathfrak{m}^n \setminus \mathfrak{m}^{n+1}$$

This is a consequence of Krull's Intersection Theorem. The map v extends to a valuation of $Q(R)$ by $v(a/b) := v(a) - v(b)$. (The reader who has read Exercise 18 of Chapter 1 will know about this already.)

Every principal domain is a (not necessarily local) regular ring. But the class of regular domains of dimension ≤ 1 is bigger than that of the principal domains. It is the class of the so called **Dedekind rings**, which we will study later in this chapter.

c) Also $k[x_1, \ldots, x_n]_{\mathfrak{m}}$ is a regular local ring of dimension n, if k is a field and $\mathfrak{m} = (x_1, \ldots, x_n)$. More generally this holds for every maximal ideal $\mathfrak{m}$ of the polynomial ring $k[x_1, \ldots, x_n]$, since $\mathfrak{m}$ is of height n and generated by n elements by Corollary 6.7.3.

Proposition 8.1.3 (Jacobian Criterion) *Let* $I = (f_1, \ldots, f_r)$ *be an ideal of* $A = k[x_1, \ldots, x_n]$ *with a field* k. *Let further* $\mathfrak{m} = (x_1 - a_1, \ldots, x_n - a_n) \supset I$ *be a maximal ideal, and write* $a := (a_1, \ldots, a_n)$. *Then* $(A/I)_{\mathfrak{m}}$ *is a regular local ring if and only if*

$$\mathrm{rk}\left(\frac{\partial f_i}{\partial x_j}(a)\right) = n - \dim(A/I)_{\mathfrak{m}}$$

Proof. Note that the canonical map $k \to A/\mathfrak{m}$ is an isomorphism, due to the special form of $\mathfrak{m}$. For every $a \in k^n$ one has a k-linear map

$$\vartheta : A \to k^n, \quad \vartheta(f) = \left(\frac{\partial f}{\partial x_1}(a), \ldots, \frac{\partial f}{\partial x_n}(a)\right)$$

Clearly, $\vartheta(x_i - a_i) = e_i$, $\vartheta(\mathfrak{m}^2) = \{0\}$, and so ϑ induces an isomorphism

$$\vartheta : \mathfrak{m}/\mathfrak{m}^2 \overset{\sim}{\to} k^n.$$

We get

$$\mathrm{rk}\left(\frac{\partial f_i}{\partial x_j}(a)\right) = \dim_k \vartheta(I) = \dim_k(I + \mathfrak{m}^2)/\mathfrak{m}^2.$$

Further

$$\dim_k(I + \mathfrak{m}^2)/\mathfrak{m}^2 + \dim_k \mathfrak{m}/(I + \mathfrak{m}^2) = \dim_k \mathfrak{m}/\mathfrak{m}^2 = n.$$

Therefore

$$\mu(\mathfrak{m}/I) = \dim_k(\mathfrak{m}/I)\big/(\mathfrak{m}/I)^2 = \dim_k \mathfrak{m}/(I + \mathfrak{m}^2) = n - \mathrm{rk}\left(\frac{\partial f_i}{\partial x_j}(a)\right)$$

and the statement follows. $\square$

J-P. Serre and also Auslander and Buchsbaum proved the following *homological characterization of regular local rings*: A Noetherian local ring $(R, \mathfrak{m})$ is regular, if and only if every R-module M has finite homological dimension, if and only if $\mathfrak{m}$ has finite homological dimension. We give a proof of this later in this chapter after defining the requisite notions.

8.2 Regular Residue Class Rings

Proposition 8.2.1 (J. Ohm) *Let $(A, \mathfrak{m})$ be a Noetherian local ring in which there is an element x, such that (x) is a prime ideal of height 1. Then A is a domain.*

Proof. Since (x) is a prime ideal of height 1, there is a minimal prime ideal $\mathfrak{p}$ of A, properly contained in (x). We will show that $\mathfrak{p} = (0)$. Let $y \in \mathfrak{p} \subset (x)$. Then $y = ax$ for some a. Since $\mathfrak{p}$ is prime, $a \in \mathfrak{p}$. So $\mathfrak{p} = \mathfrak{p}x$. Using $x \in \mathfrak{m}$ by Nakayama's Lemma we get $\mathfrak{p} = (0)$, whence A is a domain. $\qquad\square$

Proposition 8.2.2 *Every regular local ring is a domain.*

Proof. We proceed by induction on $\dim(A)$. If $\dim A = 0$, then A is a field, and so a domain. So let $\dim(A) = n > 0$. Let $\mathfrak{m}$ be the maximal ideal and $\mathfrak{p}_1, \dots, \mathfrak{p}_r$ the minimal prime ideals of A, and set $k := A/\mathfrak{m}$. We have $\mathfrak{m} \not\subset \mathfrak{m}^2 \cup \mathfrak{p}_1 \cup \cdots \cup \mathfrak{p}_r$. Otherwise, since $\mathfrak{m} \neq \mathfrak{m}^2$, we have $\mathfrak{m} \subset \mathfrak{p}_i$ for some i by the Prime Avoidance Lemma 1.1.8. This contradicts $\dim(A) > 0$.

Choose $x \in \mathfrak{m} \setminus (\mathfrak{m}^2 \cup \mathfrak{p}_1 \cup \cdots \cup \mathfrak{p}_r)$ and $x_2, \dots, x_n \in \mathfrak{m}$ so that $x, x_2, \dots, x_n$ represent a basis of $\mathfrak{m}/\mathfrak{m}^2$ over k. We know $\dim(A/(x)) \geq n - 1$. The maximal ideal $\mathfrak{n} := \mathfrak{m}/(x)$ of the ring $A/(x)$ is generated by $n - 1$ elements, namely by the classes of $x_2, \dots, x_n$. Since generally $\dim(A/(x)) \leq \mu(\mathfrak{n})$ we see that $A/(x)$ is regular of dimension $n - 1$. By induction $A/(x)$ is a domain, whence (x) is a prime ideal. By the choice of x the ideal (x) is not minimal. Proposition 8.2.1 now says that A is a domain. $\qquad\square$

Corollary 8.2.3 *A regular local ring of dimension 1 is a principal domain.*

Proof. It is a domain by the proposition. And it is factorial by Corollary 6.5.9. Let p be a generator of the maximal ideal. Since (p) is the only prime ideal of height one, p is essentially (i.e. up to multiplication with a unit) the only prime element. Every non-zero element is of the form up^n with a unit u. From this one easily derives that every non-zero ideal is generated by some p^m. $\square$

Definition 8.2.4 *Let A be a ring, M an A-module. An r-tuple $x_1, \dots, x_r$ is called an M-regular sequence (or simply an M-sequence), if the x_i belong to $\mathfrak{m}$ and every x_i is a non-zero-divisor of $M/(x_1 M + \cdots + x_{i-1} M)$. (Especially x_1 has to be a non-zero-divisor of M.) Also this applies to $M = A$.*

Proposition 8.2.5 a) *Let $(A, \mathfrak{m})$ be a local Noetherian ring, and $x_1, \dots, x_r$ an A-regular sequence in $\mathfrak{m}$. Then $\dim(A/(x_1, \dots, x_r)) = \dim(A) - r$.*

b) *Let A be a regular local ring. Every minimal generating system $x_1, \dots, x_n$ of its maximal ideal $\mathfrak{m}$ is an A-regular sequence.*

c) *Let A be a Noetherian local ring whose maximal ideal $\mathfrak{m}$ is generated by an A-regular sequence $x_1, \ldots, x_n$. Then A is regular, and $x_1, \ldots, x_n$ is a minimal generating system of $\mathfrak{m}$.*

Proof. a) Since x_1 is a non-zero-divisor, it lies in no minimal prime ideal of A. Hence $s := \dim(A/(x_1)) \leq \dim(A) - 1$. On the other hand, there are s residue classes $\overline{y_1}, \ldots, \overline{y_s}$ modulo (x_1) such that $\mathfrak{m}/(x_1)$ is a minimal prime over-ideal of $(\overline{y_1}, \ldots, \overline{y_s})$. But then $\mathfrak{m}$ is a minimal prime over ideal of $(x_1, y_1, \ldots, y_s)$, whence $\dim(A) \leq s + 1$.

In this way one shows a) inductively.

b) Induction on n, the case $n = 0$ being obvious. Let $n > 0$. As in the proof of Proposition 8.2.2 we see that $A/(x_1)$ is regular of dimension $n - 1$. By induction hypothesis the residue classes $\overline{x_2}, \ldots, \overline{x_n}$ of $x_2, \ldots, x_n$ form an $A/(x_1)$-regular sequence, and since A is a domain, x_1 is a non-zero-divisor of A. Together we see that $x_1, x_2, \ldots, x_n$ is an A-regular sequence.

c) Since $x_1 \in \mathfrak{m}$ is a non-zero-divisor, x_1 is not contained in any minimal prime ideal. Hence $\dim(A/(x_1)) = \dim(A) - 1$. Inductively we see $\dim(A/(x_1, \ldots, x_i)) = \dim(A) - i$, especially $0 = \dim(A/(x_1, \ldots, x_n)) = \dim(A) - n$, i.e. $\dim(A) = n$, which proves that A is regular. Since in an n-dimensional local ring the maximal ideal cannot generated by less then n elements, the n-tuple $x_1, \ldots, x_n$ is a minimal generating system. $\square$

Corollary 8.2.6 *Let $(A, \mathfrak{m})$ be a regular local ring, $x_1, \ldots, x_r \in \mathfrak{m}$. The following statements are equivalent:*

(i) *$x_1, \ldots, x_r$ is a part of a minimal generating system of $\mathfrak{m}$;*

(ii) *the residue classes of $x_1, \ldots, x_r$ are linear independent modulo $\mathfrak{m}^2$;*

(iii) *$A/(x_1, \ldots, x_r)$ is regular of dimension $\dim(A) - r$.*

Proof. A subset of a finitely generated module M over a local ring $(A, \mathfrak{m})$ is a minimal generating set if and only if it represents a basis of $M/\mathfrak{m}M$ over $A/\mathfrak{m}$. Therefore (i) $\Longrightarrow$ (ii) is clear.

(ii)$\Longrightarrow$ (iii). The ring $B := A/(x_1, \ldots, x_r)$ is of dimension $n - r$ where $n = \dim A$, since $x_1, \ldots, x_r$ is an A-regular sequence. And the maximal ideal of B is generated by $n - r$ elements, since $\dim_{A/\mathfrak{m}} \mathfrak{m}/(\mathfrak{m}^2, x_1, \ldots, x_r) = n - r$. Therefore B is regular.

(iii) $\Longrightarrow$ (i). Let B and n be as above. By hypothesis the maximal ideal of B is generated by $n - r$ elements, say $y_{r+1}, \ldots, y_n$. Let $x_{r+1}, \ldots, x_n \in A$ represent them. Then $x_1, \ldots, x_n$ generate $\mathfrak{m}$. Since $\dim A = n$, this is a minimal generating system of $\mathfrak{m}$ and $x_1, \ldots, x_r$ is a part of it. $\square$

Proposition 8.2.7 *Let I be an ideal of a regular local ring $(A, \mathfrak{m})$ such that A/I is regular. Then I is generated by a part of a minimal generating system of $\mathfrak{m}$.*

Proof. Choose $y_1, \ldots, y_s \in \mathfrak{m}$ so that their residue classes modulo I form a minimal generating system of $\mathfrak{m}/I$. They form a basis of $\mathfrak{m}/(\mathfrak{m}^2 + I)$ over $A/\mathfrak{m}$. So we have $(y_1, \ldots, y_s) + I = \mathfrak{m}$. Then choose $x_1, \ldots, x_r \in I$ so that their classes form a basis of $(\mathfrak{m}^2 + I)/\mathfrak{m}^2$ over $A/\mathfrak{m}$. Then $x_1, \ldots, x_r, y_1, \ldots, y_s$ is a minimal generating set of $\mathfrak{m}$. Set $I' := (x_1, \ldots, x_r)$. The proposition now follows from the

CLAIM: $I = I'$.

Clearly $I \supset I'$. Further A/I and A/I' both are regular of the same dimension s. So I and I' are prime ideals with $I \supset I'$ and $\dim(A/I) = \dim(A/I')$. This implies $I = I'$. □

Corollary 8.2.8 *Let A be a regular local ring. Then the numbers $\mu(\mathfrak{p})$ for the ideals $\mathfrak{p}$ of A with regular $A/\mathfrak{p}$ are bounded by $\dim(A)$.* □

8.3 Homological Dimension

The concept of **homological dimension** $\mathrm{hd}_R M$ of an R-module M is a measure of how far an R-module is from being projective. Therefore one takes that a projective module is the simplest type, i.e. $\mathrm{hd}_R M = 0$ if and only if M is R-projective. One studies other R-modules in terms of projective R-modules.

Any module M is the image of free module F, i.e. there is an exact sequence

$$0 \to K \to F \to M \to 0.$$

The next simplest modules are those for which K is projective. Note that if for some free module F and some epimorphism $F \to M \to 0$ the kernel K is projective, then for any other free module F' and surjection $F' \to M$, with kernel K', we will have K' projective! This is a consequence of Schanuel's Lemma:

$$
\begin{array}{ccccccccc}
0 & \longrightarrow & K & \longrightarrow & F & \longrightarrow & M & \longrightarrow & 0 \\
& & \downarrow & & \downarrow & & \| & & \\
0 & \longrightarrow & K' & \longrightarrow & F' & \longrightarrow & M & \longrightarrow & 0
\end{array}
$$

One has $F \oplus K' \simeq F' \oplus K$, and so K' is projective. We will say that $\mathrm{hd}_R M = 1$ in the above case.

Definition 8.3.1 *More generally we say that* $\mathrm{hd}_R M = n$ *if there is an exact sequence (called projective resolution of length n) with projective R-modules* $P_i, 0 \leq i \leq n$, *and maps*

$$0 \to P_n \to P_{n-1} \to \cdots \cdots \to P_0 \to M \to 0,$$

and no such sequence of smaller length. By Schanuel's Lemma it will follow that if $\mathrm{hd}_R M = n$, *and one had a sequence as above with* P_i *R-projective for* $0 \leq i \leq n-1$, *then necessarily* P_n *must be R-projective.*

If no such integer n exists we say that $\mathrm{hd}_R M$ *is infinite.*

Examples 8.3.2 (1) $\mathrm{hd}_R M = 0$ if and only if M is R-projective.

(2) $\mathrm{hd}_R M = 1$ if and only if M is not R-projective and $M \simeq P/Q$, with P, Q projective.

(3) Let $a \in R$ be neither a zero-divisor nor a unit. Then $\mathrm{hd}_R R/(a) = 1$ as one has a free resolution

$$0 \to R \xrightarrow{\cdot a} R \to R/(a) \to 0.$$

Moreover since $\mathrm{Ann}(R/(a)) = (a)$, and a is a non-zero-divisor, $R/(a)$ is not projective.

(4) Let $R = \mathbb{Z}/(4)$ and $I = (\overline{2})R \cong R/(2)$. Then $\mathrm{hd}_R I = \infty$ as one has an exact sequence

$$0 \to I \to R \to I \to 0,$$

which does not split. The statement follows by splicing this sequence with itself.

Note that we have seen

$$\mathrm{hd}_{\mathbb{Z}/(2)} \mathbb{Z}/(2) = \mathrm{hd}_{\mathbb{Z}/(6)} \mathbb{Z}/(2) = 0, \quad \mathrm{hd}_{\mathbb{Z}} \mathbb{Z}/(2) = 1, \quad \mathrm{hd}_{\mathbb{Z}/(4)} \mathbb{Z}/(2) = \infty.$$

(5) More generally, let R be an Artinian local ring. Then for any finitely generated R-module M either $\mathrm{hd}_R M = 0$ or $\mathrm{hd}_R M = \infty$. (Details are left to the reader).

The key lemma, used actually to compute homological dimension of an R-module M, is

Lemma 8.3.3 *Let* $0 \to M' \to M \to M'' \to 0$ *be an exact sequence of R-modules. Then*

$$\mathrm{hd}\, M \leq \mathrm{Max}(\mathrm{hd}\, M', \mathrm{hd}\, M'')$$

with equality except possibly when $\mathrm{hd}\, M'' = \mathrm{hd}\, M' + 1$.

Proof. We first claim that if any two of hd M, hd M', hd M'' are finite then so is the third: If M or M'' is projective, the result is easy to see. So assume hd $M > 0$, hd $M'' > 0$, and let $M = P/Q$, with P projective. Therefore, $M' = Q'/Q$ with $Q \subset Q' \subset P$, and $M'' \simeq P/Q'$. Now hd $Q' = $ hd $M'' - 1$, hd $Q = $ hd $M - 1$. One has the exact sequence

$$0 \to Q \to Q' \to Q'/Q = M' \to 0.$$

By induction on the sum of the two finite homological dimensions in an exact sequence then we prove the claim.

Now proceed by induction on the sum of all three finite homological dimension! By induction one gets

$$\text{hd } Q' \leq \text{Max(hd } Q, \text{hd } M')$$

with equality except possibly when hd $M' = $ hd $Q + 1$. Therefore,

$$\text{hd } M'' - 1 = \text{Max(hd } M - 1, \text{hd } M')$$

with equality except possibly when hd $M' = $ hd M. Thus,

$$\text{hd } M \leq \text{Max(hd } M', \text{hd } M'')$$

with equality except possibly when hd $M'' = $ hd $M' + 1$ (by considering the various possibilities which may occur).   □

Remark 8.3.4 The above lemma is most useful to compute hd M'' from hd M' and hd M. This is successful except in the case when hd $M' = $ hd M; then it is ambiguous as one only gets an inequality.

We note that homological dimension of a finitely generated module over a Noetherian ring is a 'local property' i.e. to compute $\text{hd}_R M$ one may assume that R is a local ring to begin with. This is due to

Lemma 8.3.5 *Let R be a Noetherian ring and M a finitely generated R-module. Then*

$$\text{hd } M = \sup \text{hd}_{R_\mathfrak{p}} M_\mathfrak{p}, \quad \mathfrak{p} \in \text{Spec}(R).$$

Proof. Let r.h.s. $= d$. Since localization preserves exact sequences, $\text{hd}_{R_\mathfrak{p}} M_\mathfrak{p} \leq \text{hd}_R M$. Hence $d \leq \text{hd}_R M$. So if d is infinite then $\text{hd}_R M$ is infinite. Let us consider an exact sequence

$$0 \to K_d \to F_{d-1} \to \cdots \to F_0 \to M \to 0$$

with finitely generated free R-modules F_i. Then $(K_d)_\mathfrak{m}$ is a finitely generated free $R_\mathfrak{m}$-module for every maximal ideal $\mathfrak{m}$ of R. Now K is projective by Proposition 2.2.8, and so hd $M \leq d$.   □

Lemma 8.3.6 *If in a commutative diagram of module homomorphisms of the form*

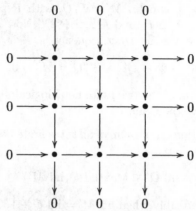

the columns and the first two rows are exact, then so is the third row.

Proof. To show the surjectivity at the south east corner is easy. For the rest use the Snake Lemma 1.2.18.   □

Proposition 8.3.7 *Let $(R, \mathfrak{m})$ be a local noetherian commutative ring, M a finitely generated R-module, and $x \in \mathfrak{m}$ a non-zero-divisor of R and M. Then*

$$\mathrm{hd}_R M = \mathrm{hd}_{R/(x)}(M/xM).$$

Proof. Choose an exact sequence

$$0 \to K \to F \to M \to 0$$

with a free R-module F. We get a commutative diagram whose columns and first two rows are exact:

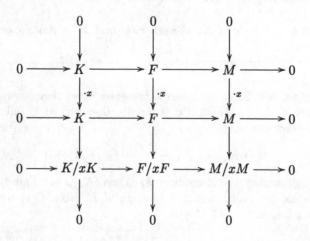

By Lemma 8.3.6 also the last row is exact.

Applying the same procedure to an exact sequence

$$0 \to K_1 \to F_1 \to K \to 0$$

with a free R-module F_1 , we get again an exact sequence

$$0 \to K_1/xK_1 \to F_1/xF_1 \to K/xK \to 0.$$

'Splicing' this with the previous one, we get the exact sequence

$$0 \to K_1/xK_1 \to F_1/xF_1 \to F/xF \to M/xM \to 0.$$

Iterating this procedure, from an R-free resolution of M

$$\ldots \to F_2 \to F_1 \to F_0 \to M \to 0$$

we get the $R/(x)$-free resolution of M/xM:

$$\ldots \to F_2/xF_2 \to F_1/xF_1 \to F_0/xF_0 \to M/xM \to 0.$$

This clearly implies $\mathrm{hd}_{R/(x)}(M/xM) \leq \mathrm{hd}_R M$. (Note that upto this point we did not need that M is finitely generated, R is local noetherian and x in its maximal ideal.)

To show the inverse inequality assume $\mathrm{hd}_{R/(x)}(M/xM) = n < \infty$ and set $L := \ker(F_{n-1} \to F_{n-2})$ in a free resolution of M. By assumption L/xL is free over $R/(x)$. So we need prove the following:

CLAIM: If M/xM is finitely generated and free over $R/(x)$, then M is free over R.

PROOF: There is an isomorphism $f' : R^n/xR^n \xrightarrow{\sim} M/xM$, which gives an epimorphism $R^n \to M/xM$. This lifts to a linear map $f : R^n \to M$, so that – writing $F := R^n$ – we have a commutative square:

$$
\begin{array}{ccc}
F & \xrightarrow{\ f\ } & M \\
\downarrow & & \downarrow \\
F/xF & \xrightarrow{\ f'\ } & M/xM
\end{array}
$$

We show that f is an isomorphism. Since $f(R^n) + xM = M$ and $x \in \mathrm{Jac}(R)$, by Nakayama's Lemma, we see that f is surjective. Now the square fits into a diagram as above with $K = \ker(f)$. The column and the first two rows are again exact. Since this implies the exactness of the third row we get $K/xK = 0$. Now K is finitely generated, since $F = R^n$ is so and R is noetherian. Again by Nakayama's Lemma we see $K = 0$, i.e. that f is an isomorphism. $\square$

8.4 Associated Prime Ideals

We will need the following device in the proof the homological characterization of regular local rings.

Definition 8.4.1 *Let M be an A-module. A prime ideal $\mathfrak{p}$ of A is called* **associated to** M, *if there is an injective A-linear map $A/\mathfrak{p} \to M$. The set of associated prime ideals is denoted by $\mathrm{Ass}_A M$ or $\mathrm{Ass}(M)$.*

Equivalently $\mathfrak{p}$ is associated to M, if and only if there is an $x \in M$ with $\mathfrak{p} = \mathrm{Ann}(x)$. (Namely, for such an x define $A/\mathfrak{p} \to M$ by $\bar{1} \mapsto x$, and conversely define x as the image of $\bar{1}$ under the injection $A/\mathfrak{p} \to M$.)

This definition formally makes sense for any ring and module. But it is of significance only if the ring A is Noetherian!

Remarks 8.4.2 a) Clearly $\mathrm{Ass}(U) \subset \mathrm{Ass}(M)$, if U is a submodule of M.

b) If $\mathfrak{p}$ is a prime ideal of A, then $\mathrm{Ass}_A(A/\mathfrak{p}) = \{\mathfrak{p}\}$. More generally, if $N \neq 0$ is a submodule of $A/\mathfrak{p}$, then $\mathrm{Ass}_A N = \{\mathfrak{p}\}$. Namely $\mathrm{Ann}(x) = \mathfrak{p}$ for any $x \in A/\mathfrak{p} \setminus \{\bar{0}\}$.

c) If $\mathfrak{p} \in \mathrm{Ass}(M)$, then $\mathfrak{p}$ is the annihilator of a submodule of M, hence $\mathfrak{p} \supset \mathrm{Ann}(M)$.

Proposition 8.4.3 *Let A be Noetherian and M any (not necessarily finitely generated) A-module. The set of the zero-divisors of M, i.e. of the $a \in A$ with $am = 0$ for some $m \in M \setminus \{0\}$, is the union of the $\mathfrak{p} \in \mathrm{Ass}_A(M)$.*

Proof. Every $\mathfrak{p} \in \mathrm{Ass}(M)$ consists of zero-divisors. Namely, if $A/\mathfrak{p} \to M$ is injective, let x be the image of $(1 \bmod \mathfrak{p})$. Then $x \neq 0$ and $\mathfrak{p}x = \{0\}$.

Conversely let $ax = 0$ with $a \in A$, $x \in M \setminus \{0\}$. Let $\mathfrak{p}$ be maximal in those ideals $J \supset (a)$ with $Jy = \{0\}$ for some $y \in M \setminus \{0\}$.

We claim that $\mathfrak{p}$ is prime. First, $\mathfrak{p} \neq A$, since $1y = y \neq 0$. Now let $bc \in \mathfrak{p}$. Then $bcy = 0$.

If $cy = 0$ then $\mathfrak{p} + (c)$ annihilates y. From the maximality condition on $\mathfrak{p}$ we derive that $c \in \mathfrak{p}$.

Now let $cx \neq 0$. Then $\mathfrak{p} + (b)$ annihilates cy. As above we derive $b \in \mathfrak{p}$.

At last define $A \to M$ by $1 \mapsto y$. Its kernel contains $\mathfrak{p}$, hence equals $\mathfrak{p}$ by its maximality property. So $\mathfrak{p} \in \mathrm{Ass}(M)$. $\square$

Corollary 8.4.4 *Under the assumptions of the proposition – namely A Noetherian, M any A-module – we have:*

$$M = 0 \iff \mathrm{Ass}_A M = \emptyset.$$

Proof. If $M = 0$, there is no zero-divisor. If $M \neq 0$ at least $0 \in A$ is a zero-divisor of M. So $\text{Ass}_A M$ cannot be empty. $\square$

Proposition 8.4.5 *Let S be a multiplicative subset of the Noetherian ring A and M an A-module. Then* $\text{Ass}_{A_S} M_S = \{\mathfrak{p}_S \mid \mathfrak{p} \in \text{Ass}(M),\ \mathfrak{p} \cap S = \emptyset\}$.

Proof. '$\supset$'. An injection $A/\mathfrak{p} \hookrightarrow M$ induces one $A_S/\mathfrak{p}_S \hookrightarrow M_S$.

'$\subset$'. Let $\mathfrak{p}_S \in \text{Ass}_{A_S} M_s$, say $\mathfrak{p}_S = \text{Ann}(x/t)$ with $x \in M$, $t \in S$ and $\{a_1, \ldots, a_n\}$ be a generating set of $\mathfrak{p}$. Then $(a_i/1)(x/t) = 0$, i.e. there are $s_i \in S$ with $s_i a_i x = 0$. So, if $s := s_1 \cdots s_n$, then $sax = 0$ for all $a \in \mathfrak{p}$, whence $\mathfrak{p} \subset \text{Ann}(sx)$. Conversely, if $b \in A$ annihilates sx, then $b/1$ annihilates $x/t \in M_S$, i.e. $b/1 \in \mathfrak{p}_S$. This implies $b \in \mathfrak{p}$, the latter being prime. So we have shown that $\mathfrak{p} = \text{Ann}(sx)$. $\square$

Most important is the *finiteness* of $\text{Ass}(M)$ for finitely generated M, which we will show in the next proposition.

Lemma 8.4.6 *Let U be a submodule of a module M. Then* $\text{Ass}(M) \subset \text{Ass}(U) \cup \text{Ass}(M/U)$.

Proof. Let $\mathfrak{p} \in \text{Ass}(M)$ and $E \subset M$ be the image of an injection $A/\mathfrak{p} \hookrightarrow M$. If $E \cap U \neq 0$, then $\{\mathfrak{p}\} = \text{Ass}(E \cap U) \subset \text{Ass}(U)$. If $E \cap U = 0$, then $E \cong (E + U)/U \subset M/U$, hence $\mathfrak{p} \in \text{Ass}(M/U)$. $\square$

Proposition 8.4.7 *Let M be a finitely generated module over a Noetherian ring A.*

a) *There are finitely many prime ideals $\mathfrak{p}_1, \ldots, \mathfrak{p}_n$ and a composition series, i.e. a finite filtration of M:*

$$0 = M_0 \subset M_1 \subset \cdots \subset M_n = M$$

with $M_i/M_{i-1} \cong A/\mathfrak{p}_i$. (There is no uniqueness statement!)

b) *For any such composition series and prime ideals as in a) we have $\text{Ass}(M) \subset \{\mathfrak{p}_1, \ldots, \mathfrak{p}_n\}$. Especially $\text{Ass}(M)$ is finite.*

c) *In the above situation for $\mathfrak{p} \in \text{Spec}(A)$ the following statements are equivalent:*

(1) $\mathfrak{p}$ *is a minimal prime over-ideal of* $\text{Ann}(M)$;
(2) $\mathfrak{p}$ *is minimal in* $\text{Ass}(M)$;
(3) $\mathfrak{p}$ *is minimal in* $\{\mathfrak{p}_1, \ldots, \mathfrak{p}_n\}$.

Proof. a) If $M = 0$ then $n = 0$ and all is trivial. If $M \neq 0$, let $\mathfrak{p}_1 \in \text{Ass}(M)$ and M_1 be the image of some injection $A/\mathfrak{p}_1 \hookrightarrow M$. Then apply the same procedure to M/M_1 to find M_2, and so on. The sequence $M_0 \subsetneq M_1 \subsetneq \cdots$ will stop, since M is Noetherian. But it does not stop as long as $M/M_i \neq 0$. So a) is established.

b) follows by the Lemma 8.4.6.

c) Let $\{x_1, \ldots, x_r\}$ be a generating set of M. Then $\text{Ann}(M) = \bigcap_{i=1}^{r} \text{Ann}(x_i)$. So, if $\mathfrak{p}$ is a minimal prime over-ideal of $\text{Ann}(M)$ it is one of some $\text{Ann}(x_i) =:$ $\mathfrak{a}$. To show that $\mathfrak{p} \in \text{Ass}(M)$, it is enough to show that $\mathfrak{p}$ is associated to the submodule $Ax_i \cong A/\mathfrak{a}$. Now $\mathfrak{p}_\mathfrak{p}$ is the only prime ideal of $A_\mathfrak{p}/\mathfrak{a}_\mathfrak{p}$, hence associated to $A_\mathfrak{p}$. By Lemma 8.4.5 we conclude that $\mathfrak{p} \in \text{Ass}(A/\mathfrak{a}) \subset \text{Ass}(M)$.

Now let $\mathfrak{p}_1, \ldots, \mathfrak{p}_n$ and $M_0 \subset \cdots \subset M_n$ be as above, then $\text{Ann}(M)$ clearly annihilates every M_i/M_{i-1}, and so $\mathfrak{p}_i \supset \text{Ann}(M)$. Therefore we have inclusions

$$\text{Ass}(M) \subset \{\mathfrak{p}_1, \ldots, \mathfrak{p}_n\} \subset \text{V}(\text{Ann}(M)).$$

Since we have shown that the minimal elements of $\text{V}(\text{Ann}(M))$ belong to $\text{Ass}(M)$, statement c) is true.   □

Corollary 8.4.8 *Let A be a Noetherian ring. There is a finite set of prime ideals, including the minimal ones, whose union is the set of zero-divisors of A.*   □

8.5 Homological Characterization of Regular Local Rings

The characterization is: *A local Noetherian ring R is regular if and only if every finitely generated R-module has a finite homological dimension, i.e. it possesses a finite free resolution.*

Definition 8.5.1 *Let M be a finitely generated module of finite homological dimension over a local Noetherian ring R. Then there is an exact sequence*

$$0 \longrightarrow R^{\beta_n} \longrightarrow R^{\beta_{n-1}} \longrightarrow \cdots \longrightarrow R^{\beta_0} \longrightarrow M \longrightarrow 0$$

with finite β_i. The **Euler characteristic** *of M is denoted by $\chi(M)$ and is defined to be $\sum_{i=1}^{n} (-1)^i \beta_i$.*

Using Schanuel's Lemma one can show that this number does not depend on the particular free resolution of M that is chosen.

Lemma 8.5.2 *Let $(R, \mathfrak{m})$ be a Noetherian local ring and M a finitely generated R-module with a finite free resolution*

$$0 \longrightarrow R^{\beta_n} \longrightarrow R^{\beta_{n-1}} \longrightarrow \cdots \longrightarrow R^{\beta_0} \longrightarrow M \longrightarrow 0.$$

Suppose that $\mathfrak{m}$ has a non-zero annihilator. Then M is free.

Proof. Using induction on n we easily see that it suffices to prove the result when $n = 1$. So assume $n = 1$ and that one has an exact sequence

$$0 \longrightarrow R^{\beta_1} \xrightarrow{g} R^{\beta_0} \xrightarrow{f} M \longrightarrow 0.$$

We may assume that f maps the canonical basis of R^{β_0} to a minimal generating set of M i.e. $g(R^{\beta_1}) \subset \mathfrak{m}R^{\beta_0}$. By the hypothesis there is a $z \in R \setminus (0)$ such that $z\mathfrak{m}R^{\beta_0} = 0$, and so $zR^{\beta_1} = 0$. Hence $R^{\beta_1} = (0)$ and $M \cong R^{\beta_0}$ is free. $\qquad\square$

Remarks 8.5.3 a) Localizing does not change the Euler characteristic: $\chi_{R_\mathfrak{p}}(M_\mathfrak{p}) = \chi_R(M)$. It follows that $\chi(M) \geq 0$ (provided it is defined). Namely localize in a minimal prime ideal $\mathfrak{p}$. Then $M_\mathfrak{p}$ is free over $R_\mathfrak{p}$ by Lemma 8.5.2, and $\chi_{R_\mathfrak{p}}(M_\mathfrak{p})$ is equal to the rank of $M_\mathfrak{p}$ over $R_\mathfrak{p}$.

b) Let I be an ideal of finite homological dimension over a Noetherian local ring R. Then also the R-module R/I has finite homological dimension and $\chi(I) + \chi(R/I) = 1$. Namely from a finite free resolution of I

$$0 \longrightarrow R^{\beta_n} \longrightarrow \cdots \longrightarrow R^{\beta_0} \longrightarrow I \longrightarrow 0$$

by finitely generated free modules we get the finite free resolution

$$0 \longrightarrow R^{\beta_n} \longrightarrow \cdots \longrightarrow R^{\beta_0} \longrightarrow R \longrightarrow R/I \longrightarrow 0$$

of R/I, and one directly observes that $\chi_R(R/I) = 1 - \chi_R(I)$.

Lemma 8.5.4 *Let R be a Noetherian ring, M a finitely generated R-module of finite homological dimension. If $\chi(M) = 0$ then $\mathrm{Ann}(M)$ contains a non-zero-divisor.*

Proof. Let $\mathrm{Ass}(R) = \{\mathfrak{p}_1, \ldots, \mathfrak{p}_n\}$, then $M_{\mathfrak{p}_i}$ is an $R_{\mathfrak{p}_i}$-module having a finite free resolution. Now $M_{\mathfrak{p}_i}$ is free by Lemma 8.5.2. But $\chi(M_{\mathfrak{p}_i}) = \chi(M) = 0$ and so $M_{\mathfrak{p}_i} = 0$. Hence $\mathrm{Ann}(M) \not\subset \mathfrak{p}_i$ for any $\mathfrak{p}_i \in \mathrm{Ass}(R)$. But then $\mathrm{Ann}(M) \not\subset \cup \mathfrak{p}_i$ and so must contain a non-zero-divisor. $\qquad\square$

Theorem 8.5.5 (Ferrand, Vasconcelos) *Let $(R, \mathfrak{m})$ be a local Noetherian ring and $I \subset \mathfrak{m}$ be an ideal of R. Assume that $\mathrm{hd}\, I < \infty$. Then I/I^2 is a free R/I-module if and only if I can be generated by an R-sequence.*

Proof. Here we prove only one implication of the equivalence and postpone the other one to Lemma 10.3.2.

So assume that I/I^2 is free over R/I. If $I = (0)$, it is generated by the empty regular sequence. So assume $I \neq (0)$. Then also $I/I^2 \neq 0$ by Nakayama's Lemma, and – since I/I^2 is free over R/I – we have $\operatorname{Ann}(I/I^2) = I$, hence $\operatorname{Ann}(I) \subset I$. By hypothesis I is of finite homological dimension. Hence R/I over R is so, too, and $\chi(I) + \chi(R/I) = 1$. We distinguish two cases.

CASE 1: $\chi(R/I) = 1$ and $\chi(I) = 0$. Then for every minimal prime ideal $\mathfrak{p}$ we have $I_\mathfrak{p} = (0)$. This implies first $I \subset \mathfrak{p}$. Otherwise there is an $a \in I \setminus \mathfrak{p}$ and for this $sa \neq 0$ for every $s \in R \setminus \mathfrak{p}$, and hence $I_\mathfrak{p} \neq (0)$. Secondly $I_\mathfrak{p} = (0)$ implies $\operatorname{Ann}(I) \not\subset \mathfrak{p}$. Namely every $a \in I$ is annihilated by some $s \in R \setminus \mathfrak{p}$, and so does a whole finite generating set.

Together this contradicts $\operatorname{Ann}(I) \subset I$.

CASE 2: $\chi(R/I) = 0$ and $\chi(I) = 1$. By Lemma 8.5.4 we see that I has a non-zero-divisor. By the Prime Avoidance Lemma 1.1.8 – applied to $\{\mathfrak{m}I\} \cup \operatorname{Ass}(R)$ – we can deduce that $I \setminus \mathfrak{m}I$ has a non-zero-divisor, say a. One can by Nakayama's Lemma complete $\{a\}$ to a set $\{a, b_2, \ldots, b_r\}$ of elements of I which represent a basis of $I/\mathfrak{m}I$ over $R/\mathfrak{m}$, hence a basis of the free R/I-module I/I^2.

It is easily verified that the conormal module $(I/(a))/(I/(a))^2 = I/(I^2 + (a))$ is a free $R/(I + (a))$-module of rank $r - 1$. Note that $\operatorname{hd}_{R/(a)}(I/aI) < \infty$ by Proposition 8.3.7.

Therefore by induction $I/(a)$ can be generated by an $R/(a)$ sequence $\overline{c_2}, \ldots, \overline{c_r}$; whence $I = (a, c_2, \ldots, c_r)$ is generated by an R-sequence. $\square$

Corollary 8.5.6 *A local Noetherian ring R whose maximal ideal $\mathfrak{m}$ has finite homological dimension is a regular local ring.* $\square$

Conversely,

Proposition 8.5.7 *Let $(R, \mathfrak{m})$ be a regular local ring of dimension d. Then any finitely generated R-module M has finite homological dimension $\leq d$.*

Proof. We may assume $d > 0$. Let $\pi \in \mathfrak{m} \setminus \mathfrak{m}^2$. Then $R/(\pi)$ is regular of dimension $d - 1$. Consider an exact sequence

$$0 \longrightarrow K \longrightarrow F \longrightarrow M \longrightarrow 0$$

with F a finitely generated free R-module. Since π is not a zero-divisor on K, by induction hypothesis $\operatorname{hd}_{R/(\pi)} K/\pi K = \operatorname{hd}_R K \leq d - 1$. Hence $\operatorname{hd} M = \operatorname{hd} K + 1 \leq d$. $\square$

We put together:

Theorem 8.5.8 *(J-P. Serre) Let $(R, \mathfrak{m})$ be a Noetherian local ring. The following properties are equivalent:*

(1) *R is regular.*

(2) *$R/\mathfrak{m}$ is of finite homological dimension.*

(3) *Every finitely generated R-module of finite homological dimension.*   □

Corollary 8.5.9 *If R is a regular local ring and $\mathfrak{p} \in \mathrm{Spec}(R)$ then $R_\mathfrak{p}$ is regular.*

Proof. Let M be a (finitely generated) $R_\mathfrak{p}$-module, generated by say $x_1, \dots, x_n$, and N be the R-submodule, generated by the same $x_1, \dots, x_n$. Then $M \cong N_\mathfrak{p}$. Any finite free resolution of N over R produces one of M over $R_\mathfrak{p}$. Hence the $R_\mathfrak{p}$-module M has finite homological dimension.   □

Remarkably, this corollary had not been proved in total generality before one knew the homological characterization of regular local rings.

Corollary 8.5.10 *A not necessarily local ring R is regular if it is Noetherian and every localization $R_\mathfrak{p}$ in any prime ideal $\mathfrak{p}$ is regular.*   □

The finiteness of the homological dimension of modules over regular local rings permits one to deduce that regular local rings are factorial, i.e. Unique Factorization Domains!

Lemma 8.5.11 (Nagata) *Let R be a Noetherian domain and $p \in R$ be a prime element. R_p is factorial if and only if R is factorial.*

Proof. It is easy to see that if R is factorial. then R_p is so. For the converse, first note that all the units of R_p are of the type up^n, for some $n \in \mathbb{Z}$ and $u \in R^*$. Moreover, any prime element in R_p is of the type uq, where u is a unit in R_p and q prime in R: If $\pi \in R_p$ is prime and $\pi = q/p^n$, then $\pi R_p = qR_p$ is a prime ideal and so $(qR_p) \cap R = qR$ is a prime ideal, i.e. q is a prime element. From this it is easy to deduce that R is factorial.   □

Theorem 8.5.12 (Auslander - Buchsbaum) *A regular local ring $(R, \mathfrak{m})$ is factorial.*

Proof. We use induction on $\dim(R)$, the case $\dim(R) = 0$, i.e. R a field, being obvious. Let $\pi \in \mathfrak{m} \setminus \mathfrak{m}^2$. Then $R/(\pi)$ is a regular local ring, and hence by Proposition 8.2.2 it is a domain. Thus π is a prime element. By Nagata's Lemma 8.5.11 it is enough to show that R_π is factorial.

We have $\dim(R_\pi) < \dim R$, since $\mathfrak{m}$ does not 'survive' in R_π. So, by induction, $(R_\pi)_\mathfrak{q}$ is factorial for every prime ideal $\mathfrak{q}$ of R. If $\mathfrak{p}$ is a prime ideal of R_π of height 1 then $\mathfrak{p}_\mathfrak{q}$ is principal for every prime ideal $\mathfrak{q}$ of R_π. Therefore, by Proposition 2.4.1 we see that $\mathfrak{p}$ is a projective R_π-module of rank 1. Now $\mathfrak{p}$ has a finite free resolution; namely every finitely generated R-module, hence every finitely generated R_π-module has one. (See the proof of Lemma 8.3.5.) Therefore $\mathfrak{p}$ is stably free, and so $\mathfrak{p}$ is free of rank 1 by Lemma 3.1.14. Hence $\mathfrak{p}$ is principal. So R_π is factorial by Corollary 6.5.9. $\qquad\square$

Corollary 8.5.13 *A regular local ring is an integrally closed domain.*

This is clear by Example 1.4.10 a).

A domain is integrally closed if the localization in every maximal ideal is integrally closed. Therefore a regular domain is integrally closed. $\qquad\square$

Corollary 8.5.14 *A regular domain A is factorial, if and only if* $\mathrm{Pic}(A) = 0$.

Proof. Recall that by our definition a regular ring is Noetherian. So A is factorial, if and only if its prime ideals of height 1 are principal. They are locally principal, hence invertible by Theorem 8.5.12. So they are principal if $\mathrm{Pic}(A) = 0$. The converse is Corollary 2.4.14. $\qquad\square$

8.6 Dedekind Rings

Dedekind rings are certain (commutative) domains which generalize slightly principal domains. They appear as integral closures of $\mathbb{Z}$ in finite extensions of $\mathbb{Q}$ and also as 'coordinate rings of nonsingular affine curves'. There are several formally different, but actually equivalent possibilities to define them.

Theorem 8.6.1 *Let A be a domain. The following properties are equivalent:*

(1) *A is Noetherian, integrally closed and of dimension ≤ 1;*

(2) *every non-zero ideal of A is invertible;*

(2') *A is regular of dimension ≤ 1, i.e. A is Noetherian and $A_\mathfrak{m}$ is a principal ideal domain for every maximal ideal $\mathfrak{m}$;*

(3) *every non-zero ideal of A is a unique product of maximal ideals. (Of course the uniqueness is meant upto the order of the factors.)*

Definition 8.6.2 *A domain A which fulfills the above equivalent properties is called a* **Dedekind ring**.

Clearly all principal domains – fields included – fulfill all above properties, so they are Dedekind rings.

The equivalence (ii) $\iff$ (ii') follows directly from the fact that invertible ideals are those which are finitely presented and locally principal.

We will prove the theorem during this section.

Without loss of generality we will assume that A is not a field.

At first we prove two lemmas.

Lemma 8.6.3 *Every minimal prime ideal $\mathfrak{p}$ of a (not necessarily Noetherian) ring A consists of zero-divisors. If additionally $\mathfrak{p}$ is finitely generated, there is a $b \in A \setminus (0)$ with $b\mathfrak{p} = (0)$.*

Proof. a) Let $a \in \mathfrak{p} \setminus (0)$ and $a_{\mathfrak{p}}$ be its canonical image in $\mathfrak{p}A_{\mathfrak{p}}$. Since $\mathfrak{p}A_{\mathfrak{p}}$ is the only prime ideal of $A_{\mathfrak{p}}$, we have $a_{\mathfrak{p}}^n = 0$ and $a_{\mathfrak{p}}^{n-1} \neq 0$ for some $n \geq 1$, hence $sa^{n-1}a = 0$ for some $s \in A \setminus \mathfrak{p}$ and $sa^{n-1} \neq 0$.

b) Since $\mathfrak{p}$ is finitely generated, there is an $n \geq 1$ with $(\mathfrak{p}A_{\mathfrak{p}})^n = (0)$. Let n be minimal with this property. Let $b \in A$ have an image in $A_{\mathfrak{p}}$ which is non-zero and lies in $\mathfrak{p}^{n-1}A_{\mathfrak{p}}$. If $a_1, \dots, a_r$ generate $\mathfrak{p}$ there are $s_i \in A \setminus \mathfrak{p}$ with $s_i b a_i = 0$. For $s := s_1 \cdots s_r$ we have $sb \neq 0$ and $sb\mathfrak{p} = (0)$. $\qquad\square$

Lemma 8.6.4 *Let A be a local Noetherian domain whose maximal ideal $\mathfrak{m}$ is non-zero and principal, say $\mathfrak{m} = (p)$. Then every element of $A \setminus (0)$ is of the form $p^n u$ with $n \in \mathbb{N}$ and $u \in A^{\times}$. Consequently $\mathfrak{m}$ is the only non-zero prime ideal of A.*

Proof. Let $a \in A \setminus (0)$. Since, by Krull's Intersection Theorem, $\bigcap_n (p^n) = \bigcap_n \mathfrak{m}^n = (0)$, there is an $n \in \mathbb{N}$ such that $a \in (p^n)$, but $a \notin (p^{n+1})$. This means $a = p^n u$ with some $u \in A \setminus (p) = A^{\times}$. Now if $\mathfrak{q}$ is any non-zero prime ideal of A and $p^n u \in \mathfrak{q} \setminus (0)$, then $u \notin \mathfrak{q}$, whence $p \in \mathfrak{q}$, whence $(p) = \mathfrak{q}$. $\qquad\square$

8.6.5 We prove $(1) \Rightarrow (2)$, (3) of Theorem 8.6.1

First we show that every maximal ideal $\mathfrak{m}$ is invertible. Since A is not a field, $\mathfrak{m} \neq (0)$. Let $a \in \mathfrak{m} \setminus (0)$ and consider $\mathfrak{m}^{-1}$. We have $A \subset \mathfrak{m}^{-1} \subset a^{-1}A$ and we will show that $A \neq \mathfrak{m}^{-1}$. Since $\mathfrak{m}/(a)$ is a minimal prime ideal of $A/(a)$ and since it is finitely generated, by Lemma 8.6.3 there is a $b \in A \setminus (a)$ with $b\mathfrak{m} \subset (a)$. Now clearly $ba^{-1}\mathfrak{m} \subset A$ and $ba^{-1} \notin A$, whence $ba^{-1} \in \mathfrak{m}^{-1} \setminus A$.

We have $\mathfrak{m} \subset \mathfrak{m}\mathfrak{m}^{-1} \subset A$. The first inclusion follows from $A \subset \mathfrak{m}^{-1}$ and the second one from the very definition of $\mathfrak{m}^{-1}$. We have to exclude $\mathfrak{m}\mathfrak{m}^{-1} = \mathfrak{m}$.

Assume it were so and $x \in \mathfrak{m}^{-1}$. Then $x\mathfrak{m} \subset \mathfrak{m}$, hence $x^{n+1}\mathfrak{m} \subset x^n\mathfrak{m}$, hence – by induction – $x^n\mathfrak{m} \subset \mathfrak{m}$ for all $n \in \mathbb{N}$. So $x^n \in \mathfrak{m}^{-1} \subset a^{-1}A$ for all n. Therefore every $x \in \mathfrak{m}^{-1}$ is integral over A. Since A is integrally closed, we get $\mathfrak{m}^{-1} = A$, which we already have disproved.

The only possibility is $\mathfrak{m}\mathfrak{m}^{-1} = A$, i.e. $\mathfrak{m}$ is invertible.

Now we show that every ideal $I \neq (0)$ in A is the product of maximal ideals. I is the empty product, if $I = A$. If not, I is contained in some maximal ideal $\mathfrak{m}$. Clearly $I \subset \mathfrak{m}^{-1}I$. But also $I \neq \mathfrak{m}^{-1}I$. Namely otherwise $\mathfrak{m}I = \mathfrak{m}\mathfrak{m}^{-1}I = I$. Localizing in $\mathfrak{m}$ and using Nakayama's Lemma, we would get $I_\mathfrak{m} = (0)$, hence $I = (0)$, since $I \hookrightarrow I_\mathfrak{m}$.

So we derive a chain of ideals in A

$$I \subsetneq \mathfrak{m}_1^{-1}I \subsetneq \mathfrak{m}_2^{-1}\mathfrak{m}_1^{-1}I \subsetneq \cdots$$

with maximal ideals $\mathfrak{m}_i$, which must eventually stop by Noetherian property, but which can be continued as long as the last ideal differs from A. Therefore

$$\mathfrak{m}_1^{-1} \cdots \mathfrak{m}_r^{-1}I = A, \quad \text{i.e.} \quad I = \mathfrak{m}_1 \cdots \mathfrak{m}_r$$

with suitable maximal ideals $\mathfrak{m}_1, \ldots, \mathfrak{m}_r$.

This implies that I is invertible, since the $\mathfrak{m}_i$ are so.

Now to the uniqueness. Let

$$I = \mathfrak{m}_1^{n_1} \cdots \mathfrak{m}_r^{n_r}$$

with different maximal ideals $\mathfrak{m}_1, \ldots, \mathfrak{m}_r$, $n_i \geq 1$ and $\mathfrak{m}$ any maximal ideal of A. Then $I_\mathfrak{m} = \mathfrak{m}^{n_i}A_\mathfrak{m}$, if $\mathfrak{m} = \mathfrak{m}_i$, and $I_\mathfrak{m} = A$ if $\mathfrak{m} \neq \mathfrak{m}_i$ for $i = 1, \ldots, r$. Since A is Noetherian, $\mathfrak{m}^nA_\mathfrak{m} = \mathfrak{m}^mA_\mathfrak{m}$ implies $n = m$ by Nakayama's Lemma. So the $\mathfrak{m}_i$ and n_i are unique.

8.6.6 We prove (3) $\Rightarrow$ (2). Let I be a non-zero ideal of A and $a \in I \setminus (0)$. We can write

$$I = \mathfrak{m}_1^{n_1} \cdots \mathfrak{m}_r^{n_r} \quad \text{and} \quad (a) = \mathfrak{m}_1^{m_1} \cdots \mathfrak{m}_r^{m_r}$$

with maximal ideals $\mathfrak{m}_i$ and $n_i, m_i \geq 0$. Since $(a)_\mathfrak{m} \subset I_\mathfrak{m}$ for every maximal ideal $\mathfrak{m}$, we get $n_i \leq m_i$ for every i. (Note that we do not need the Noetherian property for this argument.) Define

$$J := \mathfrak{m}_1^{m_1-n_1} \cdots \mathfrak{m}_r^{m_r-n_r}.$$

Then $IJ = (a)$, whence $I(a^{-1}J) = A$. So I is invertible.

8.6.7 At last we show (2) $\Rightarrow$ (1). Since every invertible ideal is finitely generated, A is Noetherian.

Now let $\mathfrak{p}$ be a non-zero prime ideal of A. Since it is invertible, its localization $\mathfrak{p}A_\mathfrak{p}$ is principal and so by Lemma 8.6.4 is the only non-zero prime ideal of $A_\mathfrak{p}$. Therefore, in view of Theorem 1.3.21, there does not exist any non-zero prime ideal properly contained in $\mathfrak{p}$.

Still we have to show that A is integrally closed. So assume that $x \in Q(A)$ is integral over A, i.e. the ring $I := A[x]$ is a finitely generated A-module. Therefore I is also a fractional ideal, isomorphic as a module to an ideal, contained in A. So I is invertible and, since I is a ring, $I^2 = I$. The invertible ideals form a group, whence $I = I^2$ implies that I is the identity element of this group, i.e. $I = A$. So $x \in A$. $\qquad\square$

Proposition 8.6.8 *Every ideal of a Dedekind ring A is generated by 2 elements. More precisely, if $I \neq (0)$ is an ideal and $c \in I \setminus (0)$, then there is an $a \in I$ with $I = (a, c)$.*

Proof. Let $\mathfrak{m}_1, \ldots, \mathfrak{m}_r$ be the maximal ideals containing c and $S = A \setminus \bigcup_i \mathfrak{m}_i$. Let $S^{-1}I$ in $S^{-1}A$ be generated by $a \in I$. Then $(a, c)_\mathfrak{m} = I_\mathfrak{m}$ for every maximal ideal $\mathfrak{m}$ of A, as one sees by considering the two cases $c \in \mathfrak{m}$ and $c \notin \mathfrak{m}$. $\qquad\square$

8.7 Examples

Proposition 8.7.1 *Let R be a Dedekind ring (e.g. $R = \mathbb{Z}$ or $R = k[X]$ with a field k), $K := Q(R)$, $K \subset L$ a finite field extension and A the integral closure of R in L. Assume further that A is finite over R. Then A is a Dedekind ring.*

Proof. We have $\dim(A) = \dim(R) \leq 1$ by Remark 1.4.4. Since R is Noetherian and A is a finitely generated R-module, every ideal of A is a finitely generated R-module as well. So, *a fortiori*, it is finitely generated as an A-module. Finally A, being an integral closure in L, is integrally closed. $\qquad\square$

Remarks 8.7.2 a) The so called Theorem of Krull-Akizuki assures that the assumption of the finiteness of A over R is not necessary. (See [10] VII.2.5.) Though its proof is much longer, and we will see that the finiteness holds in important cases.

b) In the above situation, L is the quotient field of A. Moreover $L = \{a/r \mid a \in A, \ r \in R \setminus (0)\}$. Namely every $x \in L$ fulfills an equation of the form

$$a_0 + a_1 x + \cdots + a_n x^n = 0$$

with $a_i \in R$, $a_n \neq 0$. Multiplying this equation by a_n^{n-1}, we see that $b := a_n x$ is integral over R, i.e. $b \in A$, whence $x = b/a_n$.

c) From b) we see that there exists a basis of L over K consisting of elements of A.

Proposition 8.7.3 *Let R be a Noetherian integrally closed domain with quotient field K and A the integral closure of R in a finite extension field $L \supset K$. Assume that there exists a non-zero K-linear map $\tau : L \to K$ with $\tau(A) \subset R$. Then A is finite over R. Especially this holds if L is separable over K.*

Proof. For every R-submodule M of L define

$$M^D := \{b \in L \mid \tau(ba) \in R \text{ for all } a \in A\}.$$

Then $M \subset N$ implies $M^D \supset N^D$. Especially we get $A^D \subset E^D$ if $E = Ra_1 + \cdots + Ra_n$ where $a_1, \ldots, a_n$ is a basis of L over K with $a_i \in A$. Now $\tau(A) \subset R$ implies $A \subset A^D$. Further we have an R-linear map $E^D \to R^n$ by $b \mapsto (\tau(ba_1), \ldots, \tau(ba_n))$, which is injective. Otherwise $\tau(ba_1) = \cdots = \tau(ba_n) = 0$ for some $b \in L^\times$. This would imply $\tau = 0$, since $ba_1, \ldots, ba_n$ is a basis of L over K.

In all, we have an R-linear embedding $A \hookrightarrow R^n$, which gives the result, since R is Noetherian.

If $L \supset K$ is separable, the trace $\mathrm{Tr}_{L/K}$ is known to be different from zero (see ([110], Vol I, pg. 92)) and it maps A into R. Namely, if $a \in A$, all conjugates of a also are integral over R, and so is their sum $\mathrm{Tr}_{L/K}(a)$. Since R is integrally closed, $\mathrm{Tr}_{L/K}(a) \in R$. $\qquad\square$

Corollary 8.7.4 *Let k be a field, R a k-algebra of finite type and a domain and $L \supset \mathrm{Q}(R)$ a finite field extension. Then the integral closure A of R in L is finite over R.*

Proof. According to Noether's Normalization Theorem 6.6.4, the ring R is finite over some polynomial ring $k[x_1, \ldots, x_n]$. So we may assume $R = k[x_1, \ldots, x_n]$, in which case R is integrally closed.

If char $k = 0$ then L is separable over K and the statement follows directly from Proposition 8.7.3. So assume char $k = p \neq 0$.

Let L' be the normal closure of L over K. Then – R being Noetherian – it is enough to show that the integral closure of R in L' is finite over R. So we assume the extension $L \supset K$ to be normal. In this case L is separable over the maximal purely inseparable sub-extension L_i of K. Once we have shown that the integral closure A' of R in L_i is finite over R, the corollary follows. For, then A' is integrally closed and Noetherian and $L \supset L_i = \mathrm{Q}(A')$ is separable, whence A is finite over A' and so over R by Proposition 1.4.6 b).

So finally we may assume $R = k[x_1, \ldots, x_n]$ and L purely inseparable over $K = \mathrm{Q}(R)$. We will construct a $\tau : L \to R$ which fulfills the conditions on τ in Proposition 8.7.3.

There is a power q of p with $L \subset K^{1/q}$. Then $R^{1/q} = k^{1/q}[x_1^{1/q}, \ldots, x_n^{1/q}]$ is integrally closed (being isomorphic to R) and integral over R, so is the integral closure of R in $K^{1/q}$. Then $A = R^{1/q} \cap L$.

CLAIM: $R^{1/q}$ is a free R-module.

Namely $B := k[x_1^{1/q}, \ldots, x_n^{1/q}]$ is free over R, a basis being the family of all products $x_1^{a_1/q} \cdots x_n^{a_n/q}$ with $0 \le a_i < q$. Further $R^{1/q} = k^{1/q}[x_1^{1/q}, \ldots, x_n^{1/q}]$ is free (but not necessarily finitely generated) over $B = k[x_1^{1/q}, \ldots, x_n^{1/q}]$, a basis being any basis of $k^{1/q}$ over k. Then all the products of one basis element of $R^{1/q}$ over B by one basis element of B over R form a basis of $R^{1/q}$ over R.

Since $R^{1/q}$ is free over R there is an R-linear projection $p : R^{1/q} \to R$ with $p(A) \ne 0$. Define then

$$\tau : L \to K \quad \text{by} \quad a/b \mapsto p(a)/b \quad \text{for} \quad a \in A,\ b \in R.$$

This has the required properties. □

Examples 8.7.5 Let R be a principal domain and $d \in R$ be neither a square of a unit, nor divisible by a square of a prime element. Consider $L := K(\sqrt{d})$ where $K = Q(R)$. Assume $\text{char}(K) \ne 2$, in which case especially L is separable over K. We will compute the integral closure A of R in L. Let $\alpha = a + b\sqrt{d}$ with $a, b \in K$. If $\alpha \in A$ clearly the trace $t = 2a$ and the norm $n = a^2 - b^2 d$ of α must belong to R. Conversely, if they do, α is integral over R, since then $\alpha^2 - t\alpha + n = 0$. We consider two special cases, and the reader is invited to try other ones.

1) Suppose $2 | d$. Then one easily derives $A = R + R\sqrt{d} = R \oplus R\sqrt{d}$.

This includes the case that 2 is a unit, especially the case where A is an algebra over a field of characteristic $\ne 2$. So the rings, considered in Examples 2.4.15 b) are Dedekind rings, but often not principal domains.

2) Suppose $R = \mathbb{Z}$. Then one computes

$$A = \begin{cases} R \oplus R\sqrt{d} & \text{if } d \equiv 2 \text{ or } 3 \pmod 4 \\ R \oplus R \cdot \frac{1+\sqrt{d}}{2} & \text{if } d \equiv 1 \pmod 4 \end{cases}$$

Especially the ring $\mathbb{Z} \oplus \mathbb{Z}\sqrt{5}$, considered in Examples 2.4.15 a) is a Dedekind ring, but not a principal domain.

8.8 Modules over Dedekind Rings

The following proposition will be improved later:

Proposition 8.8.1 *Let A^n be a finitely generated free module over the Dedekind ring A and $U \subset A^n$ be a submodule. Then U is the direct sum of at most n projective A-modules of rank 1.*

Proof. We show this by induction on n, the statement being true for $n \leq 1$. Let pr_n be the projection onto the last factor and denote its kernel by A^{n-1}. Then we get an exact sequence

$$0 \longrightarrow U \cap A^{n-1} \longrightarrow U \longrightarrow \mathrm{pr}_n(U) \longrightarrow 0.$$

Since $I_n := \mathrm{pr}_n(U) \subset A$ is an ideal, hence projective (maybe 0), the sequence splits. So $U \cong (U \cap A^{n-1}) \oplus I_n$. By the induction hypothesis we are done. $\square$

Remark 8.8.2 Using transfinite induction or Zorn's Lemma in a clever way, one can show the same also in the case that U is a submodule of any free module, not necessarily finitely generated.

Corollary 8.8.3 *Every finitely generated projective module over a Dedekind ring is a direct sum of projective modules of rank 1.*

As already mentioned, one can do better as we will see in the sequel.

Definitions 8.8.4 *Let R be any domain and M any R-module.*

a) *Define its* **torsion** *(or* **torsion submodule***) to be*

$$T(M) := \{x \in M \mid \text{there is an } a \in R \setminus (0) \text{ with } ax = 0\}$$

b) *M is called a* **torsion module**, *if $T(M) = M$.*

c) *M is called* **torsion free**, *if $T(M) = 0$.*

Remark 8.8.5 Clearly $T(M)$ is a submodule of M, and $T(T(M)) = T(M)$. So $T(M)$ is a torsion module. And $M/T(M)$ is torsion free.

Proposition 8.8.6 a) *Every finitely generated torsion-free module over a Dedekind ring is projective.*

b) *Consequently every finitely generated module M over a Dedekind ring is of the form*

$$M \cong T \oplus P, \quad T = T(M) \text{ a torsion module}, \quad P \cong M/T(M) \text{ projective}.$$

Note that the torsion free, but not finitely generated $\mathbb{Z}$-module $\mathbb{Q}$ is not projective.
The projective summand P above is not a canonically defined submodule of M.

Proof. a) Let M be a finitely generated torsion-free module over the Dedekind ring A and $S := A \backslash (0)$. Since M is torsion-free, $i_{M,S} : M \to S^{-1}M$ is injective. We regard M as a submodule of $S^{-1}M$. The latter is a finitely generated vector space over $K := Q(A)$, and we can choose a basis $x_1, \dots, x_n$ with $x_i \in M$. Since M is finitely generated, there is a $b \in A \backslash (0)$, with $bM \subset \sum Ax_i$. Namely if M is generated by $y_1, \dots, y_r$ and $y_j = \sum_{i=1}^{n} \frac{a_{ij}}{s_{ij}} x_i$ then one may set $b = \prod_{i,j} s_{ij}$. The map $M \to bM$, $x \mapsto bx$ is an isomorphism. Since bM is projective by Corollary 8.8.3, so is M.

b) is clear, since $M/T(M)$ is projective. $\qquad\qquad\qquad\qquad\qquad\qquad\square$

For a complete classification of all finitely generated modules over a Dedekind ring we have to analyze the finitely generated torsion modules and the finitely generated projective ones.

The following proposition is essentially a special case of Serre's Splitting Theorem. Here we give an independent elementary proof.

Proposition 8.8.7 *Let P, Q be projective R-modules of rank 1 where R is a Noetherian ring of dimension 1. Then*

$$P \oplus Q \cong (P \otimes Q) \oplus R.$$

Proof. P and Q are isomorphic to non-zero (invertible and integral) ideals I, resp. J of R.

CLAIM. There is an (integral) ideal $I' \cong I$ with $I' + J = R$.

Namely there are only finitely many prime ideals containing J, say $\mathfrak{p}_1, \dots, \mathfrak{p}_n$. Set $S := R \backslash (\bigcup_{i=1}^{n} \mathfrak{p}_i \cup \bigcup_{\mathfrak{p} \in \mathrm{Ass}(R)} \mathfrak{p}$. Then $S^{-1}(I^{-1})$ is a principal fractional ideal of the semilocal ring $S^{-1}R$. We know that every invertible fractional ideal of a semilocal ring is principal by Proposition 2.2.14. Let $z \in I^{-1}$ be a generator of $S^{-1}(I^{-1})$. Then $I' := zI \subset R$ and $S^{-1}I' = S^{-1}R$. Therefore I' is not contained in any of the prime ideals containing J, whence $I' + J = R$.

Therefore the map $I' \oplus J \to R$, defined by $(x, y) \mapsto x - y$, is surjective. And its kernel consists of the pairs (x, x) with $x \in I' \cap J$. So we have an exact sequence

$$0 \longrightarrow I' \cap J \overset{f}{\longrightarrow} I' \oplus J \overset{g}{\longrightarrow} R \longrightarrow 0,$$

with $f(a) = (a, a)$, $g(a, b) = a - b$ which clearly splits. By the Chinese Remainder Theorem and Lemma 2.4.10 we get $I' \cap J = I'J \cong P \otimes Q$. And so $(P \otimes Q) \oplus R \cong P \oplus Q$. $\qquad\qquad\qquad\qquad\qquad\square$

Corollary 8.8.8 *Any finitely generated projective module P of rank n over a Dedekind ring is of the form*

$$P = L \oplus A^{n-1},$$

where L is a rank 1 projective module. L is unique upto isomorphism.

Proof. We know already $P \cong L_1 \oplus \cdots \oplus L_n$ with L_i of rank 1. By Proposition 8.8.7 we get $P \cong (L_1 \otimes \cdots \otimes L_n) \oplus A^{n-1}$.

The uniqueness of L follows from $\bigwedge^n (L \oplus A^{n-1}) \cong L$. $\qquad\qquad\square$

Remark 8.8.9 Again we see that every ideal of a Dedekind ring is generated by at most 2 elements. Namely let $I \neq (0)$ be an ideal. Then $I \oplus I^{-1} \cong II^{-1} \oplus R \cong R^2$. Therefore there is a surjective homomorphism $R^2 \to I$.

Consequently, any projective module of rank n over a Dedekind ring is generated by a set of at most $n + 1$ elements.

Proposition 8.8.10 (Elementary Divisors for Dedekind Rings) *A finitely generated torsion module M over a Dedekind ring A is of the form*

$$M \cong (A/I_1) \oplus \cdots \oplus (A/I_n),$$

where $I_1 \supset \cdots \supset I_n$ is a descending series of non-zero ideals of A. The I_j are unique.

Proof. Let $e_1, \ldots, e_m$ form a generating set of M. For every e_i there is a $c_1 \in A \setminus (0)$ with $c_i e_i = 0$. So $cM = 0$ for the non-zero $c = c_1 \cdots c_m$. There are only finitely many maximal ideals of A containing c, say $\mathfrak{m}_1, \ldots, \mathfrak{m}_r$. Let $S = A \setminus \bigcup_i \mathfrak{m}_i$. Then $S^{-1}A$ is a semilocal Dedekind ring, hence a principal domain.

The elements of S are units modulo (c), since their residue classes are outside every maximal ideal of $A/(c)$. Therefore

$$S^{-1}A/(S^{-1}A)c \cong A/(c).$$

So M as an $A/(c)$-module is an $S^{-1}A$-module. By the Elementary Divisors Theorem 4.1.1 for principal domains we get

$$M \cong S^{-1}A/\mathfrak{a}_1 \oplus \cdots \oplus S^{-1}A/\mathfrak{a}_n \cong (A/I_1) \oplus \cdots \oplus (A/I_n)$$

where $I_j := \mathfrak{a}_j \cap A$. By an easy consideration we see that it is enough to show the uniqueness of the I_j locally. So assume that A is a local principal domain with the maximal ideal $\mathfrak{m}$ and $k = A/\mathfrak{m}$. Then $I_j = \mathfrak{m}^{r_j}$ where $r_1 \leq r_2 \leq \cdots \leq r_n$. The numbers r_j are determined by the vector space dimensions $d_i := \mathrm{rk}_k(\mathfrak{m}^{i-1}M/\mathfrak{m}^i M)$ in the following way:

$$
\begin{aligned}
d_i &= n &&\text{for } 0 < i \leq r_1 \\
d_i &= n - 1 &&\text{for } r_1 < i \leq r_2 \\
&\cdots\cdots && \cdots\cdots\cdots \\
d_i &= 1 &&\text{for } r_{n-1} < i \leq r_n \\
d_i &= 0 &&\text{for } r_n < i
\end{aligned}
$$

$\qquad\qquad\square$

8.9 The Finiteness of the Class Number

Our goal is to prove the following important classical result.

Theorem 8.9.1 *Let $R = \mathbb{Z}$ or $R = F[X]$ with a finite field F. Let further $A \supset R$ be a finite ring extension, A a domain. Then there are only finitely many isomorphism classes of ideals of A. Especially, $\mathrm{Pic}(A)$ is a finite group.*

($Q(A)$ is a so named global field and A an order in it.)

Remarks 8.9.2 a) Recall the following facts on ideals, fractional ideals and rank 1 projective modules of a domain A. Ideals I, J are isomorphic, if and only if there is an $x \in Q(A)^\times$ with $xI = J$. By Remark 2.4.13 every rank 1 projective module over the domain A is isomorphic to an invertible ideal of A. Especially $\mathrm{Pic}(A) \cong \mathrm{Inv}(A)/\mathrm{Prin}(A)$, where $\mathrm{Inv}(A)$ denotes the group of invertible fractional ideals, $\mathrm{Prin}(A)$ that of principal fractional ideals of A. Note that every fractional ideal is isomorphic to an 'integral' one, i.e. one contained in A. (If $I \subset s^{-1}A$, then $sI \subset A$.)

b) If A is the integral closure of R in a finite field extension of $Q(R)$, then A is finite over R according to Proposition 8.7.3 and Corollary 8.7.4.

8.9.3 If $f \in R \setminus (0)$, then R/fR is a finite ring. We define $|f| := \#(R/Rf)$ if $f \neq 0$ and $|0| = 0$. If $R = \mathbb{Z}$ this is the usual absolute value. If $R = F[X]$, $q := \#F$ and $f \in R \setminus (0)$ we have $|f| = q^{\deg(f)}$. In both cases:

$$|fg| = |f| \cdot |g| \quad \text{and} \quad |f + g| \le |f| + |g|. \tag{8.1}$$

(Even $|f + g| \le \mathrm{Max}\{|f|, |g|\}$ in the case $R = F[X]$.)

As an R-module the ring A is torsion-free and finitely generated, hence free. Since $K = Q(A) = \{a/s \mid a \in A, s \in R \setminus (0)\}$ by Remark 8.7.2 b), clearly A is of rank $n := [Q(A) : Q(R)]$. Fix an R-basis $\alpha_1, \dots, \alpha_n$ of A. This is also a basis of $Q(A)$ as a vector space over $Q(R)$.

Now let $I \neq (0)$ be an ideal of A. It is also a finitely generated torsion-free R-module, hence free. Let $\alpha \in I \setminus (0)$. Then the homothesy of α on the $Q(R)$-vector-space $Q(A)$ is an automorphism. Therefore $\alpha\alpha_1, \dots, \alpha\alpha_n$ are linearly independent over $Q(R)$, hence over R. It follows that I, contained in A and containing $\alpha\alpha_1, \dots, \alpha\alpha_n$, is a free R-module of rank n and that A/I is a finite ring. We write $\|I\| := \#(A/I)$.

Lemma 8.9.4 *Let $\alpha \in A$ and h_α denote the homothesy of α on A, regarded as a free R-module. Then:*

a) $\det(h_\alpha) \in A\alpha$,

b) $\|A\alpha\| = |\det(h_\alpha)|$.

Proof. The case $\alpha = 0$ being clear, assume $\alpha \neq 0$. By the Elementary Divisor Theorem 4.1.1 there are an R-basis $\alpha'_1, \ldots, \alpha'_n$ of A and $d_1, \ldots, d_n \in R$ with $\alpha \alpha'_i = d_i \alpha'_i$. So $\det(h_\alpha) = d := d_1 \cdots d_n \in A\alpha$.

Further both sides of *b)* are equal to $|d|$. $\qquad\square$

As a consequence of *a)* we see that $I \cap R \neq (0)$ for every non-zero ideal I of A, in other words, that A/I is a torsion module over R.

Lemma 8.9.5 *There is an integer $C > 0$ such that in every non-zero ideal I of A there is a $\gamma \neq 0$ with $\|A\gamma\| \leq C\|I\|$, i.e. $\#(I/A\gamma) \leq C$.*

Proof. To determine $\gamma \in I$, we distinguish two cases.

CASE 1: $R = \mathbb{Z}$. Let $m \in \mathbb{N}$ be such that $m^n \leq \|I\| < (m+1)^n$. At least two of the following $(m+1)^n$ elements

$$\sum_{j=1}^{n} b_j \alpha_j, \quad b_j \in \mathbb{Z}, \quad 0 \leq b_j \leq m$$

must be congruent modulo I, since $\#(A/I) < (m+1)^n$. Their difference will be our γ.

CASE 2: $R = F[X]$. Let $q = \#F$ and $s \in \mathbb{N}$, such that $q^{sn} \leq \|I\| < q^{(s+1)n}$. Two of the following $q^{(s+1)n}$ elements

$$\sum_{j=1}^{n} b_j \alpha_j, \quad b_j \in F[X], \quad |b_j| \leq q^s \ \text{(i.e. } \deg(b_j) \leq s)$$

must be congruent modulo I. Again call their difference γ.

In both cases γ has the properties:

(i) $\gamma \in I \setminus (0)$,

(ii) $\gamma = \sum_{j=1}^{n} m_j \alpha_j, \quad m_j \in R, \quad |m_j|^n \leq \|I\|,$

The latter means $|m_j| \leq r$, where we set $r := \sqrt[n]{\|I\|}$.

Let as above $\alpha_1, \ldots, \alpha_n$ be a basis of A over R. Then one can write $\alpha_i \alpha_j = \sum_{i,j,k} a_{ijk} \alpha_k$ with $a_{ijk} \in R$. For $\gamma = \sum_j m_j \alpha_j$ (with $m_j \in R$) we have $\gamma \alpha_i = \sum_{j,k} m_j a_{ijk} \alpha_k$. So $\det(h_\gamma)$ is a homogeneous polynomial of degree n in the m_j over R. Therefore by facts (8.1) there is a $C \in \mathbb{N}$ such that $|m_j| \leq r$ implies $|\det(h_\gamma)| \leq Cr^n = C\|I\|$. $\qquad\square$

Proof of Theorem 8.9.1: Let $c \in R$ be the product of all $a \in R \setminus (0)$ with $|a| \leq C$. (There are only finitely many of them.) We will show that every ideal $I \neq (0)$ of A is isomorphic to one between A and Ac. Since A/Ac is a finite ring there are only finitely many of the latter.

Choose γ as in Lemma 8.9.5. Then $I\gamma^{-1}/A \cong I/A\gamma$ is of order $\leq C$. There is an R-module isomorphism $I\gamma^{-1}/A \cong R/(d_1) \oplus \cdots \oplus R/(d_m)$ with suitable $d_i \in R \setminus (0)$. Then $|d_i| \leq C$, whence $d_i|c$. So $c(I\gamma^{-1}/A) = 0$, i.e. $c\gamma^{-1}I \subset A$. On the other hand $cA \subset c\gamma^{-1}I$, and we are done. $\square$

Corollary 8.9.6 *Let A be as above. For every $n \in \mathbb{N}$ there are only finitely many isomorphism classes of projective A-modules P of rank n.*

Proof. Since $\dim(A) = \dim(R) = 1$, by Proposition 8.8.7 we have an invertible ideal I with $P \cong I \oplus A^{n-1}$. $\square$

A strong generalization of the theorem and its corollary is the Jordan-Zassenhaus Theorem. See ([97] Theorem 3.9).

The finiteness of class number has other interesting consequences:

Proposition 8.9.7 *Let A be a Dedekind ring with finite Picard group. (Or more general, let A be any domain with only finitely many isomorphism classes of ideals.) Then there is an $f \in A \setminus (0)$, such that A_f is principal.*

Proof. Let the ideals $I_1, \ldots, I_n$ represent the isomorphism classes of the non-zero ideals and choose $f \in I_1 \cdots I_n \setminus (0)$. Then every non-zero ideal of A_f is isomorphic to one of the $(I_j)_f = A_f$, hence principal. $\square$

Proposition 8.9.8 *Let A be a Dedekind ring whose Picard group is a torsion group. Then for every ideal I of A there is an $f \in I$ with $\sqrt{I} = \sqrt{Af}$. In the terminology of the next chapter, every ideal is a set theoretical complete intersection.*

Proof. For $I \neq (0)$ there is an r with $I^r = Af$ for some $f \in A$. This f has the required property. $\square$

Bounds on the Number of Generators

The theory of the numbers of generators is the second main theme of our book. All rings in this chapter will be supposed to be *commutative*.

9.1 The Problems

Recall some facts of Linear Geometry (or Vector Space Theory). Let k be a field, U a linear subspace of k^n of dimension $n-r$ and $\underline{c} \in k^n$. Sets of the form $\underline{c} + U$ are sometimes called affine linear subsets of k^n. We know that $\underline{c} + U$ is the set of solutions of a system of r linear equations

$$a_{11}X_1 + \cdots + a_{1n}X_n = b_1$$
$$\cdots\cdots\cdots\cdots\cdots\cdots\cdots\cdots\cdots$$
$$a_{r1}X_1 + \cdots + a_{rn}X_n = b_r$$

where the matrix $A' := (a_{ij})_{1 \le i,j \le n}$ has rank r. With the notations we used in the section on Hilbert's Nullstellensatz

$$\underline{c} + U = \mathrm{V}(a_{11}X_1 + \cdots + a_{1n}X_n - b_1, \ldots, a_{r1}X_1 + \cdots + a_{rn}X_n - b_r)$$

is an algebraic subset of k^n.

The dimension of $\underline{c}+U$ as an algebraic subset of k^n also is $n-r$, the dimension of U as a vector space. This can for e.g be seen as follows: The matrix A' can be completed to an invertible $n \times n$-matrix $A := (a_{ij})_{1 \le i,j \le n}$. Performing the following change of variables:

$$Y_i = a_{i1}X_1 + \cdots + a_{in}X_n - b_i$$

for $i = 1, \ldots, n$ with $b_i = 0$ for $i > r$, we see

$$\mathrm{I}(\underline{c} + U) = (Y_1, \ldots, Y_r) \text{ and } k[X_1, \ldots, X_n]/\mathrm{I}(\underline{c} + U) \cong k[Y_{r+1}, \ldots, Y_n].$$

So, in algebraic terms, we get the following:

Let k be a field, $A := k[X_1, \ldots, X_n]$ and I an ideal of A which is generated by polynomials *of degree 1*. Then (I is a prime ideal and)

$$\mu(I) = \operatorname{ht}(I) = n - \dim(A/I).$$

We ask to what extent these equalities will hold if I is not necessarily generated by linear polynomials, or if A is a more general Noetherian ring. The second equality, by Proposition 6.7.5, holds for every affine domain. We will concentrate our attention on the first one. Generally for any Noetherian ring A and any ideal $I \neq A$ we only have the inequality $\mu(I) \geq \operatorname{ht}(I)$.

The case of equality deserves a name:

Definitions 9.1.1 a) *An ideal I of a Noetherian ring A is called an* **ideal theoretic complete intersection** *or simply a* **complete intersection** *if* $\mu(I) = \operatorname{ht}(I)$.

b) *The ideal I is called a* **local complete intersection** *if $I_{\mathfrak{m}}$ is a complete intersection in $A_{\mathfrak{m}}$ for every maximal ideal $\mathfrak{m}$ containing I.*

In the next section we will show that a complete intersection in the polynomial ring $A = k[X_1, \ldots, X_n]$ (more generally in any regular or 'Cohen-Macaulay' ring) can be generated by an A-regular sequence. (Note: If $I \subset A$ of height r is generated by r elements $a_1, \ldots, a_r$ then $a_1, \ldots, a_r$ need *not* be a regular sequence itself. See Remark 9.2.4.)

Examples 9.1.2 a) Every maximal ideal of $A := k[X_1, \ldots, X_n]$, where k is a (not necessarily algebraically closed) field, is of height n and generated by n elements, as we know from Corollary 6.7.3. So it is a complete intersection. (As one easily sees from the construction in the proof of Corollary 6.7.3, the maximal ideals are even generated by regular sequences.)

b) Also every prime ideal of height 1 of A is a complete intersection, since A is factorial. See Corollary 6.5.9

c) If A is a regular local ring and I an ideal with regular residue class ring A/I. Then I is a complete intersection (and prime) by Corollary 8.2.6.

d) On the other hand there are trivial counterexamples. Let for e.g. $A = k[X, Y]$ and $\mathfrak{m} := (X, Y)$. Then $\mathfrak{m}^n$ is of height 2 for every n, but $X^n, X^{n-1}Y, \ldots, XY^{n-1}, Y^n$ is a generating system of $\mathfrak{m}^n$ whose residue classes form a basis of $\mathfrak{m}^n/\mathfrak{m}^{n+1}$ over $k = A/\mathfrak{m}$. So $\mathfrak{m}^n$ cannot be generated by less than $n + 1$ elements.

e) There are also prime ideals in polynomial rings which are not complete intersections. Macaulay gave examples which show that $\mu(\mathfrak{p})$ is not bounded for prime ideals of height 2 in $k[X, Y, Z]$, where k is an infinite field. *cf.* ([52] p. 36), ([27], §4) and [59]. (Moh gave examples in the formal power series ring $k[[X, Y, Z]]$, in which Macaulay's examples do not remain prime ideals.)

In the exercises we will give an example of a curve in 3-space (i.e. a prime ideal of height 2 in $k[X, Y, Z]$) which is no complete intersection, even not locally.

f) An invertible ideal is locally a principal ideal, generated by a non-zero-divisor, hence a locally complete intersection. But in Examples 2.4.15 we found many invertible ideals which are not principal and hence not (global) complete intersections. In these cases the rings were Dedekind rings and hence the ideals generated by 2 elements.

In the Exercises of Chapter 10 we also will give an example of a regular curve in 3-space which is not a (ideal-theoretical) complete intersection, though it locally clearly is so by c).

9.1.3 Also we are interested to what extent we can bound the number of polynomials which are needed to describe an algebraic subset E of k^n, where k is an algebraically closed field. I.e. what can we say about the minimal r such that $E = V(f_1, \ldots, f_r)$ for suitable polynomials f_i? We may ask the same question if E is a closed subset of $\text{Spec}(A)$ where A is any Noetherian ring.

Since we ask, how to describe a *set* $E = V(I)$, and we have (in both cases) $V(I) = V(\sqrt{I})$, we define:

Definitions 9.1.4 a) *An ideal I of a ring A is called* **set-theoretically r-generated** *if there are r elements $a_1, \ldots, a_r \in A$ with $\sqrt{I} = \sqrt{(a_1, \ldots, a_r)}$.*

b) *An ideal I of a Noetherian ring A is called a* **set-theoretic complete intersection** *if I is set theoretically r-generated where $r = \text{ht}(I)$.*

The latter means $V(I) = V(f_1) \cap \cdots \cap V(f_r)$ with $r = \text{ht}(I)$. In geometric terms: '$V(I)$ is the intersection of the right number of hypersurfaces.'

Dimension Theory tells us that the maximal ideal of a local Noetherian ring is a set-theoretic complete intersection. Using prime avoiding techniques, one can show this for $\text{Jac}(A)$ in a semilocal Noetherian ring A. But in general one only knows the following: Given a prime ideal $\mathfrak{p}$ of height r in a Noetherian ring A, then there are r elements $a_1, \ldots, a_r$ and finitely many prime ideals $\mathfrak{p}_2, \ldots, \mathfrak{p}_s \not\subset \mathfrak{p}$, such that

$$\sqrt{(a_1, \ldots, a_r)} = \mathfrak{p} \cap \mathfrak{p}_2 \cap \cdots \cap \mathfrak{p}_s.$$

One can achieve that the $\mathfrak{p}_2, \ldots, \mathfrak{p}_s$ are different from finitely many given ones, but not always get rid of them.

If $(A, \mathfrak{m})$ is a local ring then it is often easy to compute $\mu(I)$ of a finitely generated ideal I or more general $\mu(M)$ of a finitely generated module M, since it equals the vector space dimension of $M/\mathfrak{m}M$ over $A/\mathfrak{m}$. So the question, whether some ideal is a locally complete intersection, is easy in some sense (may the answer be negative in most cases). But the question, whether some ideal might be a set-theoretic complete intersection, often is troublesome.

In Chapter 10 we will show the following:

Let I be an 'ideal of a curve', i.e. an ideal of $A := k[X_1, \ldots, X_n]$ with $\dim(A/I) = 1$.

a) If k is an arbitrary field and I locally a complete intersection, then I is a set-theoretic complete intersection.

b) If k is a field of positive characteristic, then I is a set theoretic complete intersection.

It is still open, whether the ideal of a curve is a set-theoretic complete intersection in every case.

9.2 Regular Sequences in Regular Rings

Proposition 9.2.1 *Let A be a Noetherian ring, $a_1, \ldots, a_r$ be an A-regular sequence, $\mathfrak{p}$ a minimal prime over-ideal of $I := (a_1, \ldots, a_r)$. Then $\mathrm{ht}_A(\mathfrak{p}) = r$. Consequently also $\mathrm{ht}(I) = r$.*

Proof. We know already $\mathrm{ht}(\mathfrak{p}) \leq r$ by the Small Dimension Theorem.

To show the converse, we may assume by induction that $\mathrm{ht}_{A/(a_1)}(\mathfrak{p}/(a_1)) = r - 1$. Therefore there is a series of r prime ideals

$$\mathfrak{p}_1 \subsetneq \cdots \subsetneq \mathfrak{p}_r = \mathfrak{p}$$

with $(a_1) \subset \mathfrak{p}_1$. Now a_1, as a non-zero-divisor, does not belong to any minmal prime ideal, since these are associated to A. Hence $\mathfrak{p}_1$ is not a minimal prime ideal of A. So there is a minimal prime ideal $\mathfrak{p}_0 \subsetneq \mathfrak{p}_1$, by which we conclude $\mathrm{ht}(\mathfrak{p}) \geq r$. $\qquad\square$

The goal of this section is, to show some kind of converse, namely: Let A be a (not necessarily local) regular (or more generally a 'Cohen-Macaulay') ring, and $I = (a_1, \ldots, a_r)$ with $r = \mathrm{ht}(I)$ ($< \infty$). Then there is an A-regular sequence $b_1, \ldots, b_r$, generating I. (The definition of a Cohen-Macaulay ring will be given below.)

Remarks 9.2.2 a) Let A be a Noetherian ring, M finitely generated and I an ideal with $IM \neq M$. Then for every M-regular sequence $a_1, \ldots, a_n$ in I we have $(a_1, \ldots, a_i)M \neq (a_1, \ldots, a_{i+1})M$ for $0 \leq i \leq n-1$ since a_{i+1} operates injectively on the non-zero module $M/(a_1, \ldots, a_i)M$.

Consequently every M-regular sequence in I can be extended to a maximal (finite) one.

b) If $a_1, \ldots, a_n$ is an M-regular sequence then for every multiplicative $S \subset A$ the sequence $a_1/1, \ldots, a_n/1$ is $S^{-1}M$-regular. This follows from the exactness of the localization functor and the fact that localization commutes with the forming of factor modules.

Lemma 9.2.3 a) *Let a, b be an M-regular sequence and b not a zero-divisor of M. Then b, a is an M-regular sequence, too.*

b) *If A is Noetherian, M finitely generated and $a_1, \ldots, a_n$ an M-regular sequence contained in the Jacobson radical of A, then any permutation of it is also M-regular.*

Proof. a) Assume, a were a zero-divisor modulo bM, say $ax = by$ with $x, y \in M$, $x \notin bM$. We have $y \in aM$, say $y = az$, since b is no zero divisor modulo aM. So $ax = abz$, whence $x = bz$ (a being no zero divisor of M) in spite of $x \notin bM$.

b) It is enough to show that one can interchange a_i with a_{i+1} for every i. Arguing modulo $(a_1, \ldots, a_{i-1})M$, one sees that one may assume $n = 2$. So we have to show that b, a is an M-regular sequence, if a, b is one. By a) it remains to show that b is a non-zero-divisor of M.

So assume, b were a zero-divisor of M, and choose $x \in M \setminus \{0\}$ in such a way that Ax is maximal under those monogene submodules with $bx = 0$. Since b is a non-zero-divisor modulo aM, we have $x \in aM$, i.e. $x = ay$ for some $y \in M$. So $bay = 0$, which implies $by = 0$, since a is no zero-divisor of M. From $y \neq 0$, $a \in \mathrm{Jac}(A)$, $x = ay$ we see with Nakayama's Lemma that $Ax \subsetneq Ay$, contradicting the maximality condition on Ax. $\square$

Remark 9.2.4 In general, regular sequences need not remain regular under permutation. Consider for e.g. the sequence $1, 0$, which is A-regular. If you think this is too primitive, then take a product $A = B \times C$, a B-regular sequence b_1, b_2 of B, and consider the sequence of pairs $(b_1, 1), (b_2, 0)$. And if you are not yet content, begin with an $a_1 \in A$ such that the ring $A/(a_1)$ splits. For example (if k is a field) in the ring $k[X, Y, Z]$ the sequence $X(X+1)$, $(X+1)(Z+1) - 1$, $Y(X+1)$ is regular, whereas $X(X+1)$, $Y(X+1)$, $(X+1)(Z+1) - 1$ is not so. Moreover, the latter sequence is *not* regular, but it is of length 3 and generates an ideal of height 3, namely (X, Y, Z).

Proposition 9.2.5 *Let I be an ideal of a Noetherian ring A and M a finitely generated A-module with $IM \neq M$. Then any two maximal M-regular sequences in I have the same length.*

Proof. Assume, n is the minimal length of maximal M-regular sequences in I, and let $b_1, \ldots, b_n$ such one. We use induction on n, the case $n = 0$ being obvious.

If $a_1, \ldots, a_n$ is another regular sequence in I, we have to show that it is maximal too, i.e. that I consists of zero-divisors of $M/(a_1, \ldots, a_n)M$.

Let first $n = 1$. Then I is contained in the finite union $\bigcup \mathfrak{p}$ of primes $\mathfrak{p} \in \mathrm{Ass}(M/b_1 M)$, hence $I \subset \mathfrak{p}$ for some $\mathfrak{p} \in \mathrm{Ass}(M/b_1 M)$. By the definition of Ass there is some $z \in M - b_1 M$ with $\mathfrak{p}z \subset b_1 M$, hence for such z we have $Iz \subset b_1 M$, hence $a_1 z \in b_1 M$, say $a_1 z = b_1 y$.

CLAIM: $y \notin a_1 M$ but $Iy \subset a_1 M$.

This shows that I consists of zero-divisors of $M/a_1 M$, whence also a_1 is a maximal regular sequence in I.

PROOF OF THE CLAIM: The assumption $y \in a_1 M$ would imply $a_1 z = a_1 b_1 y'$ for some y', hence $z = b_1 y' \in b_1 M$, which was excluded. For $c \in I$ we have $cb_1 y = a_1(cz) \in a_1(b_1 M)$, which implies $cy \in a_1 M$ since b_1 is no zero-divisor of M.

Now assume $n > 1$. Form the – finite – union U of all prime ideals which are associated to $M/(a_1, \ldots, a_k)M$ or $M/(b_1, \ldots, b_k)M$ for some $k < n$. Then $I \not\subset U$, since for every prime ideal $\mathfrak{p}$ of the above kind $a_n \notin \mathfrak{p}$ or $b_n \notin \mathfrak{p}$. Choose $c \in I \setminus U$. Then c is no zero-divisor of $M/(a_1, \ldots, a_k)M$ nor of $M/(b_1, \ldots, b_k)M$ for every $k = 0, \ldots, n-1$. Therefore, by Lemma 9.2.3 a), we may permute c successively with $a_{n-1}, \ldots, a_1$, resp. $b_{n-1}, \ldots, b_1$ to see that $c, a_1, \ldots, a_{n-1}$ and $c, b_1, \ldots, b_{n-1}$ are regular sequences. The latter is a maximal one. Otherwise there were a non-zero-divisor $b \in I$ of $M/(b_1, \ldots, b_{n-1}, c)M$, whence $b_1, \ldots, b_{n-1}, c, b$ would be regular. But $b_1, \ldots, b_{n-1}, c$ is already maximal by the case $n = 1$ applied to $M/(b_1, \ldots, b_{n-1})M$. So, arguing modulo cM and using the induction hypothesis, one sees that $c, a_1, \ldots, a_{n-1}$ is a maximal M-regular sequence in I. So $c, a_1, \ldots, a_{n-1}$ hence also $a_1, \ldots, a_{n-1}, c$ is maximal. Applying the case $n = 1$ again, we finally see that $a_1, \ldots, a_n$ is a maximal A-regular sequence in I. $\qquad\qquad\qquad\square$

Definition 9.2.6 *If I is an ideal of a Noetherian ring A and M a finitely generated A-module with $IM \neq M$, we denote by $\mathrm{depth}(I, M)$ the common length of the maximal M-regular sequences in I and call it the I-**depth** of M. If additionally $(A, \mathfrak{m})$ is local we call $\mathrm{depth}(\mathfrak{m}, M)$ simply the **depth** of M and denote it by $\mathrm{depth}(M)$. ($\mathrm{depth}(I, M)$ is also called **grade** of I on M.)*

Corollary 9.2.7 *In the above situation let $a_1, \ldots, a_r$ be an M-regular sequence in I. Then* $\mathrm{depth}(I, M/(a_1, \ldots, a_r)M) = \mathrm{depth}(I, M) - r$.

Corollary 9.2.8 *Let A be Noetherian, M finitely generated, the ideal I with $IM \neq M$ generated by n elements and $\mathrm{depth}(I, M) = m$. Then $m \leq n$ and there is a generating system $a_1, \ldots, a_n$ of I such that its first m elements $a_1, \ldots, a_m$ make up an M-regular sequence.*

Proof. Induction on m, the case $m = 0$ being obvious. So assume $m > 0$. Then clearly $n > 0$. Let $I = (b_1, \ldots, b_n)$. We will see that there is a non-zero-divisor a_1 of M with $I = (a_1, b_2, \ldots, b_n)$. The corollary will then follow by induction, arguing modulo $a_1 M$.

The condition $m > 0$ means $I \not\subset \bigcup \mathfrak{p}$, for $\mathfrak{p} \in \mathrm{Ass}(M)$. Hence also $b_1 + (b_2, \ldots, b_n) \not\subset \bigcup \mathfrak{p}$, for $\mathfrak{p} \in \mathrm{Ass}(M)$, by Lemma 1.1.8 a). Then there is a non-zero-divisor of M of the form $b_1 + \sum_{i=2}^{n} \lambda_i b_i$ with $\lambda_i \in A$. Clearly this may be chosen to be a_1. $\qquad\square$

Definition 9.2.9 *The dimension $\dim(M)$ of a finitely generated A-module M is defined to be $\dim(A/\mathrm{Ann}M)$.*

Proposition 9.2.10 *Let M be a non-zero finitely generated module over a local Noetherian ring A. Then*

$$\mathrm{depth}(M) \leq \mathrm{Min}_{\mathfrak{p} \in \mathrm{Ass}(M)} \dim(A/\mathfrak{p}) \leq \dim(M).$$

Proof. Since $\mathfrak{p} \in \mathrm{Ass}(M)$ implies $\mathfrak{p} \supset \mathrm{Ann}(M)$, the second inequality is clear.

We prove the first inequality by induction on $n := \mathrm{depth}(M)$, the case $n = 0$ being obvious. So let $n > 0$ and a be a non-zero-divisor of M. Since $\mathrm{depth}(M/aM) = n - 1$ we have by induction hypothesis that $n - 1 \leq \mathrm{Min}_{\mathfrak{p} \in \mathrm{Ass}(M/aM)} \dim(A/\mathfrak{p})$. Therefore it will be enough to show that for every $\mathfrak{p} \in \mathrm{Ass}(M)$ there is a $\mathfrak{p}' \in \mathrm{Ass}(M/aM)$ with $\mathfrak{p} + (a) \subset \mathfrak{p}'$. Namely then $n - 1 \leq \dim(A/\mathfrak{p}') \leq \dim(A/\mathfrak{p}) - 1$.

Consider the submodules

$$N' := \{x \in M \mid \mathfrak{p}x = 0\} \text{ and } N := \{x \in M \mid \mathfrak{p}x \subset aM\}.$$

We have $N' \subset N$, $aM \subset N$ and $N' \neq 0$, the latter since $\mathfrak{p} \in \mathrm{Ass}(M)$. Any $\mathfrak{p}' \in \mathrm{Ass}(N/aM)$ will fulfil the above statements. So we only have to show that $aM \neq N$.

Assume $aM = N$ and let $y \in N'(\subset N)$. Then $y = ax$ for some $x \in M$ and $\mathfrak{p}ax = \mathfrak{p}y = 0$, whence $\mathfrak{p}x = 0$ since a is a non-zero-divisor of M. This would mean $aN' = N'$, contradicting $N' \neq 0$ by Nakayama's Lemma. $\qquad\square$

Definition 9.2.11 *Let A be a Noetherian ring. A* **Cohen-Macaulay A-module** *is a finitely generated module M with* $\operatorname{depth}_{A_{\mathfrak{m}}} M_{\mathfrak{m}} = \dim(M_{\mathfrak{m}})$ *for every maximal ideal* $\mathfrak{m}$. *A ring A is called a* **Cohen-Macaulay ring** *if it is Noetherian and Cohen-Macaulay as an A-module.*

For us, the most interesting examples of Cohen-Macaulay rings are the regular ones, especially the polynomial rings over fields.

Corollary 9.2.12 *Let M be a Cohen-Macaulay module over a Noetherian ring.*

a) *Every element of* $\operatorname{Ass}(M)$ *is minimal in* $\operatorname{Ass}(M)$. *In other words:* $\mathfrak{p}, \mathfrak{p}' \in \operatorname{Ass}(M)$ *and* $\mathfrak{p} \subset \mathfrak{p}'$ *imply* $\mathfrak{p} = \mathfrak{p}'$.

b) *If in addition A is local, then* $\dim(M) = \dim(A/\mathfrak{p})$ *for every minimal prime over-ideal $\mathfrak{p}$ of* $\operatorname{Ann}(M)$.

Proof. a) Since Ass behaves well under localization according to Proposition 8.4.5, we may assume that A is local. Suppose $\mathfrak{p} \neq \mathfrak{p}'$. Then $\dim(M) \geq \dim(A/\mathfrak{p}) > \dim(A/\mathfrak{p}') \geq \operatorname{depth}(M)$, contrary to $\dim(M) = \operatorname{depth}(M)$.

b) $\dim(M) \geq \dim(A/\mathfrak{p}) \geq \operatorname{depth}(M)$ for every minimal prime over-ideal of $\operatorname{Ann}(M)$. $\qquad\square$

Corollary 9.2.13 *Let $(A, \mathfrak{m})$ be a local Noetherian ring, M a finitely generated Cohen-Macaulay A-module and $a_1, \dots, a_r$ an M-regular sequence in $\mathfrak{m}$. Then $M/(a_1, \dots, a_r)M$ is a Cohen-Macaulay module, too.*

Proof. By induction, we may assume $r = 1$. Clearly $\operatorname{depth}(M/a_1 M) = \operatorname{depth}(M) - 1$. On the other hand, $\dim(M/a_1 M) \leq \dim(M) - 1$. This is because $\operatorname{Ann}(M) \subset \operatorname{Ann}(M) + (a_1) \subset \operatorname{Ann}(M/a_1 M)$. Moreover no minimal prime over-ideal $\mathfrak{p}$ of $\operatorname{Ann}(M)$ contains a_1, as $\mathfrak{p} \in \operatorname{Ass}(M)$ by Proposition 8.4.7 c). $\qquad\square$

Theorem 9.2.14 *Let A be a Cohen-Macaulay ring, $I \neq A$ an ideal of height r, generated by r elements. Then there is an A-regular sequence $a_1, \dots, a_r$, generating I.*

Proof. Let $a_1, \dots, a_\nu$ be a maximal A-regular sequence in I and $J := (a_1, \dots, a_\nu)$. Then I consists of zero-divisors of A/J and so is contained in some $\mathfrak{p} \in \operatorname{Ass}(A/J)$. But since A/J is a Cohen-Macaulay module, its associated prime ideals are minimal prime over-ideals of J. By Proposition 9.2.1 we get $\operatorname{ht}(\mathfrak{p}) = \nu$. So $r \leq \nu$, whence $r = \nu$. Thus $\operatorname{depth}(I, A) = r$. Hence by Corollary 9.2.8 the r generators of I can be replaced by a regular sequence which generates I. $\qquad\square$

9.3 Forster-Swan Theorem

Let R be a commutative noetherian ring with 1, and let M be a finitely generated R-module. The Forster-Swan theorem gives a bound on the number of generators of a finitely generated R-module M in terms of its local number of generators $\mu_{\mathfrak{p}}(M)$. (Note that the local number of generators of M is always less than the number of generators of M. So the theorem gives a hold in the reverse direction.)

By Nakayama's lemma, to find the minimal local number of generators $\mu_{\mathfrak{p}}(M)$ of M is the same as to calculate the dimension of the vector space $M_{\mathfrak{p}}/\mathfrak{p}M_{\mathfrak{p}}$ over the field $k(\mathfrak{p}) = R_{\mathfrak{p}}/\mathfrak{p}R_{\mathfrak{p}}$. Thus the calculation of a bound on $\mu(M)$ is reduced to a problem in linear algebra.

Let us describe this bound. Given a prime ideal $\mathfrak{p}$ define the local $\mathfrak{p}$-bound as the number
$$b_{\mathfrak{p}}(M) = \mu_{\mathfrak{p}}(M) + \dim R/\mathfrak{p}.$$
The Forster-Swan theorem asserts that
$$\mu(M) \leq \sup {}_{\mathfrak{p}}\{b_{\mathfrak{p}}(M) \mid \mathfrak{p} \text{ prime }, M_{\mathfrak{p}} \neq 0\}.$$

The theorem was discovered by O. Forster in 1964, and shortly generalised by R. G. Swan. We give four proofs of this theorem.

- O. Forster's proof via the *Kaplansky ideals*.

- R.G. Swan's proof via *basic* elements, a concept which was introduced by him.

- R. A. Rao's proof based on classical K-theory. (Alternatively, use Quillen's Splitting Lemma as done by Plumstead.) The advantage is one can prove similar bounds, conjectured by Eisenbud-Evans, for finitely generated modules over polynomial extensions of noetherian rings.

- We deduce the Forster-Swan theorem from the Eisenbud-Evans theorem.

Schematically, the different viewpoints to approach the Forster-Swan theorem are given by the following diagram:

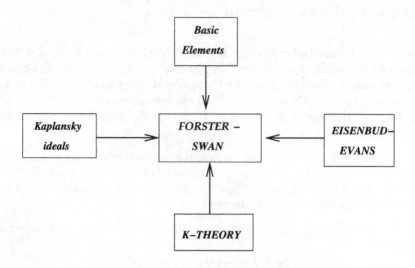

9.3.1 Let M be a finitely generated R-module. The following ideals $I_r(M)$ for every $r \geq 0$ were defined by S. Kaplansky:

$$I_r(M) = \sum \text{Ann}(M/\sum_{i=1}^{r} Rm_i)$$

where the summation is done over all possible collections of r elements $m_1, \ldots, m_r$ in M. We call them **Kaplansky ideals**.

They have the following properties:

a) $I_0(M) = \text{Ann}(M)$, $I_k(M) \subset I_{k+1}(M)$ for all k, and $I_r(M) = R$ if $r \geq \mu(M)$.

b) For any multiplicatively closed subset S of R

$$I_k(S^{-1}M) = S^{-1}I_k(M)$$

(where on the left hand side $S^{-1}M$ is considered as an $S^{-1}A$-module).

PROOF: One has

$$S^{-1}I_k(M) = S^{-1}\sum \text{Ann}_A(M/\sum_{i\leq k} Rm_i) \overset{(1)}{=} \sum S^{-1}\text{Ann}(M/\sum_{i\leq k} Rm_i)$$

$$\overset{(2)}{=} \sum \text{Ann}_{A_S}(S^{-1}(M/\sum_{i\leq k} Rm_i))$$

$$\overset{(3)}{=} \sum \text{Ann}_{A_S}(S^{-1}M/\sum S^{-1}Rm_i) = I_k(S^{-1}M).$$

To see the equalities (1) and (3), show that for every family (N_i) of submodules of any module M we have $\sum_i S^{-1}N_i = S^{-1}(\sum_i N_i)$. For equality (2) we need that M, hence $M/\sum Rm_i$ is finitely generated.

c) Recall the definition $\mu_{\mathfrak{p}}(M) := \mu_{R_{\mathfrak{p}}}(M_{\mathfrak{p}}) = \dim_{R_{\mathfrak{p}}/\mathfrak{p}R_{\mathfrak{p}}}(M_{\mathfrak{p}}/\mathfrak{p}M_{\mathfrak{p}})$.

We have $\mu_{\mathfrak{p}}(M) \geq k$ if and only if $\mathfrak{p} \supset I_{k-1}(M)$. In particular if $\mathfrak{p} \supset I_{k-1}(M)$ and $\mathfrak{p} \not\supset I_k(M)$ then $\mu_{\mathfrak{p}}(M) = k$. (This is a topological characterization of $\mu_{\mathfrak{p}}(M)$.

PROOF: Suppose that $\mu_{\mathfrak{p}}(M) \geq k$ then

$$I_{k-1}(M_{\mathfrak{p}}) = \sum_{i=1}^{k-1} \text{Ann}(M_{\mathfrak{p}}/\sum R_{\mathfrak{p}}m_i) \neq R_{\mathfrak{p}},$$

and so $I_{k-1}(M_{\mathfrak{p}}) \subset \mathfrak{p}R_{\mathfrak{p}}$. Since $I_{k-1}(M_{\mathfrak{p}}) = I_{k-1}(M)_{\mathfrak{p}}$ it follows that $I_{k-1}(M) \subset \mathfrak{p}$, i.e. $\mathfrak{p} \in V(I_{k-1}(M))$. Conversely, if $\mu_{\mathfrak{p}}(M) < k$ then $I_{k-1}(M_{\mathfrak{p}}) = R_{\mathfrak{p}}$, whence $I_{k-1}(M)_{\mathfrak{p}} = R_{\mathfrak{p}}$, and so $\mathfrak{p}R_{\mathfrak{p}} \not\supset I_{k-1}(M)_{\mathfrak{p}}$. Consequently, $\mathfrak{p} \not\supset I_{k-1}(M)$.

The third property allows us to induce a 'stratification' of $\text{Spmax}(R)$, resp. $\text{Spec}(R)$, with respect to a finitely generated R-module M:

Let $X_k(M) = \{\mathfrak{m} \in \text{Spmax}(R) \mid \mu_{\mathfrak{m}}(M) \geq k\} = V(I_{k-1}(M)) \cap \text{Spmax}(R)$ then

$$\text{Spmax}(R) = X_0(M) \supset X_1(M) \supset \cdots \supset X_{r+1}(M) = \emptyset,$$

where $r = \mu(M)$, the minimal number of generators of M.

Definition 9.3.2 *The* **Forster - Swan bound** *(on $\mu(M)$) is the number*

$$b(M) = \max_k \{k + \dim \ X_k(M) \mid k \geq 1, \ X_k(M) \neq \emptyset\}.$$

We set $b(M) = 0$ if $X_1(M) = \emptyset$, i.e. $M = 0$. Note that if N is a factor module of M then $b(N) \leq b(M)$ as $\mu_{\mathfrak{p}}(N) \leq \mu_{\mathfrak{p}}(M)$ for all primes $\mathfrak{p}$.

Remark 9.3.3 An alternative description of the Forster - Swan bound is the following: Let

$$b(M) = \max\{\mu_{\mathfrak{p}}(M) + \delta(\mathfrak{p}) \mid \mathfrak{p} \in \text{Spec}(R)\},$$

where we shortly write $\delta(\mathfrak{p}) := \dim(\text{Spmax}(R/\mathfrak{p})) = \dim(V(\mathfrak{p}) \cap \text{Spmax}(R))$. (Clearly $\delta(\mathfrak{p}) \leq \dim(R/\mathfrak{p})$.)

PROOF: Denote the right hand number by $f(M)$. Since $X_k(M) = V(I_{k-1}(M)) \cap \text{Spmax}(R)$ there is a prime ideal $\mathfrak{p}_0 \supset I_{k-1}(M)$ such that $\delta(\mathfrak{p}_0) = \dim(X_k(M))$. From $\mathfrak{p}_0 \supset I_{k-1}(M)$ we get $\mu_{\mathfrak{p}_0}(M) \geq k$. Hence $k + \dim X_k(M) \leq \mu_{\mathfrak{p}_0}(M) + \delta(\mathfrak{p}_0)$. Consequently $b(M) \leq f(M)$

To prove the converse inequality, let $\mathfrak{p} \in \text{Spec}(R)$ and $\mu_{\mathfrak{p}}(M) = k$. Then $\mathfrak{p} \supset I_{k-1}(M)$, whence $\delta(\mathfrak{p}) \leq \dim \ \text{Spmax}(R/I_{k-1}(M)) = \dim(X_k(M))$. Hence $\mu_{\mathfrak{p}}(M) + \delta(\mathfrak{p}) \leq k + \dim(X_k(M))$. Hence taking the supremum on both sides $f(M) \leq b(M)$.

Theorem 9.3.4 (Forster - Swan) *Let R be a noetherian ring and M a finitely generated R-module. Then $\mu(M) \leq b(M)$.*

Proof. We may assume $b(M) < \infty$ and argue by induction on $b(M)$. If $b(M) = 0$ then $\mathrm{Ann}(M) = R$ and $M = 0$ is generated by 0 elements!

Let $X_{kj}(M)$ be the finitely many irreducible components of $X_k(M)$. Let J be the finite set $\{(k,j) \mid k \geq 1$ and $k + \dim(X_{kj}(M)) = b(M)\}$.

If $(k,j) \in J$ the $X_{kj}(M) \not\subset X_{k+1}(M)$. Otherwise we would have $k + 1 + \dim(X_{k+1}(M)) > k + \dim(X_{kj}(M)) = b(M)$, which contradicts the definition of $b(M)$.

Let $\mathfrak{m}_{kj} \in X_{kj}(M) \setminus X_{k+1}(M)$. Then $\mu_{\mathfrak{m}_{kj}}(M) = k > 0$ for $(k,j) \in J$.

CLAIM: There is an $x \in M$ such that $x \notin \mathfrak{m}_{kj}M$ for all $(k,j) \in J$. This is equivalent to saying that there is an element $x \in M$ whose image in every $R/\mathfrak{m}_{kj}$-vector space $M/\mathfrak{m}_{kj}M$ belongs to some basis.

Let J' be the set of all $\mathfrak{m}_{kj}$ with $(k,j) \in J$. (It may happen that $\mathfrak{m}_{kj} = \mathfrak{m}_{kj'}$ for different j, j'.) By the Chinese Remainder Theorem the canonical map

$$\kappa : M \longrightarrow \bigoplus_{\mathfrak{m} \in J'} \frac{M}{\mathfrak{m}M}$$

is surjective. Hence there is an $x \in M$ whose image under κ has no zero component.

Let $N = M/Rx$. By the choice of x, $\mu_{\mathfrak{m}_{kj}}(N) = k - 1$ for all $(k,j) \in J$, i.e. $\mathfrak{m}_{kj} \notin X_{k+1}(N)$. Therefore $k + \dim(X_k(N)) < b(M)$, hence $b(N) < b(M)$. By induction hypothesis $\mu(N) \leq b(N)$. Since $b(N) < b(M)$, $\mu(M) \leq \mu(N) + 1 \leq b(M)$. $\qquad\square$

Corollary 9.3.5 *Let P be a finitely generated projective R-module of rank r over a noetherian ring whose maximal spectrum has dimension d. Then P can be generated by $d + r$ elements.*

Proof. $X_0(P) = X_1(P) = \cdots = X_r(P) \supsetneq X_{r+1}(P) = \emptyset$, so $b(P) = r + \dim(X_r(P)) \leq r + \dim(\mathrm{Spmax}(R)) = r + d$. $\qquad\square$

Corollary 9.3.6 *Let A be a Noetherian ring of dimension n and I an ideal which is a local complete intersection. Then $\mu(I) \leq n + 1$.*

Proof. Let $\mathfrak{p} \in \mathrm{Spec}(A)$. We have to show $\mu_{\mathfrak{p}}(I) + \dim(A/\mathfrak{p}) \leq n + 1$. We distinguish two cases.

CASE 1: $\mathfrak{p} \not\supset I$. Then $I_{\mathfrak{p}} = A_{\mathfrak{p}}$, whence $\mu_{\mathfrak{p}}(I) = 1$. But clearly $\delta(\mathfrak{p}) \leq \dim(A/\mathfrak{p}) \leq n$.

CASE 2: By hypothesis $I_{\mathfrak{p}}$ is generated by a regular sequence of length $\leq \dim(A_{\mathfrak{p}}) = \mathrm{ht}(\mathfrak{p})$. So $\mu_{\mathfrak{p}}(I) + \dim(A/\mathfrak{p}) \leq \mathrm{ht}(\mathfrak{p}) + \dim(A/\mathfrak{p}) \leq n$. $\qquad\square$

Corollary 9.3.7 *Let A be a regular ring of dimension n, for example $A :=$ $k[X_1, \dots, X_n]$ with a field k, and I an ideal such that A/I is regular. Then $\mu(I) \leq n + 1$.*

Indeed in this case I is a local complete intersection by Corollary 8.2.6.   □

An immediate corollary is

Corollary 9.3.8 *Let A be a residue class ring of a finite dimensional regular ring, for e.g. an affine algebra, then the set of the $\mu(I)$ where I ranges over all ideals with regular A/I is bounded. Especially the set $\{\mu(\mathfrak{m}) \mid \mathfrak{m} \in \mathrm{Spmax}(A)\}$ is bounded.*   □

Note that an algebra A of finite type over any regular ring R (for e.g. $R = \mathbb{Z}$) is of the above kind, namely a residue class ring of $R[X_1, \dots, X_m]$, which is regular as well.

Another special case is that I is invertible. Then $b(I) = 1 + \dim \mathrm{Spmax}(A)$. We get a new proof that ideals of a Dedekind ring can be generated by 2 elements. We know Dedekind rings which are not principal domains. For these the Forster-Swan bound is sharp.

Also if A is a regular domain and I a prime ideal of height 1. Then by Theorem 8.5.12 I is locally 1-generated, i.e. invertible. So the above inequality holds.

The bound $b(M)$ in the Forster-Swan Theorem is the best possible in general also in higher dimension. We recall an example of R.G. Swan which shows this:

Let A be the subring of $\mathbb{R}[x_0, \dots, x_d]/(x_0^2 + \cdots + x_d^2 - 1)$ consisting of all polynomials all of whose terms have even (total) degree. Let P be the A-submodule of $\mathbb{R}[x_0, \dots, x_d]/(x_0^2 + \cdots + x_d^2 - 1)$ generated by $x_0, \dots, x_d$. Thus P consists of the classes of those polynomials all of whose terms have odd (total) degree. It was shown by R.G. Swan in [96] that P is projective (of rank 1) and that if $P_r = P \oplus A^{r-1}$ then rank $P_r = r$ and P_r cannot be generated by fewer than $d + r$ elements, *cf.* Proposition 5.6.10.

We end this section by a weak, but very general generalization of Corollary 6.7.3.

There is a wide class of Noetherian rings which have the following property. **(Closedness of the singular locus)**

Definition 9.3.9 *We say that a Noetherian ring A has a **closed singular locus**, if the set $\{\mathfrak{p} \in \mathrm{Spec}(A) \mid A_{\mathfrak{p}} \text{ is not regular}\}$ is closed in $\mathrm{Spec}(A)$.*

One can derive from the Jacobian Criterion 8.1.3 that every affine algebra has a closed singular locus. But of course the following proposition is of value, only if the class of rings with this property is much wider. This is the case, but unfortunately we cannot show it here.

Proposition 9.3.10 *Let A be a Noetherian ring of dimension $n < \infty$. Assume that $A/\mathfrak{p}$ has a closed singular locus for all $\mathfrak{p} \in \mathrm{Spec}(A)$. The the set $\{\mu(\mathfrak{m}) \mid \mathfrak{m} \in \mathrm{Spmax}(A)\}$ is bounded.*

Proof. Induction on $n = \dim(A)$. If $n = 0$ the ring A has only finitely many prime ideals. So let $n > 0$. Since there are only finitely many minimal prime ideals, by reduction modulo these, we may assume that A is a domain. By hypothesis, the singular locus of A is closed, say it is $\mathrm{V}(I)$. Since the localization $A_{(0)}$ of A in the zero ideal is a field, $(0) \notin \mathrm{V}(I)$, i.e. $I \neq (0)$. So $\dim(A/I) < n$. By induction hypothesis the set of the $\mu(\mathfrak{m}/I)$ is bounded where $\mathfrak{m}$ ranges over the maximal ideals containing I. Hence also the set of the $\mu(\mathfrak{m})$ is bounded for these $\mathfrak{m}$. The other maximal ideals $\mathfrak{m} \notin \mathrm{V}(I)$ are locally complete intersections, hence, by Corollary 9.3.6 can be generated by $\leq n + 1$ elements. $\square$

9.3.1 Basic elements

Let R be a commutative ring with 1, and let M be a finitely generated R-module. An element $m \in M$ is said to be **unimodular** in M if Rm is a direct summand of M, i.e. there is a (finitely generated) module M' such that $M \simeq Rm \oplus M'$. For instance, any non-zero vector v in a vector space V is a unimodular element. For another example, Serre's Splitting Theorem can be restated as asserting that a finitely generated projective A-modules of rank $> \dim(A)$ has a unimodular element.

One can attach an ideal $O_M(m)$ of R with an element $m \in M$, called the *order ideal* of m, viz.,

$$\mathcal{O}_R(m) = \{f(m) \mid f \in M^* = Hom_R(M, R)\}.$$

Clearly, m is unimodular if and only if $\mathcal{O}_R(m) = R$. $\mathrm{Um}_R(M)$ will denote the set of all unimodular elements of M.

How far does the analogy of being "part of a basis" go in the ring-theoretic setup? Let us consider the (commutative) case closest to the vector space situation: This is the case when R is a local ring with maximal ideal $\mathfrak{m}$. Are you surprised? Recall Nakayama's lemma, and its consequence,

$m_1, \ldots, m_r \in M$ generate M if and only if $\overline{m}_1, \ldots, \overline{m}_r$ is a basis of the $R/\mathfrak{m}$-space $M/\mathfrak{m}M$.

This gives a clue. We can expect an element $m \in M$ to be "good" if $m \notin \mathfrak{m}M$; for then $\overline{m} \in M/\mathfrak{m}M$ is part of a basis. Of course, in the case when R is not local, we would wish to have this property at every localization of R at its maximal ideals. Such an element is called a **basic** element of M. Formally,

Let $m \in M$. We say m is basic at $\mathfrak{p}$ if $m \notin M_\mathfrak{p} \equiv m$ is part of basis of the $k(\mathfrak{p})$ $(= R_\mathfrak{p}/\mathfrak{p}R_\mathfrak{p})$-vector space $M_\mathfrak{p}/\mathfrak{p}M_\mathfrak{p} \equiv m \neq 0$ in $M \otimes_R k(\mathfrak{p})$. We say m is a *basic element* of M if m is basic at $\mathfrak{p}$, for every prime ideal $\mathfrak{p}$ of R.

Let us relate this definition to the "local number of generators of M" $\mu_\mathfrak{p}(M) =$ the minimal number of generators of the $R_\mathfrak{p}$-module $M_\mathfrak{p}$. By Nakayama's lemma, $\mu_\mathfrak{p}(M) = \dim M_\mathfrak{p}/\mathfrak{p}M_\mathfrak{p}$. Observe that

$$\dim((M_\mathfrak{p}/R_\mathfrak{p}m/\mathfrak{p}M_\mathfrak{p}+R_\mathfrak{p}m/R_\mathfrak{p}m) = \dim(M_\mathfrak{p}/\mathfrak{p}M_\mathfrak{p}+R_\mathfrak{p}m) = \dim(M_\mathfrak{p}/\mathfrak{p}M_\mathfrak{p})-1$$

if and only if $m \notin \mathfrak{p}M_\mathfrak{p}$. Thus, m is basic in M at $\mathfrak{p}$ if and only if $\mu_\mathfrak{p}(M/Rm) = \mu_\mathfrak{p}(M) - 1$.

Of course, any unimodular element $m \in M$ is a basic element in M. The converse is true when M is projective. There are several ways to see this. For instance, consider the case when M is free of rank n, say $M \simeq R^n$. If $m = (a_1, \ldots, a_n)$. Then m basic means that the ideal $(a_1, \ldots, a_n) \not\subset \mathfrak{m}$, for any maximal ideal $\mathfrak{m}$ of M. Hence, $(a_1, \ldots, a_n) = R$, whence m is unimodular. For the general case, note that any finitely generated projective module is a summand of a free module R^n, for some n. So m is unimodular in A^n, by above; whence there is a linear map $\psi : A^n \longrightarrow R$ for which $\psi(m) = 1$. Composing with the inclusion map $P \longrightarrow A^n$ yields a linear map $\theta : P \longrightarrow R$, with $\theta(m) = 1$. This means that m is unimodular in M.

9.3.2 Basic elements and the Forster-Swan theorem

We begin with a basic lemma.

Lemma 9.3.11 *Let M be a non-zero finitely generated R-module. Let $\mathfrak{p}_1, \ldots, \mathfrak{p}_r \in \operatorname{Supp}(M)$. Then there exists $m \in M$ which is basic at $\mathfrak{p}_1, \ldots, \mathfrak{p}_r$.*

PROOF: By a reordering we may assume that $\mathfrak{p}_i$ is maximal among $\{\mathfrak{p}_i, \mathfrak{p}_{i+1}, \ldots, \mathfrak{p}_r\}$, for each i.

We induct on r, the result being clear for $r = 1$. Let m' be basic at $\mathfrak{p}_1, \ldots, \mathfrak{p}_{r-1}$. If m' is basic at $\mathfrak{p}_r$, then there is nothing more to be done. If not, choose $n' \in M$ which is basic at $\mathfrak{p}_r$.

Let $r \in \mathfrak{p}_1 \cdots \mathfrak{p}_{r-1} \setminus \mathfrak{p}_r$. (Such a r clearly exists.) Let $m = m' + rn'$. Then m is basic at $\mathfrak{p}_1, \cdots, \mathfrak{p}_r$. $\square$

Swan's proof of Forster-Swan theorem:

$$\mu(M) \leq \sup {}_{\mathfrak{p}}\{b_{\mathfrak{p}}(M) \mid \mathfrak{p} \in \mathrm{Supp}(M)\}.$$

Let $f(M) = \sup {}_{\mathfrak{p}}\{b_{\mathfrak{p}}(M) \mid \mathfrak{p} \text{ prime}, M_{\mathfrak{p}} \neq 0\}$. CLAIM : We show that the bound $f(M)$ can be attained at atmost finitely many prime ideals.

Note that $U_r = \{\mathfrak{p} \mid \mu_{\mathfrak{p}}(M) = r\}$ is an open subset of $\mathrm{Spec}(R)$. This is because if the images of $m_1, \dots, m_r \in M$, generate $M_{\mathfrak{p}}$, then the images of $m_1, \dots, m_r \in M$ will also generate $M_{\mathfrak{q}}$, for all $\mathfrak{q} \in D(s)$, for some $s \notin \mathfrak{p}$. (The existence of the $m_1, \dots, m_r \in M$, as above, will determine a s.)

Let $X_r = \{\mathfrak{p} \mid \mu_{\mathfrak{p}}(M) \geq n\}$. Then $X_r = (\cup_{i=1}^{n-1} U_i)^c$ is a closed set. Hence, there is an ideal I_r such that $X_r = V(I_r)$. Let $\sqrt{I_r} = \cap_{i=1}^{n_r}\mathfrak{p}_{ri}$ be its primary decomposition, so that the irreducible decomposition of the closed set is $V(I_r) = \cup_{i=1}^{n_r}V(\mathfrak{p}_{ri})$.

Let $E = \{\mathfrak{p}_{ri} \mid r \geq 0, 1 \leq i \leq n_r\}$. Note that E is a finite set as $X_r = \emptyset$, for $r > \mu(M)$. We show that if $f(M) = b_{\mathfrak{p}}(M)$ then $\mathfrak{p} \in E$.

Let $\mu_{\mathfrak{p}}(M) = n$. Then $\mathfrak{p} \in X_n$. Hence $\sqrt{I_n} \subset \mathfrak{p}$. Hence $\mathfrak{p}_{ni} \subset \mathfrak{p}$, for some i. Note that $\mu_{\mathfrak{p}}(M) = \mu_{\mathfrak{p}_{ni}}(M)$. Therefore, if coht $(\mathfrak{p}) <$ coht $(\mathfrak{p}_{ni})$, then $b_{\mathfrak{p}}(M) < b_{\mathfrak{p}_{ni}}(M) = b_{\mathfrak{p}}(M)!$. Therefore, $\mathfrak{p} = \mathfrak{p}_{ni} \in E$. This settles the claim.

By above lemma, choose a $m \in M$ which is basic at all primes of E. Let $\overline{M} = M/Rm$. Then $f(\overline{M}) \leq f(M) - 1$: This is clear if $\mathfrak{p} \notin E$, and also if $\mathfrak{p} \in E$ as then m is basic at E.

By induction, $\mu(\overline{M}) \leq f(\overline{M})$. Hence $\mu(M) \leq \mu(\overline{M}) + 1 \leq f(M)$, as required to be shown. $\square$

9.3.3 Forster-Swan theorem via K-theory

We first explain the notion of generalized dimension of a commutative noetherian ring R. This notion evolved from the work of J-P. Serre-H.Bass (who began by concentrating on the Maximum spectrum), then in the work of R.G.Swan (who started working with j-primes), and then by B. Plumstead who axiomatised the notion in his doctoral thesis [70].

Given a set $\mathcal{P}$ of prime ideals of R and a function $\delta : \mathcal{P} \to \mathbb{N} \cup \{0\}$, define a partial order $<<$ on $\mathcal{P}$ by setting $\mathfrak{p} << \mathfrak{q}$ if $\mathfrak{p} \subset \mathfrak{q}$ and $\delta(\mathfrak{p}) > \delta(\mathfrak{q})$.

Definition 1. *The function* $\delta : \mathcal{P} \to \mathbb{N} \cup \{0\}$ *is called a* **generalised dimension function** *on* $\mathcal{P}$ *(g.d.f. in short) if for any ideal I of R, $V(I) \cap \mathcal{P}$ has only a finite number of minimal elements with respect to* $<<$.

We say R has **generalised dimension** $\delta(R)$ if

$$\delta(R) = \min_{\delta}(\max_{\mathfrak{p} \in \operatorname{Spec} R} \delta(\mathfrak{p})).$$

For instance, the standard dimension function $\delta(\mathfrak{p}) = $ coheight $\mathfrak{p}$ is a g.d.f. Thus g.d.$R \le \dim R$.

We recollect an example of B. Plumstead in [70] of a ring having g.d. $A < \dim A$.

First observe that if s is an element of R such that $R/(s)$ and R_s have generalised dimension $\le d$, then g.d. $R \le d$. In fact, if δ_1, δ_2 are g.d.f's on $R/(s)$, R_s respectively withh $\delta_i \le d$ for $i = 1, 2$, then define $\delta : \operatorname{Spec}(R) \to \mathbb{N} \cup \{0\}$ as follows

$$\delta(\mathfrak{p}) = \begin{cases} \delta_1(\mathfrak{p}) \text{ if } s \in \mathfrak{p} \\ \\ \delta_2(\mathfrak{p}) \text{ if } s \notin \mathfrak{p} \end{cases}$$

Clearly, δ is a g.d.f. on R with $\delta(\mathfrak{p}) \le d$ for all $\mathfrak{p} \in \operatorname{Spec}(R)$.

Plumstead's Example: Take $A = R[x]$, where R is a ring having an element $s \in$ radical of R with $\dim(R/(s)) < \dim R$. The above reasoning shows g.d. $A < \dim A$.

We begin a proof of the Forster-Swan theorem via K-theory due to R.A. Rao.

Theorem 9.3.12 (Forster-Swan)

Let A be a noetherian ring of dimension d, and let δ be a generalized dimension function on A. Then an A-module M is generated by

$$f(M) = \sup_{\mathfrak{p}}\{\mu_{\mathfrak{p}}(M) + \delta(\mathfrak{p}) \mid \mathfrak{p} \in \operatorname{Supp}(M)\}$$

elements.

PROOF: Let $f = f(M)$; we show that there is a free A-module of rank f mapping onto M. We prove this by induction on $(\dim(A), \delta(A))$; it being clear when $d = 0$.

Consider a surjection $A^n \to M \to 0$, with $n > f$, and let K be the kernel of this map.

Find a (n.z.d.) $s \in R$, such that

1. $s \notin \mathfrak{p}$, if $\mathfrak{p}$ is minimal with respect to the generalized dimension function δ. (Note that there are only finitely many primes, including the minimal ones, for which $\mathfrak{p}$ is minimal with respect to δ.

2. such that K_s is free.

Choose $x_1 \in K_s$, with x_1 is unimodular in A_s^n. Let $T = 1 + sR$. Then g.d.$A_T \leq \delta(A)$, if $\delta(A) > 0$, and so $(d, \delta(A_T)) < (d, \delta(A))$.

For any $\mathfrak{p} \in \mathrm{Spec}(A_T)$,

$$\mu_{\mathfrak{p}}(F_T) - \mu_{\mathfrak{p}}(F_T/K_T) = \mu_{\mathfrak{p}}(F_T) - \mu_{\mathfrak{p}}(M_s) > \dim(A/\mathfrak{p}) \geq \delta(\mathfrak{p}).$$

Hence K_T is $(\delta(\mathfrak{p}) + 1)$-fold basic in A_T^n at $\mathfrak{p}$ for every $\mathfrak{p} \in \mathrm{Spec}(A_T)$; by induction there is a $x_2 \in K_T$ which is unimodular in A_T^n.

Since $n > f, n \geq \delta(A) + 1$. Note that $\delta(A_{Ts}) \leq \delta(A) - 1$, and so $n \geq \delta(A_{Ts}) + 2$. By Proposition 7.1.2 one can find $\varepsilon \in E_n(A_{Ts})$ such that $\varepsilon(x_1)_T = (x_2)_s$. Let $x = x_1 \times_\varepsilon x_2 \in A_s^n \times_\varepsilon A_T^n$. Since x is "locally" unimodular, it is unimodular.

Any elementary matrix is homotopic to the identity matrix: if $\varepsilon = \prod E_{ij}^\lambda$ then $\varepsilon(T) = \prod E_{ij}^{\lambda T}$ is a homotopy. By Quillen's Splitting Lemma 4.3.8, ε splits as $(\varepsilon_1)_s(\varepsilon_2)_t$, for some $\varepsilon_1 \in SL_n(A_s)$, $\varepsilon_2 \in SL_n(A_T)$.

In view of the fact that elementary matrices split, $A_s^n \times_\varepsilon A_T^n \overset{\sim}{\to} A^n$; identify them. Let $Ax \oplus P = A^n$,

We are now in the situation

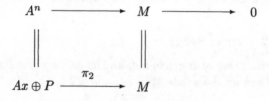

Since $x_s = s_1 \in K_s$ and $x_T = x_2 \in K_T$, the second projection map $\pi_2 : P \to M$ is "locally" surjective, and so surjective.

But P is stably free of rank $n - 1 \geq f \geq \delta(A)$. By the Bass Cancellation Theorem 7.1.11, P is free. $\qquad \square$

9.4 Eisenbud-Evans theorem

The Eisenbud-Evans theorem is the ultimate result known today by general position arguments regarding the number of generators of a finitely generated module M over a commutative noetherian ring R. (There are results also know over non-commutative noetherian rings due to R. Warfield, etc., but we shall not dwell on these here.) It can be used to derive all the earlier theorems: Serre's Splitting Theorem, Bass Cancellation Theorem, and the Forster-Swan Theorem; and in fact provides more information on the latter. These theorems play an important role in showing that curves in n-space are set-theoretic complete intersections. Today, one has more refined arguments, which combine general position arguments, and some Quillen-Suslin patching theory, (via K-theory), to derive these consequences for curves directly; without recourse to the Eisenbud-Evans theorem *per se*. A schematic picture is given by

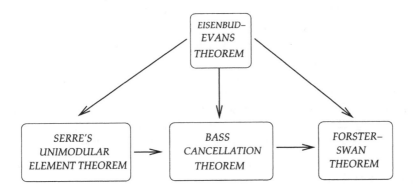

What are the Eisenbud-Evans theorem all about? One can ask under what condition can you guarantee that a submodule M' of a module M has a basic element? Moreover, in this case how will you go about finding such an element? Essentially, the condition should depend on M' being large enough, rather than on M. For instance, consider the case when $M' = Q$ is a projective R-module. If $\mathrm{rank}(Q) > \dim(R)$, then by Serre's Splitting Theorem Q will have a unimodular element $q \in Q$. q will also be basic in Q, as Q is projective. Here $\mu_{\mathfrak{p}}(Q) > \dim(R)$, for all primes ideals $\mathfrak{p}$ of R. A similar situation holds in the general case. We have to first figure out a notion of "largeness" of a module. For this we define the notion of when a submodule M' is t-fold basic in M.

If M' is a sub-module of M we say M' is t-fold basic in M at $\mathfrak{p}$ if $\mu_{\mathfrak{p}}(M/M') \le \mu_{\mathfrak{p}}(M) - t$. M' is t-fold basic in M if M' is t-fold basic in M at $\mathfrak{p}$ for every prime ideal $\mathfrak{p}$ of R. Clearly, basic is 1-basic.

Let us do a little linear algebra to understand this definition. One has

$$\mu_{\mathfrak{p}}(M/M') = \dim_{k(\mathfrak{p})} M_{\mathfrak{p}}/M'_{\mathfrak{p}}/\mathfrak{p}M_{\mathfrak{p}}/M'_{\mathfrak{p}}$$
$$= \dim_{k(\mathfrak{p})} M_{\mathfrak{p}}/M'_{\mathfrak{p}} + \mathfrak{p}M_{\mathfrak{p}}$$
$$= \dim_{k(\mathfrak{p})} M_{\mathfrak{p}}/\mathfrak{p}M_{\mathfrak{p}}/(M'_{\mathfrak{p}} + \mathfrak{p}M_{\mathfrak{p}})/\mathfrak{p}M_{\mathfrak{p}}$$
$$= \dim_{k(\mathfrak{p})} M_{\mathfrak{p}}/\mathfrak{p}M_{\mathfrak{p}} - \dim_{k(\mathfrak{p})}(M'_{\mathfrak{p}} + \mathfrak{p}M_{\mathfrak{p}})/\mathfrak{p}M_{\mathfrak{p}}.$$

This is now a vector space dimension calculation. We are essentially doing linear algebra in commutative rings - the door being opened by Nakayama's Lemma.

Thus $M' = \sum_{i=1}^{l} Rm_i$ is t-fold basic in M at $\mathfrak{p}$ if and only if $\dim\{m_i + \mathfrak{p}M_{\mathfrak{p}}\} \geq t$.

In particular, note that M' is t-fold basic in M at $\mathfrak{p}$ implies M' is t'-fold basic in M at $\mathfrak{p}$ for $t' \leq t$.

Having defined the appropriate notion of "largeness" let us describe what Eisenbud-Evans found: The Eisenbud-Evans theorem is that *if M' is (coht($\mathfrak{p}$) + 1)-fold basic in M at $\mathfrak{p}$, for every prime ideal $\mathfrak{p}$ of R, then M' has a basic element.*

The proof of the theorem is in the spirit in which the key starting Prime Avoidance Lemma 1.1.8 was proved. The technique used to prove it is called *general position argument*. One takes a set of generators $m_1, \ldots, m_t$ of M', and consider submodules $\sum_{i=1}^{t} R(m_i + a_i m_t)$ of M', for $a_i, \ldots, a_{t-1} \in R$. One tries to ensures that the choice of the submodule is such that it has similar properties as the original M' one started with. This requires a suitable choice of $a_1, \ldots, a_r$. The situation is then inductive and leads to a basic element having a specific type, as was expected by the proof of our key lemma.

We schematically describe the different types of Mathematical thinking which help in general position arguments of different types:

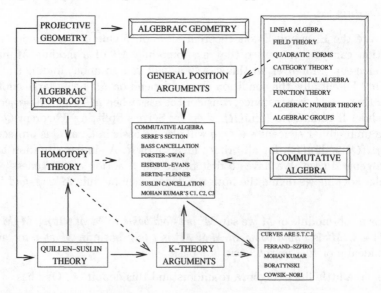

We now go to the proof of the Eisenbud-Evans theorem. Let us, as Eisenbud-Evans did, first just state the two general position argument lemmas which will be needed, and prove the theorem assuming them.

Lemma 9.4.1 *Let R be a commutative noetherian ring with 1. Let M be a finitely generated R-module, and let M' be a submodule of M Let $\mathcal{P}$ be a subset of $\mathrm{Spec}(R)$ having the following property: For every prime ideal $\mathfrak{p} \in \mathrm{Spec}(R)$, for which there is a prime ideal $\mathfrak{q} \in \mathcal{P}$ with $\mathfrak{q} \subset \mathfrak{p}$, $\mathfrak{q} \neq \mathfrak{p}$, one has $\mu_{\mathfrak{q}}(M/M') \leq \mu_{\mathfrak{p}}(M) - w$, i.e. M' is w-fold basic in M at $\mathfrak{p}$.*

Then M' is w-fold basic in M at $\mathfrak{q}$, for all but finitely many prime ideals $\mathfrak{q} \in \mathcal{P}$. (These are atmost the prime ideals which are minimal over the Kaplansky ideals $I_r(M/M')$, for some r, and which belong to $\mathcal{P}$.) $\square$

Lemma 9.4.2 *Let R be a commutative noetherian ring, M a finitely generated R- module, M' a submodule of M. Let $\mathfrak{p}_1,\ldots,\mathfrak{p}_n \in \mathrm{Spec}(R)$ be a finite set of prime ideals. Assume that M' is w_i-fold basic in M at $\mathfrak{p}_i$, with $w_i > 0$, for all i. Let $M' = \sum_{i=1}^{t} Rm_i$, for some $m_i \in M$.*

If $a \in R$ is such that $(a, m_1) \in R \oplus M$ is basic at $\mathfrak{p}_1,\ldots,\mathfrak{p}_n$, then there exist elements $a_i \in R$ such that $(a, m_1 + aa_1 m_t) \in R \oplus M$ is basic at $\mathfrak{p}_1,\ldots,\mathfrak{p}_n$. Moreover, the submodule $R(m_1 + aa_1 m_t) + \sum_{i=2}^{t-1} R(m_i + a_i m_t)$ is $\min\{t-1, w_i\}$-fold basic in M at $\mathfrak{p}_i$, for all i. $\square$

Let us now proceed to formally state and prove the famous Eisenbud-Evans theorem.

Theorem 9.4.3 *Let R be a commutative noetherian ring of dimension d, M a finitely generated R-module, M' a sumbodule of M. Let δ be a generalized dimension function on R.*

If M' is $(\delta(\mathfrak{p}) + 1)$-fold basic in M at $\mathfrak{p}$, for every prime ideal $\mathfrak{p}$ of R, then M' has a basic element.

Furthermore, let $m_1,\ldots,m_t$ generate M'. Let $a \in R$ such that $(a, m_1) \in R \oplus M$ is basic. Then there is a basic element of M of type $m_1 + a\sum_{i=2}^{t} a_i m_i$.

PROOF: A subset $\{n_1,\ldots,n_k\}$ is called a *basic* set if the R-submodule $N' = \sum_{i=1}^{k} Rn_i$ is $\min\{k, \delta(\mathfrak{p}) + 1\}$-fold basic in M at $\mathfrak{p}$, for all prime ideals $\mathfrak{p}$ of R. (Note that if $\{m\}$ is a basic set, then m is 1-basic in M.)

By hypothesis, $\{m_1,\ldots,m_t\}$ is a basic set. The idea employed by Eisenbud-Evans is to "cut down" on number of generators occuring in a basic set, retaining its basicity. This is done by means of "elementary operations": More precisely, given a basic set $\{n_1,\ldots,n_t\}$, $t > 1$, one finds $a_1,\ldots,a_{t-1} \in R$, such that the set

$$\{n_1 + aa_1n_t, n_2 + a_2n_t, \ldots, n_{t-1} + a_{t-1}n_t\}$$

is a basic set. If we do this process $(t-1)$ times to $\{m_1, \ldots, m_t\}$ we will land up with a basic element of the required type.

CLAIM: There are atmost finitely many prime ideals for which N' is **not** $\{k, \delta(\mathfrak{p}) + 2\}$-fold basic in M at $\mathfrak{p}$. We prove this.

Since R has a finite (generalized) dimension, it suffices to show, for fixed $\delta(\mathfrak{p}) = s$, say, that there are only finitely many primes for which N' is not $\{k, \delta(\mathfrak{p}) + 2\}$-fold basic in M at $\mathfrak{p}$.

Let $\mathcal{P} = \{\mathfrak{q} \mid \delta(\mathfrak{q}) = s\}$. If $\mathfrak{p} \subset \mathfrak{q} \in \mathcal{P}$, $\mathfrak{p} \neq \mathfrak{q}$, then $\delta(\mathfrak{p}) > \delta(\mathfrak{q}) = s$. Hence, $\min\{k, \delta(\mathfrak{p}) + 1\} \geq \min\{k, \delta(\mathfrak{q}) + 2\} = w$. Therefore, M' is w-fold basic in M at $\mathfrak{p}$. By Lemma 9.4.1, M' is w-fold basic in M at $\mathfrak{q}$, for all but finitely many $\mathfrak{q} \in \mathcal{P}$. Let E be this finite set of primes.

By Lemma 9.4.2, there are $a_1, \ldots, a_{t-1} \in A$ such that $N'' = A(n_1 + aa_tn_t) + \sum_{i=2}^{t-1} A(n_i + a_in_t)$ is $\min\{t-1, \delta(\mathfrak{p}) + 1\}$-fold basic in M at $\mathfrak{p} \in E$. However, if $\mathfrak{p} \notin E$, the N'' is $\min\{t-1, \delta(\mathfrak{p}) + 1\}$-fold basic in M at $\mathfrak{p}$, for any choice of a_i's. Thus N'' is a basic set as required. $\qquad\square$

We now prove the two lemma which were mentioned before the theorem and used in its proof.

PROOF OF LEMMA 9.4.1: Let $\mathfrak{p} \in \mathcal{P}$. Let $\mu_{\mathfrak{p}}(M/M') = r+1$. This is if and only if $I_r(M/M') \subset \mathfrak{p}$, and $I_{r+l}(M/M') \not\subset \mathfrak{p}$, for all $l > 0$. Suppose, in addition, that $\mathfrak{p}$ is not minimial over $I_r(M/M')$. By hypothesis, $\mu_{\mathfrak{q}}(M/M') = w \ (= r)$. Therefore, $\mu_{\mathfrak{p}}(M/M') = \mu_{\mathfrak{q}}(M/M') \leq \mu_{\mathfrak{q}}(M) - w \leq \mu_{\mathfrak{p}}(M) - w$. Thus, M' is w-fold basic in M at $\mathfrak{p}$. $\qquad\square$

PROOF OF LEMMA 9.4.2: Of course, $(a, m_1 + aa_tm_t)$ is basic at $\mathfrak{p}_1, \ldots, \mathfrak{p}_n$, for any choice of $a_1 \in R$. This is because it is the image of the basic element (a, m_1) by a flip of $R \oplus M$.

If $w_{i_0} \geq t = \mu(M')$, for some i_0, then for any choice of a_i, the submodule

$$R(m_1 + aa_1m_t) + \sum_{i \geq 2} R(m_i + a_im_t)$$

is $(t-1)$-fold basic in M at $\mathfrak{p}_{i_0}$, as required. Therefore, we may assume that $w_i < t$, for all i, by simply omitting all the primes ideals $\mathfrak{p}_i$ for which the corresponding $w_i \geq t$.

We will induct on n. We start the induction with the vacuous case $n = 0$.

Reorder the $\mathfrak{p}_i$, and assume $\mathfrak{p}_t$ is minimal amongst $\mathfrak{p}_1, \ldots, \mathfrak{p}_t$, i.e. $\mathfrak{p}_t \not\subset \cap_{i=1}^{t-1}\mathfrak{p}_i$.

By the induction hypothesis we can choose $a'_1, \ldots, a'_{t-1} \in R$ so that

$$m'_1 = m_1 + aa'_1m_t, m'_i = m_i + a'_im_t, 2 \leq i \leq t-1,$$

generate a sub-module of M which is w_i-fold basic at $\mathfrak{p}_i$, for $i < t$.

CLAIM: We can choose $a_1'', \ldots, a_{t-1}'' \in R$, so that for any $r \in R \setminus \mathfrak{p}_t$,

$$m_1'' = m_1' + raa_1''m_t, m_2'' = m_2' + ra_2''m_t, \ldots, m_{t-1}'' = m_{t-1}' + a_{t-1}''rm_t,$$

is w_t-fold basic at $\mathfrak{p}_t$.

(Therefore, if $r \in (\cap_{j<t}\mathfrak{p}_j) \setminus \mathfrak{p}_t$, then $m_1'', \ldots, m_{t-1}''$ is w_I-fold basic at $\mathfrak{p}_i$, for $i \leq t$! This is because, for $i < t$, $r \in \mathfrak{p}_i$, and $m_1', \ldots, m_{t-1}'$, are w_i-fold basic in M at $\mathfrak{p}_i$. For $i = t$, this is guaranteed by the claim.)

PROOF OF CLAIM: We may localize at $\mathfrak{p}_t$. Therefore, assume R is a local ring with maximal ideal $\mathfrak{p}_t$. Since "basicness" can be checked after "going modulo $\mathfrak{p}_t$" we assume R is a field.

If $\sum_{i=1}^{t-1} Rm_i'$ is w_t-fold basic, then choose $a_i = 0$, for all i. If not, let $k < t$, be the largest integer such that $m_k' \in \sum_{i=1}^{k-1} Rm_i'$. (Note that k exists as $\sum_{i=1}^{t-1} Rm_i'$ is a vector space of dimension $\leq w_t - 1 < t - 1$).

If $k \neq 1$, the choose $a_i'' = 0$, for $i \neq k$, and choose $a_k'' = 1$. If $k = 1$, then $m_1' = 0$. Since (a, m_1') is basic in $R \oplus M$, $a \neq 0$. Therefore, take $a_1'' = 1$, and $a_i'' = 0$, for all $i > 1$. Then we can check that, $m_1'' = m_1' = aa_1''m_t, m_i'' = m_i' = m_i + a_i'm_t$. Therefore, $\sum_{i=1}^{t} Rm_i'' = Raa_1'm_t + \sum_i R(m_i + a_i'm_t)$ is w_t-basic at $\mathfrak{p}_t$, by hypothesis. $\square$

The Eisenbud-Evan's Theorem can be used to deduce the famous Serre Splitting Theorem, the Bass Cancellation Theorem, the Forster-Swan Theorem, amongst others. We give these three applications.

Corollary 9.4.4 *(Serre's Unimodular Element Theorem)*

Let P be a finitely generated projective R-module of rank $\geq \dim(R) + 1$. Then P has a unimodular element, i.e. $P \equiv P' \oplus R$, for some finitely generated projective R-module P'.

In fact, if P is generated by $m_1, \ldots, m_t$, then one can find $a_2, \ldots, a_t \in A$, so that $m = m_1 + a_2m_2 + \ldots + a_tm_t$ is unimodular.

PROOF: Let $M = P \oplus Q$ be a free R-module. Let $M' = P$, in the Eisenbud-Evans theorem. Then M' is $(d + 1)$-fold basic in M. Now $(1, m_1) \in R \oplus M$ is basic. By the Eisenbud-Evans theorem there is $a_2, \ldots, a_t \in R$, so that $m = m_1 + a_2m_2 + \ldots + a_tm_t$ is basic in M. But then, since M is projective, m is unimodular in M. Since $m \in P$, m is unimodular in P, and splits off a summand. $\square$

Corollary 9.4.5 *(Bass Cancellation Theorem)*

Let P be a projective R-module of rank $> \dim(R)$. Then P is "cancellative", i.e. $P \oplus Q \equiv P' \oplus Q$ implies $P \equiv P'$, or equivalently, $P \oplus R \equiv P' \oplus R$ implies that $P \equiv P'$.

PROOF: Let $\sigma : P' \oplus R \equiv P \oplus R$ be an isomorphism, and let $\sigma(0.1) = (p, a)$. Then $P' \equiv \mathrm{coker}(0,1) \equiv \mathrm{coker}(p, a)$. It suffices to show that (p, a) can be mapped to $(0, 1) \in P \oplus R$ by some automorphism of $P \oplus R$. For this would show that $\mathrm{coker}(p, a) \equiv \mathrm{coker}(0, 1) \equiv P$. We shall actually find a product of flips which maps $(p.a)$ to $(0, 1)$.

Let $p_1, \ldots, p_n$ generate P. P is $(d + 1)$-fold basic in a finitely generated free module R^n, for some n. Therefore, by Eisenbud-Evans theorem, there is an element $p = p_1 + a(\sum_{i=2}^{n} a_i p_i) \in P$ which is basic in R^n. This element will be unimodular in R^n, and thereby in P. Now one can construct flips which map

$$(p_1, a) \longrightarrow (p, a) \longrightarrow (p, 1) \longrightarrow (0, 1)$$

The penultimate arrow exists as p is unimodular.   □

Corollary 9.4.6 *(Forster-Swan)* $\mu(M) \leq f(M)$.

PROOF: We have a surjection $R^n \longrightarrow M$, for some n, with kernel K. Choose n least. If $n > f(M)$, then $\mu_{\mathfrak{p}}(K) \geq n - \mu_{\mathfrak{p}}(M) > \delta(\mathfrak{p})$, for all prime ideals $\mathfrak{p}$. By Eisenbud-Evans theorem, K has an element k, which is basic (hence unimodular) in R^n. Hence $R^n \equiv Rk \oplus P'$. Now $\mathrm{rank}(P') = n-1 \geq \dim(R)+1$. By Bass Cancellation Theorem, P' is cancellative, and so P' is free of rank $n - 1$. But one can construct an onto map from P' onto M, as $k \in K$. A contradiction.   □

9.5 Any Variety in the n-Dimensional Space is an Intersection of n Hypersurfaces

In 1882 Kronecker stated without proof that if I is an ideal in the polynomial ring $k[x_1, \ldots, x_n]$ over a field k then $\sqrt{I} = \sqrt{(f_1, \ldots, f_{n+1})}$ for some $f_1, \ldots, f_{n+1} \in k[x_1, \ldots, x_n]$. If k is algebraically closed, by Hilbert's Nullstellensatz this is equivalent to saying that the variety

$$V(I) = \{x \in k^n \mid f(x) = 0 \text{ for all } f \in I\} \subset k^n = A_k^n$$

is the "set-theoretic" intersection $\bigcap_{i=1}^{n+1} V(f_i)$ of $n + 1$ hypersurfaces $V(f_i)$. (This was implicitly proved by König, Perron and van der Waerden.) We give here an algebraic version for general Noetherian rings.

Theorem 9.5.1 (Kronecker) *Let I be an ideal of a Noetherian ring A of dimension n. Then there are $f_1, \ldots, f_{n+1} \in A$ such that $\sqrt{I} = \sqrt{(f_1, \ldots, f_{n+1})}$.*

Proof. We prove, by induction on r, that there are $f_1, \ldots, f_r$, such that $V(f_1, \ldots, f_r) = V(I) \cup Z_r$, where Z_r is a closed subset of $\mathrm{Spec}(A)$ of codimension r. By the **codimension** of a closed subset Z of $\mathrm{Spec}(A)$ we mean

$$\mathrm{codim}(Z) := \mathrm{Min}\{\mathrm{ht}\, \mathfrak{p} \mid \mathfrak{p} \in Z\}.$$

The case $r = 0$ being trivial, assume the result for an $r \geq 0$ and let $Z_r = Z_r^1 \cup \ldots \cup Z_r^s$ be the decomposition of Z_r into irreducible components. We may assume, $Z_r^i \not\subset V(I)$. Let $\mathfrak{p}_i \in Z_r^i \setminus V(I)$ and so $I \not\subset \mathfrak{p}_i$ for $i = 1, \ldots, s$. By Lemma 1.1.8 there is an $f_{r+1} \in I \setminus \bigcup_i \mathfrak{p}_i$. Then $V(f_1, \ldots, f_{r+1}) = V(I) \cup Z_{r+1}$ with some closed set Z_{r+1} everyone of whose irreducible components is properly contained in one of the irreducible components of Z_r. Therefore $\mathrm{codim}(Z_{r+1}) \geq r + 1$, which means $Z_{r+1} = \emptyset$. $\qquad\square$

In 1972, the above result, restricted to *polynomial* rings, was improved upon (to the best possible in general) by U. Storch [91] and D. Eisenbud - E. Evans independently in [19] who showed that n hypersurfaces in the affine n-space suffice.

Theorem 9.5.2 (U. Storch; D. Eisenbud-E. Evans) *Let R be a Noetherian commutative ring of finite Krull dimension d and let I be an ideal of the polynomial ring $R[x]$. Then there are $d + 1$ elements $f_1, \ldots, f_{d+1} \in I$ with $\sqrt{I} = \sqrt{(f_1, \ldots, f_{d+1})}$.*

Proof. We may assume that R is a *reduced* ring, since radical ideals of R and R_{red} are in bijective, inclusion preserving correspondence. We use induction on $\dim(R)$, the case $\dim(R) = 0$ being obvious, since then R – as a Noetherian reduced ring of dimension 0 – is a product of finitely many fields, hence $R[x]$ is a finite direct product of principal domains, whence all of its ideals are monogene. Let S be the multiplicatively closed subset $R \setminus \bigcup_{i=1}^r \mathfrak{p}_i$ where $\mathfrak{p}_1, \ldots, \mathfrak{p}_r$ are all the minimal prime ideals of R. By Corollary 6.4.6 $S^{-1}R$ is a finite product of fields. Hence – as above – every ideal of $S^{-1}R[x]$ is generated by a single element. Let $S^{-1}I = (f_1(x))$ for some $f_1 \in I$. Then there is an $a \in S$, such that $aI \subset (f_1) \subset I$. Hence $V(I) \subset V(f_1) \subset V(aI) = V(a) \cup V(I)$.

Since a is not in any minimal prime ideal of R, we have $\dim(R/(a)) < d$. Let the 'overbar' denote 'modulo (a)'. By the induction hypothesis there are $f_2, \ldots, f_{d+1} \in I$ with

$$\sqrt{I + (a)/(a)} = \sqrt{(\overline{f}_2, \ldots, \overline{f}_{d+1})}.$$

Therefore, $\sqrt{I + (a)} = \sqrt{(a, f_2, \ldots, f_{d+1})}$. Hence

$$V(f_1, \ldots, f_{d+1}) = V(f_1) \cap V(f_2, \ldots, f_{d+1})$$

$$\subset (V(a) \cup V(I)) \cap V(f_2, \ldots, f_{d+1})$$

$$= V(a, f_2, \ldots, f_{d+1}) \cup V(I) = V(I).$$

Since on the other hand $(f_1, \ldots, f_{d+1}) \subset I$, we get $V(I) = V(f_1, \ldots, f_{d+1})$ or equivalently $\sqrt{I} = \sqrt{(f_1, \ldots, f_{d+1})}$ as required.   □

Corollary 9.5.3 *let I be an ideal of the polynomial ring $k[x_1, \ldots, x_n]$ over a field k (with $n > 0$). Then $\sqrt{I} = \sqrt{(f_1, \ldots, f_n)}$, for some $f_1, \ldots, f_n \in k[x_1, \ldots, x_n]$. Equivalently, any variety $V(I)$ in k^n is the set theoretic intersection of n hypersurfaces $V(f_i)$, $1 \le i \le n$.*   □

Remark 9.5.4 The above theorem had the effect of raising expectations that estimates for $\mu(M)$ would improve by 1 in the case of finitely generated modules M over a polynomial rings $R[x]$. Eisenbud - Evans conjectured that if M is a finitely generated module over a polynomial ring $R[x]$ then $\mu(M) \le \sup\{\mu_\mathfrak{p}(M) + \dim R[x]/\mathfrak{p} \mid \mathrm{ht}(\mathfrak{p}) < \dim R[x]\}$ or in O. Forster's notation, let

$$
b_k^*(M) = \begin{cases} 0 & \text{if } X_k(M) = \emptyset \\ k + \dim(X_k(M)) & \text{if } 0 \le \dim(X_k(M)) < \dim(X) \\ k + \dim(X_k(M)) - 1 & \text{if } \dim(X_k(M)) = \dim(X). \end{cases}
$$

Then M can be generated by $b^*(M) = \sup\{b_x^*(M) : k \ge 1\}$ elements. This conjecture of Eisenbud-Evans was settled by N. Mohan Kumar in [60]; we give a proof of it in the next section.

Remark 9.5.5 Note that one *cannot* conclude that every prime ideal $\mathfrak{p}$ in $k[x_1, \ldots, x_n]$ is generated by n elements! One only has $\mathfrak{p} = \sqrt{(f_1, \ldots, f_n)}$ for some polynomials f_i. In fact Macaulay has constructed examples of a sequence of prime ideals $\mathfrak{p}_r$ in $k[x, y, z]$ with $\mu(\mathfrak{p}_r) = r$ for every $r \in \mathbb{N}$.

9.6 The Eisenbud-Evans conjectures

Eisenbud-Evans made some conjectures regarding a bound on the number of generators needed for a finitely generated module over a polynomial ring $R[X]$, when R is a noetherian ring. They also proposed a plan of action by which these conjectures should be solved - essentially, establish that Serre's Unimodular Element Theorem, and Bass Cancellation Theorem hold over polynomial ring, for projective modules of rank bigger than the dimension of *the base ring*. Then deduce a similar estimate for the bound as that for the base ring.

The EE-conjecture for the bound were established for affine domains by A. Sathaye in [82], and for a general noetherian ring by N. Mohan Kumar in [60]. Later, B. Plumstead gave an argument in [70] modeled on the suggested method. We give a variant of this below.

Theorem 9.6.1 *Let R be a Noetherian ring of dimension d, and $A = R[x]$. Then an A-module M is generated by $e(M)$ elements, where*

$$e(M) = \sup_{\mathfrak{p}, \dim A/\mathfrak{p} \leq d} \{\mu_{\mathfrak{p}}(M) + \dim\ A/\mathfrak{p}\}.$$

Proof. Let $e = e(M)$; we show that there is a free A-module of rank e mapping onto M.

Consider a surjection $A^n \to M \to 0$, with $n > e$, and let K be the kernel of this map.

Find a non-zero-divisor $s' \in R, x_1 \in K_s$, such that K_s is free and x_1 is unimodular in A^n_s. Let $s = f_0 s$, $T = 1 + sR$. Let δ on $\mathrm{Spec}(A_T)$ be a g.d.f. $\leq d$, as defined by Plumstead, *cf.* §9.3.3.

For any $\mathfrak{p} \in \mathrm{Spec}(A_T)$,

$$\mu_{\mathfrak{p}}(F_T) - \mu_{\mathfrak{p}}(F_T/K_T) = \mu_{\mathfrak{p}}(F_T) - \mu_{\mathfrak{p}}(M_s) > \dim\ A/\mathfrak{p} \geq \delta(\mathfrak{p}).$$

Hence K_T is $(\delta(\mathfrak{p}) + 1)$-fold basic in A^n_T at $\mathfrak{p}$ for every $\mathfrak{p} \in \mathrm{Spec}(A_T)$; by Theorem 7.1.8 there is an $x_2 \in K_T$ which is unimodular in A^n_T.

By Proposition 7.1.2 one can find an $\varepsilon \in E_n(A_{Ts})$ such that $\varepsilon(x_1)_T = (x_2)_s$, as, clearly, $e \geq \dim\ A$.

Let $x = x_1 \times_\varepsilon x_2 \in A^n_s \times_\varepsilon A^n_T$. By Theorem 7.1.11 we see $A^n_s \times_\varepsilon A^n_T \overset{\sim}{\to} A^n$; identify them, Let $Ax \oplus P = A^n$.

We are now in the situation

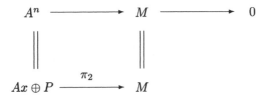

Since $x_s = s_1 \in K_s$ and $x_T = x_2 \in K_T$, the second projection map $\pi_2 : P \to M$ is 'locally' surjective, and so surjective.

But P is stably free of rank $n - 1 \geq e \geq \dim(A)$. We show that P is free, by showing that the associated unimodular row $v(x) = (v_1(x), \dots, v_n(x))$ can be completed to an invertible matrix.

Using Prime Avoidance Lemma, the reader will be easily able to show that a unimodular row of length $\geq \delta(A) + 2$, can always be completed to an elementary matrix. (Here δ will denote a g.d.f. on $\mathrm{Spec}(A)$.)

In view of this, for any $\mathfrak{p} \in \mathrm{Spec}(R_{\mathfrak{p}}[x])$, $v(x)$ can be completed over $R_{\mathfrak{p}}[x]$, whence the associated projective module $P_{\mathfrak{p}}$ is free. By Quillen's Local Global Principle, P is extended from R, i.e. $P \simeq P/xP \otimes R[x]$. By Theorem 7.1.11, the stably free module P/xP is actually free. Hence, P is free. $\square$

Curves as Complete Intersection

J-P. Serre studied the codimension 2 complete intersection problem; in particular, how does one recognize a height 2 complete intersection ideal in $R = k[X_1, \dots, X_n]$. He pointed out necessary conditions are that it should locally be a complete intersection, all its associated primes should have the same height, have homological dimension 1, and also that the module $\text{Ext}^1(I, R)$ should be monogene. These conditions are also sufficient. This is our initial objective. In the latter part we prove known results of (local complete intersection) curves in n-space which are set-theoretical complete intersections.

10.1 A Motivation of Serre's Conjecture

Hilbert's epoch theorem established the finiteness of the number of generators needed to generate an ideal I in a polynomial ring $k[x_1, \dots, x_n]$, with $k = \mathbb{Z}$ or k a field. The novelty and beauty of his proof was in its existential content: A natural outgrowth was to persevere with the old approach of finding actual generators perhaps tempered by the new methods of homological algebra which had been so grandly started by Hilbert.

An ideal I of R is generated by n elements if and only if there is a surjection $R^n \to I \to 0$. J-P. Serre's reasoning was that it would be possibly easier to show that I is locally generated by n elements, i.e. there is a cover

$$\text{Spec}(R) = D(s_1) \cup \dots \cup D(s_r) \text{ i.e. } \sum Rs_i = R$$

and surjections

$$R_{s_i}^n \longrightarrow I_{s_i}.$$

One could then endeavour to 'patch' this local information. The global information he expected was to get a finitely generated projective R-module P of rank n and a surjection

$$P \longrightarrow I.$$

If, say $R = k[x_1, \ldots, x_m]$ – with k being a field – then Serre's Conjecture was that all finitely generated projective R-module were free. Thus, P would be free and I would be generated by n elements! J-P. Serre proceeded to show how to 'patch the local information' in a special case : When I was an ideal of height 2 in $k[x_1, \ldots, x_n]$.

We need here (but not in later sections) some interesting additional facts of homological algebra especially about the functors Ext^i, which remedy the non-exactness of the functor Hom. The group (for commutative R it is an R-module) $\text{Ext}^1(M, N) = \text{Ext}^1_R(M, N)$ parameterizes the extensions of M by N, i.e. the short exact sequences

$$0 \longrightarrow N \longrightarrow E \longrightarrow M \longrightarrow 0.$$

This is shown in Appendix B.

The homological dimension $\text{hd}_R M$ of an R-module M can also be defined as the supremum of the n such that $\text{Ext}^n_R(M, N) \neq 0$ for at least one R-module N. If one knows the definition of the $\text{Ext}^n_R(M, N)$ as homology groups of a complex which is built, starting with a projective resolution of M, it is easy to show that the just given definition of the homological dimension coincides with our old one Definition 8.3.1. (Some details of this can be found in Appendix C.)

The crucial step was

Lemma 10.1.1 (J-P. Serre) *Let R be a Noetherian ring and M a finitely generated R-module with hd $M \leq 1$. Suppose that $\text{Ext}^1(M, R)$ is a cyclic (i.e. monogene) R-module. Then there is an exact sequence*

$$0 \longrightarrow R \longrightarrow P \longrightarrow M \longrightarrow 0$$

with projective P.

Proof. Let η generate $\text{Ext}^1(M, R)$ and let

$$0 \longrightarrow R \longrightarrow P \longrightarrow M \longrightarrow 0 \tag{10.1}$$

be the extension corresponding to η. Then hd $P \leq \text{Max}(\text{hd } M, \text{hd } R) = 1$. We show P is projective. Look at a part of the long exact homology sequence, derived from the sequence 10.1

$$\text{Hom}(R, R) \xrightarrow{\Delta} \text{Ext}^1(M, R) \longrightarrow \text{Ext}^1(P, R) \longrightarrow \text{Ext}^1(R, R).$$

The last term of this sequence vanishes since R is projective. Now Δ is surjective, since by Appendix B.1.13 we see that $\Delta(\text{id}_R) = \eta$. Therefore,

$\operatorname{Ext}^1(P, R) = 0$, whence $\operatorname{Ext}^1(P, R^n) = 0$ whence $\operatorname{Ext}^1(P, Q) = 0$ for every finitely generated projective Q. Since hd $P \leq 1$ we have a resolution

$$0 \longrightarrow P_1 \longrightarrow P_0 \longrightarrow P \longrightarrow 0 \tag{10.2}$$

with P_1, P_0 finitely generated projective. Since $\operatorname{Ext}^1(P, P_1) = 0$, the sequence 10.2 splits by extension theory, i.e. $P \oplus P_1 = P_0$, whence P is projective. □

10.2 Projective Generation of I

Let R be a local ring with maximal ideal $\mathfrak{m}$. For any ideal $I \subset \mathfrak{m}$ we have

$$\mu(I) = \mu(I/\mathfrak{m}I) = \mu(I/I^2)$$

since $I^2 \subset \mathfrak{m}I$. For a non local ring the numbers $\mu(I)$ and $\mu(I/I^2)$ can differ, but at most by 1, as the following lemma will show.

Lemma 10.2.1 *Let R be a ring and I a proper finitely generated ideal. Let further $a_1, \dots, a_m \in I$, such that their residue classes $\bar{a}_1, \dots, \bar{a}_m$ generate I/I^2. Then there is an $s \in I$ with $(1-s)I \subset (a_1, \dots, a_m)$. Hence $I = (a_1, \dots, a_m, s)$ and so*

$$\mu(I/I^2) \leq \mu(I) \leq \mu(I/I^2) + 1.$$

If R is Noetherian, then
$$\operatorname{ht}(I) \leq \mu(I/I^2).$$

Proof. The first inequality is obvious. Let $J := (a_1, \dots, a_m)$. If $S = 1 + I$, the ideal I_S is contained in the Jacobson radical of R_S. Therefore by Nakayama's Lemma, elements whose residue classes generate I_S/I_S^2 already generate I_S. So $J_S = I_S$, whence there is an $s \in I$ with $(1-s)I \subset J$.

For every $a \in I$ we derive $a - sa = \sum_{i=1}^m \lambda_i a_i$ for some $\lambda_i \in R$, i.e. $a \in (a_1, \dots, a_m, s)$.

The height of I can be checked in the localizations with respect to the prime ideals containing I. For such a prime ideal $\mathfrak{p}$ we have $\mu_\mathfrak{p}(I) = \mu_\mathfrak{p}(I/I^2)$, since $I_\mathfrak{p}$ is in the Jacobson radical of $R_\mathfrak{p}$. So

$$\operatorname{ht}(I_\mathfrak{p}) \leq \mu_\mathfrak{p}(I) = \mu_\mathfrak{p}(I/I^2) \leq \mu_R(I/I^2).$$

By Lemma 10.2.1 there are two possibilities for $\mu(I)$ – either it equals $\mu(I/I^2)$ or $\mu(I/I^2) + 1$. Besides, the proof of Lemma 10.2.1 also shows that $\operatorname{Spec}(R)$ is covered by two open sets $\mathrm{D}(s)$ and $\mathrm{D}(t)$, on which I is generated by $\leq \mu(I/I^2)$ elements. (Namely set $t = 1 - s$.) In other words one has surjections

$$R_s^\mu \to I_s \to 0 \qquad R_t^\mu \to I_t,$$

where $\mu = \mu(I/I^2)$ and with $(s,t) = R$. $\square$

Lemma 10.2.1 shows that the study of $\mu(I)$ is closely linked to the study of the minimal number of generators of the **conormal R/I-module I/I^2**.

In view of Serre's motivation it is natural to query, whether there is a projective module P of rank μ which maps surjectively onto I. (i.e. we are trying to 'globalize the local' information.) This was done by M. Boratynski (see [9]) and his proof provides also a non-homological proof of Serre's result mentioned in the introduction of the last section.

Theorem 10.2.2 (M. Boratynski) *Let I be a finitely generated ideal of a ring R. Then there is an ideal $J \subset I$ with $\sqrt{J} = \sqrt{I}$ and a projective R-module P of rank $\mu = \mu(I/I^2)$ which maps surjectively onto J.*

Proof. As in the proof of Lemma 10.2.1 there are $s, a_1, \ldots, a_\mu \in I$, such that $(1-s)I \subset (a_1, \ldots, a_\mu)$ and there are surjections

$$R_s^\mu \xrightarrow{\;(1,0,\ldots,0)\;} I_s(= R_s) \to 0 \qquad R_{1-s}^\mu \xrightarrow{\;(a_1,\ldots,a_\mu)\;} I_{1-s} \to 0.$$

Since $I_{s(s-1)} = R_{s(s-1)}$, the vector $v := (a_1/1, \ldots, a_\mu/1)$ is unimodular over $R_{s(s-1)}$.

Suppose that v is completable and let

$$(a_1, \ldots, a_\mu) = e_1 \sigma^{-1} \text{ for some } \sigma \in \mathrm{GL}_\mu(R_{s(s-1)}).$$

Then one has a commutative diagram

$$
\begin{array}{ccccc}
R_{s(s-1)}^\mu & \xrightarrow{\;f\;} & I_{s(s-1)} & \longrightarrow & 0 \\
\sigma \downarrow & & \| & & \\
R_{s(s-1)}^\mu & \xrightarrow{\;g\;} & I_{s(s-1)} & \longrightarrow & 0
\end{array}
$$

with $f := (1, 0, \ldots, 0)$, $g := (a_1, \ldots, a_\mu)$. Let $P := R_s^\mu \times_\sigma R_{1-s}^\mu = \{(x,y) \in R_s^\mu \times R_{1-s}^\mu \mid \sigma x_{1-s} = y_s\}$ be the so called 'fibre product' of R_s^μ and R_{s-1}^μ over σ. We have $P_s \cong R_s^\mu$ and $P_{1-s} \cong R_{1-s}^\mu$. So P is projective of rank μ. We can define an epimorphism $\varphi : P \to I$ by

$$\varphi(x,y) := (f(x), g(y)) \in I_s \times_\sigma I_{1-s} = I.$$

If $\mu = 2$ then v is always completable. So we have shown that, if $\mu(I/I^2) = 2$ there is a projective module P of rank 2 which maps epimorphically to I. (This provides a non-homological proof of Serre's theorem.)

In general one has to modify the ideal I. Let

$$J := (a_1, a_2^1, a_3^2, \dots, a_\mu^{\mu-1}) + I^{\mu-1} \subset I,$$

where $a_1, \dots, a_\mu$ are as above.

Then $\sqrt{J} = \sqrt{I}$ and $J_s = R_s$, the latter since $V(J_s) = V(I_s) = \emptyset$. Further $J_{1-s} = (a_1, a_2^1, \dots, a_\mu^{\mu-1})_{1-s}$; namely $(1-s)I^{\mu-1} \subset (a_1, a_2^1, \dots, a_\mu^{\mu-1})$, and so $I_{1-s}^{\mu-1} \subset (a_1, a_2^1, \dots, a_\mu^{\mu-1})_{1-s}$.

By Suslin's Theorem 3.5.3 the unimodular row $(a_1/1, a_2^1/1, \dots, a_\mu^{\mu-1}/1)$ over $R_{s(1-s)}$ is completable. Hence as above we get an epimorphism $P \to J$ for some projective module P of rank μ.　□

Corollary 10.2.3 *Let k be a field or a principal ideal domain and I an ideal of the polynomial ring $k[x_1, \dots, x_n]$ and $\mu = \mu(I/I^2)$. Then*

$$\sqrt{I} = \sqrt{(f_1, \dots, f_\mu)}$$

for suitable $f_1, \dots, f_\mu$.

Proof. By Theorem 10.2.2 there are an ideal J with $\sqrt{J} = \sqrt{I}$, a projective module P of rank μ and an epimorphism $P \to J$. But P is free by the Quillen-Suslin Theorem.　□

10.3 Local Complete Intersection Curves in n-Space are Set-Theoretical Complete Intersections

A classical open problem asks whether every curve in the affine n-space is a set-theoretical complete intersection. Initial progress in Serre's Conjecture led Ferrand and Szpiro to independently affirm that a curve in 3-space which is a local complete intersection is a set-theoretic complete intersection. Due to Serre's Conjecture's solution and M. Boratynski's results we can now establish this by a similar argument in n-space. (These results were conjectured by O. Forster and were first settled for $n > 3$ by N. Mohan Kumar in [60]).

We need few preliminary observations on rank 1 projective modules and on regular sequences.

For a rank 1 projective R-module L we denote by

$$L^{\otimes n} = \begin{cases} L \otimes \cdots \otimes L & n \text{ times}, & \text{if } n \geq 1 \\ L^* \otimes \cdots \otimes L^* & |n| \text{ times}, & \text{if } n \leq -1 \\ R & & \text{if } n = 0 \end{cases}$$

It is easily verified that $L^{\otimes(m+n)} = L^{\otimes n} \otimes L^{\otimes m}$ for all $n, m \in \mathbb{Z}$.

Lemma 10.3.1 *Let R be a one-dimensional Noetherian ring and P a finitely generated projective R-module of constant rank $r \geq 2$. Then every projective R-module Q of rank 1 is a direct summand of P.*

Proof. By Proposition 8.8.7 we see $P \cong P' \oplus R^{r-1}$ with some P' of rank 1. Further $r - 1 \geq 1$. Then

$$R^{r-2} \oplus (P' \otimes Q^*) \oplus Q \cong R^{r-2} \oplus (P' \otimes Q^* \otimes Q) \oplus R \cong P' \oplus R^{r-1} \cong P. \quad \square$$

Lemma 10.3.2 *Let R be a local Noetherian ring and $I \neq R$ be an ideal, generated by an R-regular sequence $x_1, \dots, x_n$. Then I/I^2 is a free R/I-module.*

Proof. We show that the residue classes $\overline{x_1}, \dots, \overline{x_n}$ form a basis. They clearly generate I/I^2. Assume they were linearly dependent, say $\sum_i a_i x_i \in I^2$ with some $a_j \notin I$. By Lemma 9.2.3 we may assume $j = n$, i.e. $a_n \notin I$. Then $a_n x_n \in (x_1, \dots, x_{n-1}) + I^2 = (x_1, \dots, x_{n-1}) + x_n I$, i.e. $(a_n + b)x_n \in (x_1, \dots, x_{n-1})$ for some $b \in I$. Since $a_n \notin I$, also $a_n + b \notin I$, whence *a fortiori* $a_n + b \notin (x_1, \dots, x_{n-1})$. But this means that x_n is a zero-divisor modulo $(x_1, \dots, x_{n-1})$ – a contradiction. $\square$

Theorem 10.3.3 (Ferrand) *Let I be an ideal of a Noetherian ring R of Krull dimension $d \geq 3$. Assume further that I is a local complete intersection ideal of height $(d-1)$, i.e. $I_{\mathfrak{m}} = (x_1, \dots, x_{d-1})$ for some regular sequence $x_1, \dots, x_{d-1}$ in $R_{\mathfrak{m}}$, for every maximal ideal $\mathfrak{m}$ of R containing I. Then there is an ideal J of R such that*

a) $I^2 \subset J \subset I$

b) *J is a local complete intersection of height $(d-1)$*

c) *J/J^2 is a free R/J-module of rank $(d-1)$.*

Proof. By Lemma 10.3.2 the R/I-module I/I^2 is locally free, i.e. projective of rank $(d-1)$. Since $\dim(R/I) = 1$ and $\bigwedge^{d-1}(I/I^2)^*$ is a projective R/I-module of rank 1, by Lemma 10.3.1 there is an ideal J, containing I^2, such that for $K := \bigwedge^{d-1}(I/I^2)^*$ we have

$$J/I^2 \oplus K = I/I^2 \tag{10.3}$$

We show that the ideal J has the required properties. We only have to prove b) and c).

b) To show that J is a local complete intersection, we may assume that R (hence also R/I) is local. Then there is a basis $x_1, \dots, x_{d-1}$ of I/I^2 such that $x_1, \dots, x_{d-2}$ is a basis of J/I^2 and x_{d-1} one of K/I^2. Then

$$J = (x_1, \dots, x_{d-2}) + I^2 = (x_1, \dots, x_{d-2}) + (x_i x_j)_{1 \leq i \leq j \leq d-1}$$

$$= (x_1, \ldots, x_{d-2}, x_{d-1}^2).$$

So $\mu(J) \le d - 1$. On the other hand $\mathrm{ht}(J) = \mathrm{ht}(I) = d - 1$, since $I^2 \subset J \subset I$.

c) J/J^2 is a projective R/J-module of rank $(d-1)$. We show that J/J^2 is generated by $(d-1)$ elements; this will suffice to show that J/J^2 is free of rank $(d-1)$.

Since $(IJ/J^2)^2 = 0$ in the ring R/J^2, by Nakayama's Lemma we have $\mu(J/IJ) = \mu(J/J^2)$. So we need only show $\mu(J/IJ) = d - 1$.

Let us analyse the R/I-module J/IJ. Notice that $J/IJ = J \otimes_R R/I = J \otimes_R R/J \otimes_{R/J} R/I = J/J^2 \otimes_{R/J} R/I$ and so is a projective R/I-module of rank $(d-1)$. We will show that it is actually a free R/I-module of rank $(d-1)$. By Serre's splitting Theorem 7.1.8 it suffices to show that $\bigwedge^{d-1}(J/IJ) \overset{\sim}{\to} R/I$.

We have an exact sequence of R/I-modules

$$0 \to I^2/IJ \to J/IJ \to J/I^2 \to 0. \tag{10.4}$$

Since J/IJ and J/I^2 both are projective over R/I, this sequence splits and I^2/IJ is also projective over R/I.

By the equality (10.3) we have $\mathrm{rk}(J/I^2) = d - 2$, and so $\mathrm{rk}(I^2/IJ) = 1$ by the exactness of (10.4) . Since the sequence (10.4) splits, $J/IJ \overset{\sim}{\to} I^2/IJ \oplus J/I^2$. Taking exterior powers gives

$$\overset{d-1}{\bigwedge}(J/IJ) = \overset{d-1}{\bigwedge}(I^2/IJ \oplus J/I^2) \overset{\sim}{\to} \overset{d-2}{\bigwedge}(J/I^2) \otimes I^2/IJ.$$

Recall that $K = \mathrm{Hom}(\bigwedge^{d-1}(I/I^2), R/I) \cong I/J$ via the equality (10.3). In particular, since $\bigwedge^{d-1}(I/I^2) \cong K^{\otimes -1}$, we get $K^{\otimes -1} \cong \bigwedge^{d-1}(K \oplus J/I^2) \cong \bigwedge^{d-2}(J/I^2) \otimes K$, hence $\bigwedge^{d-2}(J/I^2) \cong K^{\otimes -2}$.

Multiplication in I gives us a natural (R/I)-bilinear map

$$I/J \times I/J \to I^2/IJ, \quad (x + J , y + J) \mapsto xy + IJ,$$

which induces an R/I-homomorphism $I/J \otimes_{R/I} I/J \to I^2/IJ$. This map is locally an isomorphism as $I/J \cong K$ is projective of rank 1 over R/I. Hence it is an isomorphism and so $I^2/IJ \overset{\sim}{\to} K \otimes K$. Thus as required, $\bigwedge^{d-1}(J/IJ) \overset{\sim}{\to} \bigwedge^{d-2}(J/I^2) \otimes I^2/IJ = K^{\otimes -2} \otimes K^{\otimes 2} \overset{\sim}{\to} R/I$. □

Corollary 10.3.4 (Ferrand - Szpiro, $n = 3$; Mohan Kumar) *Let I be a local complete intersection ideal of $k[x_1, \ldots, x_n]$ of height $(n-1)$. Then $\sqrt{I} = \sqrt{(f_1, \ldots, f_{n-1})}$ for suitable $f_1, \ldots, f_{n-1}$.*

Proof. In view of Ferrand's construction $\sqrt{I} = \sqrt{J}$ for some local complete intersection J with J/J^2 free of rank $n-1$. By M. Boratynski's theorem, there is an ideal $J' \subset J$ with $\sqrt{J'} = \sqrt{J}$ and a projective $k[x_1, \ldots, x_n]$-module P of rank $n-1$, mapping onto J'. By Quillen-Suslin's theorem P is free of rank $n-1$, and so J' is generated by $n-1$ elements. □

10.4 The Theorem of Cowsik and Nori

The Theorem of Cowsik and Nori says that curves in the affine n-space over a field of *positive characteristic* are set-theoretical complete intersections. We need a preparation.

10.4.1 A Projection Lemma

The following statement seems intuitively plausible: A curve C in n-space can be projected into a plane in such a way, that this projection maps C isomorphically to its image outside a finite set of points. And a proof along this intuitive idea can be given, if the ground field is algebraically closed (or at least infinite) (*cf.* [30], Chapter IV).

But if this field is finite, there are only finitely many directions along which to project. And if the curve is the union of lines in every possible direction, the above statement fails to hold. So one has to admit "nonlinear projections", e.g. Nagata transformations. We will formulate and prove the correct statement in purely algebraic terms. We follow the ideas of [53].

Definition 10.4.1 *A homomorphism $A \to B$ of reduced Noetherian rings is called birational, if it is injective and induces an isomorphism of the total quotient rings $Q(A) \to Q(B)$.*

We fix a prime number p and a perfect field k of characteristic p.

Proposition 10.4.2 (Projection Lemma) *Let I be a purely 1-codimensional radical ideal of $A = k[X_1, \ldots, X_n]$, i.e. $I = \mathfrak{p}_1 \cap \cdots \cap \mathfrak{p}_r$ with $\dim(A/\mathfrak{p}_i) = 1$. Then there is a change of variables, such that after this change the ring extension $k[X_1, X_2]/(k[X_1, X_2] \cap I) \hookrightarrow A/I$ is finite and birational.*

The proof requires some preparations. First recall that every $f \in A$ with $\partial f/\partial X_i = 0$ for all i is of the form g^p with some $g \in A$, since k is perfect. So in this case f is not irreducible. As a consequence we get

Lemma 10.4.3 *Let $f \in k[X, Y]$ be irreducible. Then for big enough m, not divisible by p, we have that $f(X + Y^m, Y)$ is monic in Y upto a unit and*

$$\frac{\partial f(X + Y^m, Y)}{\partial Y} \neq 0.$$

Proof. Write $F := f(X + Y^m, Y)$. We know that F is monic in Y upto a unit for large m. The chain rule for differentiation, valid also in this formal situation, yields

$$\frac{\partial F}{\partial Y} = mY^{m-1}\frac{\partial f}{\partial X}(X + Y^m, Y) + \frac{\partial f}{\partial Y}(X + Y^m, Y). \qquad (10.5)$$

(Distinguish well between

$$\frac{\partial f(X + Y^m, Y)}{\partial Y} \quad \text{and} \quad \frac{\partial f}{\partial Y}(X + Y^m, Y).$$

The first means: first replace X by $X + Y^m$, then differentiate w.r.t. Y; the second means: first differentiate w.r.t. Y, then replace X by $X + Y^m$.)

If $\partial f/\partial Y \neq 0$, then $\partial f/\partial Y \notin (Y^r)$ for some r. Then also $\partial f/\partial Y(X + Y^m, Y) \notin (Y^r)$. Since for $m > r$ the first summand of the right hand side of (10.5) belongs to (Y^r), the left hand side does not, and so even more does not vanish.

If on the other hand $\partial f/\partial Y = 0$, then, according to the above remark, $\partial f/\partial X \neq 0$. And so if $p \nmid m$.

$$\frac{\partial F}{\partial Y} = mY^{m-1}\frac{\partial f}{\partial X}(X + Y^m, Y) \neq 0,$$

□

The following lemma is well known.

Lemma 10.4.4 Let $k \subset k(x,y)$ be a finite field extension and y separable over k. Then $k(x,y) = k(x + ay)$ for all but finitely many $a \in k$. □

Lemma 10.4.5 Let $\mathfrak{p}_1, \mathfrak{p}_2$ be two distinct maximal ideals of $K[X,Y]$. Then $\mathfrak{p}_1 \cap K[X + aY] \neq \mathfrak{p}_2 \cap K[X + aY]$ for all but finitely many $a \in K$.

Proof. By Hilbert's Nullstellensatz the fields $L_i := K[X,Y]/\mathfrak{p}_i$ are finite over K. Fix K-embeddings (i.e. K-algebra-homomorphisms) $L_i \hookrightarrow \overline{K}$, the algebraic closure of K, and call x_i, y_i the images of the residue classes of X, resp. Y modulo $\mathfrak{p}_i$ under these iembeddings The set E of all K-iembeddings $L_1 \hookrightarrow \overline{K}$ is finite. For every $\sigma \in E$ there is at most one $b \in K$ with $b(\sigma(y_1) - y_2) = x_2 - \sigma(x_1)$. Otherwise we had $x_2 = \sigma(x_1)$ and $y_2 = \sigma(y_2)$ for some σ and hence $\mathfrak{p}_1 = \mathfrak{p}_2$ – as kernels of 'isomorphic' homomorphisms. Therefore the set

$$S := \{b \in K \mid b(\sigma(y_1) - y_2) = x_2 - \sigma(x_1) \text{ for some } \sigma \in E\}$$

is finite.

We claim that $\mathfrak{p}_1 \cap K[X + aY] \neq \mathfrak{p}_2 \cap K[X + aY]$ for every $a \in K \setminus S$. By definition of S we have $\sigma(x_1 + ay_1) \neq x_2 + ay_2 \ \forall \ \sigma \in E$, whence $x_1 + ay_1$ and $x_2 + ay_2$ are not conjugate to each other. So, if $f_i(T)$ is the minimal (monic) polynomial of $x_i + ay_i$, we have $f_1 \neq f_2$. Now it is clear that $\mathfrak{p}_i \cap K[X + aY]$ is generated by $f_i(X + aY)$. So $\mathfrak{p}_1 \cap K[X + aY] \neq \mathfrak{p}_2 \cap K[X + aY]$. □

Proof of the Projection Lemma: First we handle the case $n = 3$ and call the indeterminates X, Y, Z. Let (as above) $I = \mathfrak{p}_1 \cap \cdots \cap \mathfrak{p}_r$, where $\mathfrak{p}_i$ are prime ideals with $\dim(k[X, Y, Z]/\mathfrak{p}_i) = 1$, i.e. ht $\mathfrak{p}_i = 2$ by Proposition 6.7.5. According to Theorem 6.9.3 we have

$$1 \leq \mathrm{ht}(\mathfrak{p}_i \cap k[X, Y]) \leq 2.$$

The case $\mathrm{ht}(\mathfrak{p}_i \cap k[X, Y]) = 1$ is the more general one: A curve projects to a curve. But the other case may also happen, e.g. if $\mathfrak{p}_i = (X, Y)$.

To escape this difficulty, let $g_i(X, Y) \in \mathfrak{p}_i \cap k[X, Y] \setminus \{0\}$. For big enough m, $n-m$ all $g_i(X + Z^m, Y + Z^n)$ are unital in Z. Hence $k[X, Y, Z]/\mathfrak{p}_i$ is integral over $k[X', Y', Z]/\mathfrak{p}_i \cap k[X', Y']$ for new variables $X' = X + Z^m$, $Y' = Y + Z^n$. And therefore $\dim(k[X', Y', Z]/\mathfrak{p}_i \cap k[X', Y']) = 1$.

So we may assume that $\mathrm{ht}(\mathfrak{p}_i \cap k[X, Y]) = 1$. In this case $\mathfrak{p}_i \cap k[X, Y] = f_i k[X, Y]$ for certain polynomials $f_i \in k[X, Y]$, since prime ideals of height 1 in factorial rings are principal.

Using Lemma 10.4.3, after a change of variables, we may assume that the f_i are monic upto a constant factor in Y and $\partial f_i / \partial Y \neq 0$.

Set $K_i := Q(k[X, Y, Z]/\mathfrak{p}_i)$ and denote by x_i, y_i, z_i the images of X, Y, Z in K_i. Then the field extensions $k(X) \hookrightarrow K_i$ are finite and the y_i are separable over $k(X)$. Since $k(X)$ is an infinite field, by Lemmas 10.4.4 and 10.4.5, there is an $a \in k(X)$ such that first $K_i = k(X)(ay_i + z_i)$ for $i = 1, \ldots, r$ and secondly the $k(X)[aY + Z] \cap \mathfrak{p}_i$ are distinct. Then $k(X)[Y, Z]/Ik(X)[Y, Z]$ is generated by the class of $aY + Z$ over $k(X)$.

Now write $a = c/d$ with coprime $c, d \in k[X]$ and let $\gamma, \delta \in k[X]$ be chosen such that $\gamma c - \delta d = 1$. Then $k(X)[Y, Z]/Ik(X)$ is generated by the class of $cY + dZ$ over $k(X)$.

Replacing $cY + dZ$ by Y and $\delta Y + \gamma Z$ by Z we reach the situation that the ring $k(X)[Y, Z]/Ik(X)[Y, Z]$ is generated by Y over $k(X)$. Hence

$$k[X, Y]/(I \cap k[X, Y]) \longrightarrow k[X, Y, Z]/I$$

becomes an isomorphism after tensoring with $k(X)$, whence it is birational. And it is finite since I contains a polynomial which is monic in Z.

Now assume $n > 3$. Since, by the first case, after a change of variables the extension $k[X_1, X_2]/I \cap k[X_1, X_2] \hookrightarrow k[X_1, X_2, X_n]/I \cap k[X_1, X_2, X_n]$ is finite and birational, the same holds for $k[X_1, \ldots, X_{n-1}]/I \cap k[X_1, \ldots, X_{n-1}] \hookrightarrow k[X_1, \ldots, X_n]/I$. By induction we are through. $\qquad\square$

10.4.2 The Proof of the Theorem of Cowsik and Nori

We need the following notion:

Definition 10.4.6 *Let $A \subset B$ be a ring extension. The conductor $\mathfrak{C}$ of this extension is defined by*

$$\mathfrak{C} := \mathrm{Ann}_A(B/A),$$

where B/A denotes the residue class module of the A-module B by its submodule A.

Remarks 10.4.7 a) The conductor $\mathfrak{C}$ is an ideal of B contained in A. Moreover it is the biggest common ideal of A and B.

b) Assume that the extension $A \subset B$ is a birational finite extension of reduced rings. Then $\mathfrak{C}$ contains a non-zero-divisor. Namely write the elements of a finite generating system of the A-module B as fractions over A. Then every common denominator will belong to $\mathfrak{C}$.

Theorem 10.4.8 *(Cowsik-Nori)*
Every curve in the affine n-space over a field k of positive characteristic $p > 0$, is a set-theoretic complete intersection.

Proof. Let $\mathfrak{p} \subset B := k[x_1, \ldots, x_1]$ be the radical ideal defining the curve C. If the extension $\mathfrak{p}^e$ of $\mathfrak{p}$ in $k^{\frac{1}{p^\infty}}[x_1, \ldots, x_n]$ is generated upto radical by $(n-1)$ elements $f_1, \ldots, f_{n-1}$, then $f_1^{p^N}, \ldots, f_{n-1}^{p^N} \in B$ for some N and $\mathfrak{p} = \sqrt{(f_1^{p^N}, \ldots, f_{n-1}^{p^N})}$. Therefore, we may assume $k = k^{1/p^\infty}$, i.e. that k is perfect.

By Proposition 10.4.2 after a change of variables we may assume that $k[x_1, x_2]/\mathfrak{p} \cap k[x_1, x_2] \hookrightarrow B/\mathfrak{p}$ is a finite and birational extension. Therefore the conductor $\mathfrak{C}$ of this extension contains a non-zero-divisor. Since $\dim(B/\mathfrak{p}) = 1$, we have $\dim(B/\mathfrak{p})/\mathfrak{C} = 0$, and so $(B/\mathfrak{p})/\mathfrak{C}$ is Artinian. It therefore is a finite dimensional vector space over k by Lemma 6.5.1.

Let $L_{i,r}$ (for $3 \leq i \leq n$) be the k-vector space, generated by $\overline{x}_i^{p^r}, \overline{x}_i^{p^{r+1}}, \ldots$ in $(B/\mathfrak{p})/\mathfrak{C}$. Then $L_{i,0} \supset L_{i,1} \supset \ldots \supset L_{i,r} \supset \ldots$. This series must eventually stabilize: say $L_{i,r} = L_{i,r+1} = \ldots$. Thus there exists an r such that $\overline{x}_i^{p^r} \in L_{i,r+1}$ for all i, i.e.

$$\overline{x}_i^{p^r} - s_{i1}\overline{x}_i^{p^{r+1}} - \cdots - s_{it_i}\overline{x}_i^{p^{r+t_i}} = 0,$$

with some $s_{ij} \in k$. Thus

$$f_i := x_i^{p^r} - s_{i1}x_i^{p^{r+1}} - \cdots - s_{it_i}x_i^{p^{r+t_i}} - c_i \in \mathfrak{p},$$

for suitable c_i, whose residue classes modulo $\mathfrak{p}$ belong to $\mathfrak{C}$, hence to $k[x_1, x_2]/\mathfrak{p} \cap k[x_1, x_2]$. Therefore we may and will choose c_i in $k[x_1, x_2]$.

Let $y_i := x_i^{p^r}$, $B' = k[x_1, x_2, y_3, \ldots, y_n]$ and $\mathfrak{p}' = \mathfrak{p} \cap B'$. Note that $f_i \in \mathfrak{p}'$.

Since $\mathfrak{p}'B \subset \mathfrak{p}$ and every element of $\mathfrak{p}$ raised to its p^r-th power belongs to $\mathfrak{p}'$, we see that $\sqrt{\mathfrak{p}'B} = \mathfrak{p}$.

CLAIM: $B'' = B'/(f_3, \ldots, f_n)$ is a regular ring.

Namely observe that $\partial f_i/\partial y_j = \delta_{ij}$ for $i, j \geq 3$. By the Jacobian Criterion 8.1.3, $B'/(f_3, \ldots, f_n)$ is regular of dimension 2.

Note that $\overline{\mathfrak{p}'} = \mathfrak{p}'/(f_3, \ldots, f_n)$ has height one. For any maximal ideal $\mathfrak{m} \supset \mathfrak{p}'$, $B''_{\mathfrak{m}}$ is regular and local, whence factorial. Consequently $\overline{\mathfrak{p}'}_{\overline{\mathfrak{m}}}$, being an intersection of finitely many prime ideals, is a principal ideal, i.e. $\overline{\mathfrak{p}'}$ is locally 1-generated. Therefore, $\mathfrak{p}'$ is locally $(n-1)$-generated. By Corollary 10.3.4 up to radical $\mathfrak{p}'$ is generated by $(n-1)$ elements. $\square$

A

Normality of the Elementary Subgroup

In 1964 P.M. Cohn had shown in [13] that $E_2(A)$ was not always a normal subgroup of $SL_2(A)$. For example, he showed that

$$\begin{pmatrix} 1+2x & -x^2 \\ 4 & 1-2x \end{pmatrix} \in SL_2(\mathbb{Z}[x]) \setminus E_2(\mathbb{Z}[x])$$

$$\begin{pmatrix} 1+xy & x^2 \\ -y^2 & 1-xy \end{pmatrix} \in SL_2(\mathbb{Z}[x,y]) \setminus E_2(\mathbb{Z}[x,y])$$

It came as a surprise when A. Suslin announced in mid 1976 that $E_n(A)$ is a normal subgroup of $GL_n(A)$ for $n \geq 3$ and any commutative ring A.

We have to show that *if $\gamma \in GL_n(A)$ then $\gamma E_{ij}^\lambda \gamma^{-1} \in E_n(A)$.*

Now

$$\gamma E_{ij}^\lambda \gamma^{-1} = \gamma(I_n + \lambda e_i^t e_j)\gamma^{-1} = I_n + \lambda \gamma e_i^t e_j \gamma^{-1}.$$

Let $v^t = \gamma e_i^t$, $w = e_j \gamma^{-1}$. Then $v, w \in Um_n(A)$ with $wv^t = 0$ and $\gamma E_{ij}(\lambda)\gamma^{-1} = I_n + v^t w$.

First we will see that $I_n + v^t w$ is 1-stably elementary, i.e. $\begin{pmatrix} I_n + v^t w & 0 \\ 0 & 1 \end{pmatrix} \in$ $E_{n+1}(A)$. In other words, in $(n+1)$-space $I_n + v^t w$ is elementary.

We prove this as follows: Let $wv^t = 0$ and consider $I_n + v^t w$. Since $(v^t w)^2 = 0$, we see that $I_n + v^t w$ is invertible with $I_n - v^t w$ as its inverse. Now

$$\begin{pmatrix} I_n & 0 \\ w & 1 \end{pmatrix} \begin{pmatrix} I_n & v^t \\ 0 & 1 \end{pmatrix} \begin{pmatrix} I_n + v^t w & 0 \\ 0 & 1 \end{pmatrix} \begin{pmatrix} I_n & 0 \\ -w & 1 \end{pmatrix} \begin{pmatrix} I_n & -v^t \\ 0 & 1 \end{pmatrix} = I_{n+1}.$$

It follows that $\begin{pmatrix} I_n + v^t w & 0 \\ 0 & 1 \end{pmatrix} \in E_{n+1}(A)$.

Now we consider the special case that some coordinate v_i of $v = (v_1, \cdots, v_n)$, or some coordinate w_j of w is zero. Then we will conclude from above that $I_n + v^t w \in E_n(A)$.

Let, for example, $w_j = 0$. By permuting coordinates we may assume $w_n = 0$. Let $v' = (v_1, \cdots, v_{n-1})$, $w' = (w_1, \cdots, w_{n-1})$, then $w'v'^t = 0$, and so $I_{n-1} + v'^t w' \in E_n(A)$ by above. Now

$$I_n + v^t w = \begin{pmatrix} I_{n-1} + v'^t w' & 0 \\ \star & 1 \end{pmatrix} = \begin{pmatrix} I_{n-1} + v'^t w' & 0 \\ 0 & 1 \end{pmatrix} \begin{pmatrix} I_{n-1} & 0 \\ \star & 1 \end{pmatrix} \in E_n(A)$$

(This observation was made by L.N. Vaserstein in [104] in 1973.)

We are now ready to prove the general case by reducing it to the above case. For this we need some linear algebra :

Lemma A.0.1 *Let $v = (v_1, \ldots, v_n) \in Um_n(A)$. Let further $f : A^n \to A$ be an A-linear map given by $e_i \mapsto v_i$ where $\{e_i\}_{1 \le i \le n}$ is the canonical basis of A^n. Then $\ker(f) = \{(w_1, \ldots, w_n) \in A^n \mid \sum_{i=1}^{n} w_i v_i = 0\}$ is generated by the elements $(v_j e_i - v_i e_j)$, $1 \le i < j \le n$.*

Proof. Since v unimodular, f is surjective. Let K be the submodule of $\ker(f)$ generated by $\{(v_j e_i - v_i e_j) : 1 \le i \ne j \le n\}$. We check "locally" that $K_\mathfrak{p} = (\ker(f))_\mathfrak{p} = \ker(f_\mathfrak{p})$ for all $\mathfrak{p} \in \operatorname{Spec}(A)$. This is easily done in the case $v = e_1$; and the general case can be easily reduced to this case as v can be completed to an invertible matrix over $R_\mathfrak{p}$. $\qquad\square$

Proposition A.0.2 *Let $n \ge 3$, $v = (v_1, \ldots, v_n) \in Um_n(A)$ and $w = (w_1, \ldots, w_n) \in M_{1,n}(A)$ be such that $wv^t = 0$. Then $I_n + v^t w \in E_n(A)$.*

Proof. Let $(e_i)_{1 \le i \le n}$ be the canonical basis for A^n. Define an A-linear map $f : A^n \to A$ by $f(e_i) = v_i$. Since $wv^t = 0$, we have $w \in \ker(f)$ and hence by Lemma A.0.1, there exist $a_{ij} \in A$ such that

$$w = \sum_{1 \le i < j \le n} a_{ij}(v_j e_i - v_i e_j) = \sum_{i < j} w_{ij}$$

where $w_{ij} = a_{ij}(v_j e_i - v_i e_j)$. Since $wv^t = 0$ and $w = \sum_{i<j} w_{ij}, w_{ij}v^t = 0$. Note that $n \ge 3$. By Vaserstein's observation $I_n + v^t w_{ij} \in E_n(A)$. Now

$$I_n + v^t w = I_n + v^t \sum_{i<j} v^t w_{ij}.$$

CLAIM: $I_n + \sum_{i<j} v^t w_{ij} = \prod_{i<j} (I_n + v^t w_{ij}) \in E_n(A)$.

We prove the claim by induction on the number of terms in the summation, say r. Suppose $r = 1$, then the result is clear. Suppose $r = 2$, then

$$(I_n + v^t w_1)(I_n + v^t w_2) = I_n + v^t w_1 + v^t w_2 + v^t w_1 v^t w_2 = I_n + v^t w_1 + v^t w_2$$

as $w_1 v^t = 0$.

Suppose $r \geq 3$ and assume that the result is true for $r - 1$, then,

$$\begin{aligned}
I_n + v^t w_1 + \cdots + v^t w_r &= I_n + (v^t w_1 + \cdots + v^t w_{r-1}) + v^t w_r \\
&= I_n + v^t(w_1 + \cdots + w_{r-1}) + v^t w_r \\
&= (I_n + v^t(w_1 + \cdots + w_{r-1}))(I_n + v^t w_r) \\
&= \prod_{j=1}^{r}(I_n + v^t w_j)
\end{aligned}$$

(The middle step is by the case $r = 2$, and the last step is by the inductive assumption.)

Thus by induction

$$I_n + \sum v^t w_{ij} = \prod(I_n + v^t w_{ij}) \in E_n(A)$$

as required. $\square$

Open Question: Can one drop the condition that v is unimodular in Proposition A.0.2?

This is possible when R is a regular ring containing a field.

B

Some Homological Algebra

In this Appendix we shall do the following:

* $\star$ Extension theory. We introduce extensions and Ext^1 and their relationship.

* $\star$ Derived functors, and higher Ext^n.

* $\star$ Rees' homological characterization of grade.

B.1 Extensions and Ext^1

Definition B.1.1 *Let R be a ring, M, N, X be R-modules and $f : M \to X$, $g : N \to X$ homomorphisms. A **pull-back** of (f, g) is an R-module Y together with a pair of homomorphisms $f' : Y \to M$, $g' : Y \to N$ such that $f f' = g g'$ and the following universal property holds:*

If $f'' : Z \to M$, $g'' : Z \to N$ are homomorphisms such that $f f'' = g g''$, then there exists a unique $h : Z \to Y$ such that $f' h = f''$ and $g' h = g''$.

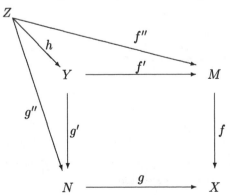

An obvious candidate for Y is $Y = \{(m,n) \in M \oplus N \mid f(m) = g(n)\}$, i.e. the kernel of the map $M \oplus N \to X$, $(m,n) \mapsto f(m) - g(n)$ with the canonical maps $f' : Y \to M$, $g' : Y \to N$.

The following definition is dual to the above one.

Definition B.1.2 *Let R be a ring, M, N, X be R-modules and $f : X \to M$, $g : X \to N$ module homomorphisms. A **push-out** of (f, g) is a module Y together with a pair of homomorphisms $f' : M \to Y$, $g' : N \to Y$ such that $f'f = g'g$ and the following universal property holds:*

If $f'' : M \to Z$, $g'' : N \to Z$ are homomorphisms such that $f''f = g''g$, then there exists a unique $h : Y \to Z$ such that $hf' = f''$ and $hg' = g''$.

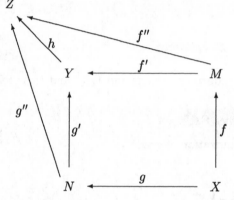

Here the obvious candidate is $Y = M \oplus N / \{(f(x), -g(x)) \mid x \in X\}$, i.e. the cokernel of $X \to M \oplus N$, $x \mapsto (f(x), -g(x))$ with the canonical $f' : M \to Y$, $g' : N \to Y$.

Proposition B.1.3 *In the category of R-modules pull backs and push outs do exist and are unique upto unique isomorphism.*   □

We leave this to the reader.

Lemma B.1.4 *If the square*

$$\begin{array}{ccc} Y & \xrightarrow{f'} & M \\ \downarrow{g'} & & \downarrow{f} \\ N & \xrightarrow{g} & X \end{array}$$

is a pull-back diagram, then g' maps every fibre $f'^{-1}(m)$ of f' bijectively to the corresponding fibre $g^{-1}(f(m))$ of g. Especially

(1) *g' induces* $\ker(f') \cong \ker(g)$;

(2) *If g is an epimorphism, then so is f'.*

Proof. We may assume $Y = \{(m,n) \in M \oplus N \mid f(m) = g(n)\}$ and $f'(m,n) = m$, $g'(m,n) = n$. Then $f'^{-1}(m) = \{(m,n) \mid n \in g^{-1}(f(m))\}$, which clearly is mapped by g' bijectively to $g^{-1}(f(m))\}$. The consequences (1) and (2) are immediate. (Note that surjectivity means: all fibres are nonempty.)   □

Dually we have

Lemma B.1.5 *Let*

$$
\begin{array}{ccc}
X & \xrightarrow{\;f\;} & M \\
{\scriptstyle g}\big\downarrow & & \big\downarrow{\scriptstyle f'} \\
N & \xrightarrow{\;g'\;} & Y
\end{array}
$$

be a push-out sqare. Then

(1) *f' induces an isomorphism* $\operatorname{coker}(f) \cong \operatorname{coker}(g')$;

(2) *f injective $\Longrightarrow$ g' injective.*

This is also easily established if one identifies Y with $M \oplus N/\{(f(x), -g(x)) \mid x \in X\}$.   □

B.1.6 Let R be a ring and M, N be two R-modules. A short exact sequence

$$0 \to M \to E \to N \to 0$$

of R modules is called an **extension** of N by M.

Two extensions

$$0 \to M \to E \to N \to 0$$
$$0 \to M \to E' \to N \to 0$$

are said to be **equivalent** if there is an isomorphism $\varepsilon : E \to E'$ such that the diagram

$$
\begin{array}{ccccccccc}
0 & \longrightarrow & M & \longrightarrow & E & \longrightarrow & N & \longrightarrow & 0 \\
 & & \big\| & & \big\downarrow{\scriptstyle \varepsilon} & & \big\| & & \\
0 & \longrightarrow & M & \longrightarrow & E' & \longrightarrow & N & \longrightarrow & 0
\end{array}
$$

is commutative. (One need only require, ε to be a homomorphism, since then bijectivity automatically holds.) This is an equivalence relation on the class of all extensions. The equivalence classes w.r.t. equivalence make up a set. (The scrupulous reader should convince himself that any E as above can be identified as a **set** with $M \times N$ – equipped with some 'skew' algebraic structure.) We denote the set of the equivalence classes of extensions by $E(N, M)$. Clearly $E(N, M) \neq \emptyset$ since $M \oplus N$, together with the canonical inclusion map i and the canonical projection π, yields an extension

$$0 \to M \xrightarrow{i} M \oplus N \xrightarrow{\pi} N \to 0 \qquad (*)$$

Any extension equivalent to $(*)$ is called a **split extension** of N by M.   □

B.1.7 From homomorphisms $\varphi : N' \to N$, resp. $\psi : M \to M'$ we deduce maps $\varphi^* : E(N, M) \to E(N', M)$, resp. $\psi_* : E(N, M) \to E(N, M')$. (Note the contravariance in the first and the covariance in the second variable!)

We proceed as follows: Let $\alpha \in E(N, M)$ given by $0 \to M \to E \to N \to 0$; then form the pull-back square

$$
\begin{array}{ccc}
E' & \longrightarrow & N' \\
\downarrow & & \downarrow{\scriptstyle \varphi} \\
E & \longrightarrow & N
\end{array}
$$

resp. the push out square

$$
\begin{array}{ccc}
M & \longrightarrow & E \\
{\scriptstyle \psi}\downarrow & & \downarrow \\
M' & \longrightarrow & E''
\end{array}
$$

these fit into the following diagrams with exact rows

$$
\begin{array}{ccccccccc}
0 & \longrightarrow & M & \longrightarrow & E' & \longrightarrow & N' & \longrightarrow & 0 \\
& & \| & & \downarrow & & \downarrow{\scriptstyle \varphi} & & \\
0 & \longrightarrow & M & \longrightarrow & E & \longrightarrow & N & \longrightarrow & 0
\end{array}
$$

respectively

$$
\begin{array}{ccccccccc}
0 & \longrightarrow & M & \longrightarrow & E & \longrightarrow & N & \longrightarrow & 0 \\
& & {\scriptstyle \psi}\downarrow & & \downarrow & & \| & & \\
0 & \longrightarrow & M' & \longrightarrow & E'' & \longrightarrow & N & \longrightarrow & 0
\end{array}
$$

Then the first row in the next to the last diagram represents $\varphi^*(\alpha)$, whereas the second row in the last diagram represents $\psi_*(\alpha)$.

There are several properties which the reader is invited to prove:

a) Functoriality: If $N'' \xrightarrow{\varphi_1} N' \xrightarrow{\varphi_2} N$, respectively $M \xrightarrow{\psi_1} M' \xrightarrow{\psi_2} M''$ are homomorphisms, then $(\varphi_2 \circ \varphi_1)^*(\alpha) \cong \varphi_1^*(\varphi_2^*(\alpha))$, resp. $(\psi_2 \circ \psi_1)_*(\alpha) \cong \psi_{2*}(\psi_{1*}(\alpha))$. Here and in the following we write '$\cong$' for equivalence.

b) $\psi_* \varphi^*(\alpha) \cong \varphi^* \psi_*(\alpha)$.

c) If 0 is the zero endomorphism of N (resp. of M) then $0^*(\alpha)$ (resp. $0_*(\alpha)$) is equivalent to the split extension of N by M.   □

B.1.8 We can now define the sum of two extensions in $E(N, M)$:

Let $0 \to M \to E \to N \to 0$, $\quad 0 \to M \to E' \to N \to 0$ be representatives of two elements α, β in $E(M, N)$. First form the direct sum $\alpha \oplus \beta$, i.e.

$$0 \to M \oplus M \to E \oplus E' \to N \oplus N \to 0.$$

Define

$$\nabla : M \oplus M \to M \text{ by } (m, m') \mapsto m + m' \text{ and } \Delta : N \to N \oplus N \text{ by } n \mapsto (n, n).$$

Then define: $\alpha + \beta := \Delta^*(\nabla_*(\alpha \oplus \beta)) = \nabla_*(\Delta^*(\alpha \oplus \beta))$.

This composition is known as the **Baer sum** .

It is related to the sum of homomorphisms in the following way. Let $\varphi_j : N' \to N$, resp. $\psi_j : M \to M'$ with $j = 1, 2$ be homomorphisms. Then $(\varphi_1 + \varphi_2)^*(\alpha) = \varphi_1^*(\alpha) + \varphi_2^*(\alpha)$ and $(\psi_1 + \psi_2)_*(\alpha) = \psi_{1*}(\alpha) + \psi_{2*}(\alpha)$.

By the Baer sum $E(N, M)$ becomes an abelian group. The class of split extensions is the zero element of this group. The inverse of the class of $\alpha : 0 \to M \xrightarrow{f} E \xrightarrow{g} N \to 0$ is the class of $0 \to M \xrightarrow{-f} E \xrightarrow{g} N \to 0$, which is equivalent to $0 \to M \xrightarrow{f} E \xrightarrow{-g} N \to 0$ and can also be described by $(-1)_*(\alpha)$, resp. $(-1)^*(\alpha)$, where (-1) denotes the homothesy of -1 on M, resp. N.

The reader may prove these statements, using a), b), c) above.   □

Lemma B.1.9 *Let*

$$
\begin{array}{ccccccccc}
0 & \longrightarrow & M & \xrightarrow{f'} & E' & \xrightarrow{g'} & M' & \longrightarrow & 0 \\
& & \| & & \downarrow{\varepsilon} & & \downarrow{\alpha} & & \\
0 & \longrightarrow & M & \xrightarrow{f} & E & \xrightarrow{g} & N & \longrightarrow & 0
\end{array}
$$

be a commutative diagram with exact rows, then the right-hand square is a pull-back diagram.

Proof. Let

$$
\begin{CD}
Z @>\gamma>> M' \\
@V\beta VV @VV\alpha V \\
E @>g>> N
\end{CD}
$$

be a pull-back diagram. By (B.1.4) γ is epimorphic and β induces an isomorphism $\ker(\gamma) \cong M$. Hence we obtain an extension

$$ 0 \to M \overset{\varphi}{\to} Z \overset{\gamma}{\to} M' \to 0 $$

By the universal property of Z there exists a map $\psi : E' \to Z$, such that $\beta\psi = \varepsilon$, $\gamma\psi = g'$. Since ψ induces the identity both in M' and M, it is an isomorphism.   □

B.1.10 Let R be a ring. A short exact sequence $0 \overset{f}{\to} N' \overset{g}{\to} P \to N \to 0$ of R-modules with projective P is called a **projective presentation** of N. Such a presentation induces for an R-module M an exact sequence

$$ 0 \to \operatorname{Hom}(N, M) \overset{g^*}{\to} \operatorname{Hom}(P, M) \overset{f^*}{\to} \operatorname{Hom}(N', M) $$

To the modules M and N and to the chosen projective presentation of N we can associate the abelian group

$$ \operatorname{Ext}^1(N, M) = \operatorname{coker}(f^* : \operatorname{Hom}(P, M) \to \operatorname{Hom}(N', M)). $$

An element in $\operatorname{Ext}^1(N, M)$ may be represented by an element φ in $\operatorname{Hom}(N', M)$. The element represented by $\varphi : N' \to M$ will be denoted by $[\varphi] \in \operatorname{Ext}^1(M, N)$.   □

Theorem B.1.11 *There is a natural isomorphism*

$$ \eta : E(N, M) \overset{\sim}{\to} \operatorname{Ext}^1(N, M) $$

This shows by the way that $\operatorname{Ext}^1(N, M)$ is independent from the chosen projective presentation of N. By the term 'natural' we especially mean 'natural' in M, i.e. a homomorphism $f : M \to M'$ induces a commutative diagram

$$
\begin{CD}
E(N, M) @>\eta>> \operatorname{Ext}^1(N, M) \\
@Vf_* VV @VVf_* V \\
E(N, M') @>\eta>> \operatorname{Ext}^1(N, M')
\end{CD}
$$

The map η is also natural in N; but to show this is a trifle more difficult.

Proof. We first define an isomorphism of sets $\eta : E(N, M) \overset{\sim}{\to} \text{Ext}^1(N, M)$.

Given an element in $E(N, M)$, represented by the extension $0 \to M \overset{h}{\to} E \overset{k}{\to} N \to 0$ we may form a diagram

$$
\begin{array}{ccccccccc}
0 & \longrightarrow & N' & \overset{f}{\longrightarrow} & P & \overset{g}{\longrightarrow} & N & \longrightarrow & 0 \\
& & \downarrow{\psi} & & \downarrow{\varphi} & & \| & & \\
0 & \longrightarrow & M & \overset{h}{\longrightarrow} & E & \overset{k}{\longrightarrow} & N & \longrightarrow & 0
\end{array}
$$

Namely φ exists (but is not unique) since P is projective; and ψ can be regarded as a restriction of φ since $\varphi(\ker g) \subset \ker k$. The homomorphism $\psi : N' \to M$ defines an element $[\psi] \in \text{Ext}^1(N, M) = \text{coker}(f^* : \text{Hom}(P, N) \to \text{Hom}(N', N))$. We claim that this element does not depend on the particular $\varphi : P \to N$ chosen. Thus let $\varphi_i : P \to N$, $i = 1, 2$ be two maps inducing $\psi_i : N' \to M$.Then $\varphi_1 - \varphi_2 = h \circ \tau$ for some $\tau : P \to M$. It follows that $\psi_1 - \psi_2 = \tau \circ f$. Therefore $[\psi_1] = [\psi_2 + \tau \circ f] = [\psi_2]$ since $\tau \circ f$ is mapped to 0 by f^*.

Since two representatives of the same element in $E(N, M)$ obviously induce the same element in $\text{Ext}^1(N, M)$, we have defined a map $\eta : E(N, M) \to \text{Ext}^1(N, M)$.

Conversely given an element in $\text{Ext}^1(N, M)$ we represent this element by a homomorphism $\psi : N' \to M$. Taking the push-out of (ψ, f) we obtain the diagram

$$
\begin{array}{ccccccccc}
0 & \longrightarrow & N' & \overset{f}{\longrightarrow} & P & \overset{g}{\longrightarrow} & N & \longrightarrow & 0 \\
& & \downarrow{\psi} & & \downarrow{\varphi} & & \| & & \\
& & M & \overset{h}{\longrightarrow} & E & \overset{k}{\longrightarrow} & M & &
\end{array}
$$

By Lemma B.1.5, the bottom row $0 \to M \overset{h}{\to} E \overset{k}{\to} N \to 0$ is an extension.

We claim that the equivalence class of this extension is independent of the particular representative $\psi : N' \to M$. Any other has the form $\psi' = \psi + \tau \circ f$ with some $\tau : P \to M$. Then the diagram

$$
\begin{array}{ccccccccc}
0 & \longrightarrow & N' & \overset{f}{\longrightarrow} & P & \overset{g}{\longrightarrow} & N & \longrightarrow & 0 \\
& & \downarrow{\psi'} & & \downarrow{\varphi'} & & \| & & \\
0 & \longrightarrow & M & \overset{h}{\longrightarrow} & E & \overset{k}{\longrightarrow} & N & \longrightarrow & 0
\end{array}
$$

with $\varphi' = \varphi + h \circ \tau$ is commutative. It follows that the extension we arrive at does not depend on the representative. We thus have defined a map

$$\vartheta : \mathrm{Ext}^1(N, M) \to E(N, M)$$

which is easily seen to be natural in M. By Lemma B.1.9 η and ϑ are inverse to each other. Thus we have an isomorphism $\eta : E(N, M) \xrightarrow{\sim} \mathrm{Ext}^1(N, M)$. □

B.1.12 Let (∗) $0 \to A \to B \to C \to 0$ be a short exact sequence of R-modules and N another R-module. By functoriality we get sequences

$$0 \to \mathrm{Hom}(C, N) \to \mathrm{Hom}(B, N) \to \mathrm{Hom}(A, N)$$

and

$$\mathrm{Ext}^1(C, N) \to \mathrm{Ext}^1(B, N) \to \mathrm{Ext}^1(A, N)$$

We know that the first of them is exact. It is left to the reader to show that also the second one is so.

If we want to splice these two exact sequences, there is an obvious way to do it – provided $\mathrm{Ext}^1(C, N)$ is interpreted as $E(C, N)$. Namely to $f : A \to N$ assign $f_*(\alpha)$ where α is given by the extension (∗). This assignment defines a map $\Delta : \mathrm{Hom}(A, N) \to \mathrm{Ext}^1(C, N)$. We leave it to the reader to show that Δ is a homomorphism and that the resulting long sequence

$$\begin{array}{ccccccc}
0 & \longrightarrow & \mathrm{Hom}(C, N) & \longrightarrow & \mathrm{Hom}(B, N) & \longrightarrow & \mathrm{Hom}(A, N) \qquad \text{(B.1)} \\
& & & & & & \downarrow {\scriptstyle \Delta} \\
\cdots & \longleftarrow & \mathrm{Ext}^1(A, N) & \longleftarrow & \mathrm{Ext}^1(B, N) & \longleftarrow & \mathrm{Ext}^1(C, N)
\end{array}$$

is exact.   □

B.1.13 In the above situation assume that $N = A$. Then $\delta(\mathrm{id}_A) = \alpha$ where α is given by the extension (∗).   □

B.2 Derived functors

In this section we go to the heart of homological algebra, by briefly studying the theory of derived functors, in special cases, and which we shall use to introduce the derived functors of $\mathrm{Ext}(M, N)$. This theory historically arose as a generalization of the theory of Ext described in the previous Appendix C. The methods there to define Ext^1 in terms of extensions has also been done by Yoneda for the higher Ext^n, which are equivalent to n-extensions.

The reader will find more details of this, and about our brief sketch below, in ([28], Chapter 7) or ([32], Chapter IV). The classic foundational book in homological algebra is Cartan-Eilenberg's Homological algebra [12].

An **additive contravariant** (covariant) **functor** is a correspondence which associates to each R-module M a R-module $T(M)$, and to each R-linear map $f : M \to M'$, a R-linear map $T(f) : T(M') \to T(M)$ (resp. $T(f) : T(M) \to T(M'))$ satisfying

(1) $T(Id.) = Id.$

(2) $T(g \circ f) = T(f) \circ T(g)$ (resp. $T(f \circ g) = T(f) \circ T(g)$).

(3) $T(f + g) = T(f) + T(g)$.

Example: If N is a R-module then the functor $T(M) = M \otimes_R N$, and $T(f) = f \otimes_R I_N$, for $f : M \to M'$, is a covariant additive functor. The functor $T(M) = \mathrm{Hom}_R(M, N)$ and $T(f) = \mathrm{Hom}(f, I_N)$, for $f : M \to M'$, is an additive contravariant functor.

We now define the n-th **right derived functors** $R^n T$ $(n \geq 0)$ of an additive contravariant functor. The reader should be able to similarly define the left derived functors $L_n(T)$ $(n \geq 0)$ of an additive covariant functor.

Let M be an R-module and $\mathcal{P}. \to M$ be a projective resolution of M. Applying T to the complex

$$\mathcal{P}. : \quad \cdots \to P_n \overset{d_n}{\to} P_{n-1} \overset{d_{n-1}}{\to} \cdots \to P_1 \to P_0$$

we have the complex $T(\mathcal{P})$

$$T(\mathcal{P}). : \quad \cdots \to T(P_0) \overset{T(d_1)}{\to} T(P_1) \overset{d_2}{\to} \cdots \overset{T(d_n)}{\to} P_{n-1} \overset{d_n}{\to} P_n \cdots$$

The n-the **homology** of the complex $T(\mathcal{P})$ is as usual the subquotients at the n-th stage, i.e. $\ker(T(d_{n+1}))/\Im(T(d_n))$. This is defined to be $R^n T(M)$, the n-th right derived functor of T.

To justify it is a functor, we must still define $R^n T(f)$, for any map $f : M \to M'$. Let $\mathcal{P}, \mathcal{P}'$ be projective resolutions of M, M' respectively. Then f can be lifted to a map F of complexes $\mathcal{P} \to \mathcal{P}'$. This induces a map $T(f) : T(\mathcal{P}.) \to T(\mathcal{P}'.)$. Since $T(F)$ is also a mapping of complexes it induces mappings on the homology

$$R^n T(f) : R^n T(M') \to R^n T(M'),$$

for $n \geq 0$. It can be shown that

(1) $R^n T(M)$, $R^n T(f)$ are well defined and independent of the projective resolution. (It is interesting to note that one cannot say this if one had only restricted oneself to free resolutions.)

Moreover, $R^n T$, $n \geq 0$, are additive contravariant functors.

(2) $R^n T(M) = 0$, $n \geq 1$, if M is projective.

(3) $R^0(T) \simeq T$, if R is left exact.

(4) For any exact sequence

$$0 \to M' \xrightarrow{f} M \xrightarrow{f} M'' \to 0$$

of R-modules, there is a long exact sequence

$$\cdots \to R^n T(M'') \xrightarrow{R^n T(g)} R^n T(M) \xrightarrow{R^n T(f)} R^n T(M') \to R^{n+1} T(M'') \to \cdots$$

which satisfies the naturality conditions.

The n-th right derived functor $R^n T$ of the contravariant additive functor $T = \mathrm{Hom}(-, N)$ is denoted by $\mathrm{Ext}^n(-, N)$.

The derived functors of Hom, $\otimes_R$ play a vital role in the homological dimensions (also called projective dimension) of rings and modules; which in turn plays a crucial role in the study of dimension of rings and modules.

With this in mind, it is natural to first how Ext^n can be used to compute (theoretically) the grade of a module w.r.t. and ideal I.

Theorem B.2.1 *(Rees)*
Let A be a Noetherian ring, M a finitely generated A-module and I an ideal. Let $n > 0$ be an integer. Then the following are equivalent:

(1) $\mathrm{Ext}_A^i(N, M) = 0$ for any finitely generated A-module N such that $\mathrm{Supp}(N) \subseteq V(I)$ and for any $i < n$;

(2) $\mathrm{Ext}_A^i(A/I, M) = 0$ for any $i < n$;

(3) there exists an M-regular sequence $f_1, \ldots, f_n$ in I.

Proof. (1) $\Rightarrow$ (2) is trivial. (2) $\Rightarrow$ (3): We have $\mathrm{Ext}_A^0(A/I, M) = \mathrm{Hom}_A(A/I, M) = 0$. If no elements of I are M-regular, then $I \subseteq \bigcup_{\mathfrak{p} \in \mathrm{Ass}(M)} \mathfrak{p}$ so that $I \subseteq \mathfrak{p}$ for some $\mathfrak{p} \in \mathrm{Ass}(M)$. Then there exists an injection $A/\mathfrak{p} \to M$, and by composing it with the natural map $A/I \to A/\mathfrak{p}$ we obtain a non-zero homomorphism $A/I \to M$, which is a contradiction.

Therefore there is an element f_1 of I which is M-regular. If $n > 1$, put $M_1 = M/f_1 M$ and consider the exact sequence $0 \to M \xrightarrow{f_1} M \to M_1 \to 0$. We then get a long exact sequence.

$$\cdots \to \mathrm{Ext}^i(A/I, M) \to \mathrm{Ext}^i(A/I, M_1) \to \mathrm{Ext}^{i+1}(A/I, M) \to \cdots$$

As $\mathrm{Ext}^i(A/I, M) = 0$ for $i < n$ by assumption, we have $\mathrm{Ext}^i(A/I, M_1) = 0$ for $i < n - 1$. By induction on n there exists a M_1-regular sequence $f_2, \ldots, f_n$ in I. Then $f_1, f_2, \ldots, f_n$ is an M-regular sequence in I.

(3) $\Rightarrow$ (1): Put $M_1 = M/f_1 M$. By induction on n we have $\mathrm{Ext}^i(N, M_1) = 0$
for $i < n - 1$. Considering the long exact sequence

$$\cdots \to \mathrm{Ext}^{i-1}(N, M_1) \to \mathrm{Ext}^i(N, M) \xrightarrow{f_1} \mathrm{Ext}^i(N, M) \to \cdots$$

derived from $0 \to M \xrightarrow{f_1} M \to M_1 \to 0$, we see that

$$0 \to \mathrm{Ext}^i(N, M) \xrightarrow{f_1} \mathrm{Ext}^i(N, M)$$

is exact for each $i < n$. Since $\mathrm{Supp}(N) = V(\mathrm{Ann}(N)) \subseteq V(I)$ we have $I \subseteq$
radical of $\mathrm{Ann}(M)$. In particular, $f_1^r \in \mathrm{Ann}(N)$ for some $r > 0$. Therefore f_1^r
annihilates $\mathrm{Ext}^i(N, M)$ also. Thus we have $\mathrm{Ext}^i(N, M) = 0$ for $i < n$. $\square$

Corollary B.2.2 *The length of a maximal M-regular sequence in I depends
on M and I only, and is equal to the number n such that $\mathrm{Ext}_A^i(A/I, M) = 0$,
for $i < n$, $\mathrm{Ext}_A^n(A/I, M) \neq 0$.*

Proof. It remains to prove that, if $f_1, \ldots, f_p$ is a maximal M-regular sequence
in I, then $\mathrm{Ext}_A^p(A/I, M) \neq 0$. By induction on p, this is an easy consequence
of the exact sequence $0 \to M \to M \to M/f_1 M \to 0$. $\square$

C

Complete intersections and Connectedness

In this Appendix we shall prove a famous theorem of Robin Hartshorn. This theorem gives a topological characterisation of a set theoretic complete intersection ideal of dimension > 1 in a local ring in terms of the connectedness of the punctured spectrum $\operatorname{Spec} R \setminus \{\mathfrak{m}\}$.

The notion and existence of a primary decomposition of an ideal in a Noetherian ring, and the notion of a completion $\widehat{R}$ of a local ring R w.r.t. its maximal ideal $\mathfrak{m}$, and the notion of a Cohen-Macaulay ring will be assumed in this section. The reader can refer to the standard texts like [1], [20], [54] for these topics.

Theorem C.0.1 *(Hartshorne)*
Let R be a local ring with maximal ideal $\mathfrak{m}$. Assume that $\operatorname{Spec}(R) \setminus \{\mathfrak{m}\}$ is disconnected in the Zariski topology. Then $\operatorname{depth}(R) \leq 1$.

Proof. Since $\operatorname{Spec}(R) \setminus \{\mathfrak{m}\}$ is disconnected all the minimal prime ideals cannot be in one connected component.

Let $(0) = \cap_{i=1}^{n} \mathfrak{q}_i$ be the primary decomposition of (0) in R, and let $\sqrt{\mathfrak{q}_i} = \mathfrak{p}_i$ be the associated prime ideals of R. If $\mathfrak{m}$ is an associated prime then depth $R = 0$; so we exclude this possibility. Let $\mathfrak{p}_1, \ldots, \mathfrak{p}_t$ be in one connected component of $\operatorname{Spec}(R) \setminus \{\mathfrak{m}\}$, and let $\mathfrak{p}_{t+1}, \ldots, \mathfrak{p}_m$ be in the other. Let $I = \cap_{i=1}^{t} \mathfrak{q}_i$, $J = \cap_{j=t+1}^{n} \mathfrak{q}_j$. Then $IJ = (0)$.

CLAIM: $\operatorname{Spec}(R) = V(I) \cup V(J)$, $V(I) \cap V(J) = \{\mathfrak{m}\}$. In fact, $\sqrt{\mathfrak{q}_i + \mathfrak{q}_j} = \mathfrak{m}$, for $i \leq t$, $j > t$: For if $\operatorname{Spec}(R) \setminus \{\mathfrak{m}\} = V(I_1) \cup V(I_2)$ with $V(I_1) \cap V(I_2) = \emptyset$, and with $V(I_1)$ a connected component, then $\sqrt{I_1 + I_2} = \mathfrak{m}$. Moreover, say $\mathfrak{q}_i \supseteq I_1$, then $\mathfrak{q}_j \supset I_2$, and $\mathfrak{p}_i + \mathfrak{p}_j \supset I_1 + I_2$, whence $\mathfrak{m}$.

Thus, $\mathfrak{m} = \sqrt{I + J}$. By general position argument, choose $a \in I \setminus \cup_{i \leq t} \mathfrak{p}_j$ so that $a + b \notin \mathfrak{p}_k$, for $1 \leq k \leq m_i$ whence, is a non-zero-divisor. Hence, $\operatorname{depth}(R) \geq 1$.

Let $\bar{R} = R/(a+b)$. Then $\bar{a} \neq 0$: If not, $a = x(a+b)$, for some $x \in R$, whence $(1-x)a = xb \in J \Rightarrow x \in \mathfrak{q}_i$ for $1 \leq i \leq t$. Hence $xb \in IJ = 0$, and so $(1-x)a = u \Rightarrow a = 0$, as $1-x$ is a unit in the local ring R.

We show that $\bar{\mathfrak{m}} = \mathfrak{m}/(a+b)$ consists of zero-divisors.

Since $\mathfrak{m} = \sqrt{I+J}$, for every $z \in \mathfrak{m}$ there is least integer $n \geq 1$ such that $z^n = p+q$, $p \in I$, $q \in J$. Now $az^n = ap = p(a+b)$, and so $\bar{a}\bar{z}^n = 0$. Since $\bar{a}\bar{z}^{n-1} \neq 0$, $\bar{z}$ is a zero-divisor in $\bar{R}$. Thus, $a+b$ is a maximal R-sequence, whence depth$(R) = 1$. $\square$

Corollary C.0.2 *(Hartshorne's Connectedness Theorem)*
Let $(R, \mathfrak{m})$ be a Cohen-Macaulay local ring. Let $V = V(I)$ be a set-theoretical complete intersection in $\mathrm{Spec}(R)$ of dimension > 1. Then $V \setminus \{\mathfrak{m}\}$ is a connected space.

Proof. Since V is a set-theoretic complete intersection, R/I is a Cohen-Macaulay local ring of dimension > 1. Hence depth$(R/I) > 1$. By the previous theorem $\mathrm{Spec}(R/I) \setminus \{\bar{\mathfrak{m}}\}$ is connected. $\square$

The natural map $i : R \to \hat{R}$ is a flat map, i.e. if one has an exact sequence $0 \to M' \to M \to M'' \to 0$ of R-modules, then the sequence $0 \to M' \otimes_R \hat{R} \to M'' \otimes_R \hat{R} \to 0$ is also exact. From this one can easily see that if $a \in R$ is a non-zero-divisor then its image $\hat{a} = i(a)$ is a non-zero-divisor in $\hat{R}$. More generally, if I is an ideal of R, then one has $(\widehat{R/I}) \simeq \hat{R}/\hat{I}$. Using this, one can show that if $a_1, \dots, a_r \in R$ is a regular sequence, then $\hat{a}_1, \dots, \hat{a}_r$ is a regular sequence in $\hat{R}$.

The map $i : R \to \hat{R}$ is, in fact faithfully flat, i.e. if $0 \to M' \to M \to M'' \to 0$ is a sequence of finitely generated R-modules, with the induced sequence of modules (and maps) $0 \to M' \otimes_R \hat{R} \to M \otimes_R \hat{R} \to M'' \otimes_r \hat{R} \to 0$ exact, then the first sequence is also exact.

Using this, one can show that if R/I is a Cohen-Macaulay local ring then $\hat{R}/\hat{I} \simeq \hat{R}/\hat{I}$ is also a Cohen-Macaulay ring.

The proof of Corollary C.0.2 shows us that:

Corollary C.0.3 *(Formal connectedness)*
If $\hat{V} = \mathrm{Spec}(\hat{R}/\hat{I})$ then $\hat{V} \setminus \{\hat{\mathfrak{m}}\}$ is also connected. (We say $V \setminus \{\mathfrak{m}\}$ is formally connected.)

Proof. $(\hat{R}/I) \simeq \hat{R}/\hat{I}$ is Cohen-Macaulay of dimension $R/I = $ depth$(R/I) > 1$. Apply Hartshorne's Connectendess theorem to deduce the result. $\square$

D

Odds and Ends

Here is a 'local-global' property which can be proved using direct sums:

Proposition D.0.1 *Let U, V be submodules of an A-module M.*

a) For every multiplicative set $S \subset A$ we have $(S^{-1}U) \cap (S^{-1}V) = S^{-1}(U \cap V)$ and $S^{-1}U + S^{-1}V = S^{-1}(U + V)$.

b) $U = V \iff U_\mathfrak{m} = V_\mathfrak{m}$ for all $\mathfrak{m} \in \mathrm{Spmax}(A)$.

Proof. a) We have a commutative diagram with exact rows

$$
\begin{array}{ccccccc}
0 & \longrightarrow & S^{-1}(U \cap V) & \longrightarrow & S^{-1}M & \longrightarrow & S^{-1}(M/U \oplus M/V) \\
& & \downarrow & & \| & & \| \\
0 & \longrightarrow & (S^{-1}U) \cap (S^{-1}V) & \longrightarrow & S^{-1}M & \longrightarrow & S^{-1}(M/U) \oplus S^{-1}(M/V).
\end{array}
$$

$\square$

Chapter 2

Let A be a principal ring, F a finitely generated free A-module and $U \subset F$ a submodule. Remember that F possesses a basis $x_1, \ldots, x_n$ such that $U = Aa_1x_1 + \ldots + Aa_nx_n$ with $a_i \in A$ and $a_i | a_{i+1}$ for $i = 1, \ldots, n-1$. Clearly U is free itself with basis $a_1x_1, \ldots, a_rx_r$ where r is maximal with $a_r \neq 0$.

Let A, F, U be as above and assume $a_1 \neq 0$. Further let $\varphi : F \to A$ be an epimorphism with $\varphi(U) = (a_1)$. Then every section (i.e. right inverse)

$\sigma' : (a_1) \to U$ of $\varphi' := \varphi|_U : U \to (a_1)$ extends uniquely to a section $\sigma : A \to F$ of φ. If B is any domain, $b \in B \setminus (0)$ and $x \in bB^n$. Then there is exactly one $y \in B^n$ with $by = x$. (If there were another one, say y', we would have $b(y - y') = 0$, which clearly cannot happen.) We denote this y by $b^{-1}x$.

Since $\sigma'(a_1) \in U \subset a_1F$, we may define σ by $\sigma(1) := a_1^{-1}\sigma'(a_1)$. And there is no other possibility to define a σ which extends σ', since $\sigma'(a_1) = \sigma(a_1) = a_1\sigma(1)$. Now we have to show $\varphi \circ \sigma = \mathrm{id}_A$. But $a_1 \varphi \circ \sigma(1) = \varphi \circ \sigma(a_1) = \varphi' \circ \sigma'(a_1) = a_1$, hence $\varphi \circ \sigma(1) = 1$. $\square$

Chapter 4

Lemma D.0.2 *Let A be a commutative ring and let $s, t \in A$ such that $As + At = A$. Let M, M' be two A-modules such that $\alpha : M_s \xrightarrow{\sim} M'_s$ and $\beta : M_t \xrightarrow{\sim} M'_t$ with $\alpha_t = \beta_s$. Then $M \cong M'$.*

Proof. Take any $m \in M$. Then $\alpha_t(m/1) = \beta_s(m/1)$. This gives $\alpha(m/1)_t = \beta(m/1)_s$. If $\alpha(m/1) = x \in M'_s$ and $\beta(m/1) = y \in M'_t$, then $x_t = y_s$. Since s and t are comaximal by the usual componendo and dividendo trick we can find an element $m' \in M'$ such that $m'_s = x$ and $m'_t = y$. Since $sA + tA = A$ the element m' is uniquely determined. Define $\vartheta : M \to M'$ by $\vartheta(m) = m'$. Then ϑ is A-linear and since $\vartheta_s = \alpha$, $\vartheta_t = \beta$, ϑ is locally an isomorphism. By (1.3.16) ϑ is an isomorphism. $\square$

Remark D.0.3 In view of Corollary 4.4.8, any unimodular vector $v \in \mathrm{Um}_r(k[x_1, \cdots, x_n])$, $r \geq 3$, can be completed to an elementary matrix $\varepsilon \in \mathrm{E}_r(k[x_1, \cdots, x_n])$.

Consequently, if $\sigma \in \mathrm{SL}_r(k[x_1, \cdots, x_n])$, then using above and the normality of $\mathrm{E}_r(k[x_1, \ldots, x_n])$ one has $\varepsilon \in \mathrm{E}_r(k[x_1, \ldots, x_n])$ such that $\sigma\varepsilon = (I_{r-2} \perp \delta)$ for some $\delta \in \mathrm{SL}_2(k[x_1, \ldots, x_n])$. Thus, $\mathrm{SL}_r(k[x_1, \cdots, x_n]) = \mathrm{SL}_2(k[x_1, \cdots, x_n])\mathrm{E}_r(k[x_1, \cdots, x_n])$ for $r \geq 3$, for all n.

In [93] A. Suslin established that $\mathrm{SL}_r(k[x_1, \cdots, x_n]) = \mathrm{E}_r(k[x_1, \ldots, x_n])$ for $r \geq 3$, for all n. This has been posed as Exercise 15 to Appendix A. We have indicated a proof of this due to M. P. Murthy with Exercises 12-14 acting as a hint. A. Suslin's proves it by establishing a Monic Inversion Principle for the Elementary group $\mathrm{E}_r(R[X])$, for $r \geq 3$, for any commutative ring R. $\square$

We give a quick bibliographic survey of the known applications of the L-G Principle in the literature, and some other directions which we feel are related.

$\star$ Extending to linear and classical groups (A.Suslin, et. al. [93], [Su][10], [Ko][2]).

⋆ Rigidity interpretation for Principle G-bundles (M.S. Raghunathan [Ra][1]).

⋆ Constructive approach to Quillen-Suslin theorems (Bose [B][][1], Laubenbacher [La][1], Fitchas [F][1], Sturmfels [St][1], etc.).

⋆ Projective generation of ideals in algebras, commutative rings (M. Boratýnski [Bt][1-10], N. Mohan Kumar [MK][2, 4, 9], S.M. Bhatwadekar, Raja Sridharan [Bh][7-10]).

⋆ Locally polynomial algebras (Bass-Connel-Wright [B][2], A. Suslin [Su][2]).

⋆ Normality of elementary subgroups, sandwich theorem for normal subgroups (L.N. Vaserstein [Va][6]).

⋆ Solvability of the quotient by the elementary subgroup (A. Bak [Bk][1]), (cf. Appendix A, Exercise 4).

⋆ Stabilization and Prestabilization for linear groups (van der Kallen [K][1-2], R.A. Rao [RR][8]).

⋆ Determining whether an element in a polynomial ring is a "variable" (A.Sathaye [Sa][2,3], T.Asanuma [A][1-3]), S.M. Bhatwadekar [Bh][13-15, 21].

⋆ Efficient generation upto radicals of ideals in polynomial rings (M. Boratýnski [Bt][1], N. Mohan Kumar [MK][7], S. Mandal [Ma][9]), (cf. Chapter 10, Exercise 16-19).

⋆ Lifting generators of the conormal module (S. Mandal [Ma,2, 3], S.M. Bhatwadekar & M. Keshari [Bt][10]).

⋆ Euler classes and complete intersections (S. Mandal [Ma][1, 2], S.M. Bhatwadekar [Bt][4], Raja Sridharan [RS][5, 7], M. K. Das [D][2]).

⋆ Orbits of linear and elementary subgroups (L. N. Vaserstein [Va][6], N. Mohan Kumar [MK 1], M. Roitman [Rt][1, 2], J. Gubeladze [G][1, 3, 4], R.A. Rao [RR][2]).

⋆ Cancellation technique for projective modules, monoid rings (A. Suslin [Su][6-9], R. A. Rao [72], [RR][1], H. Lindel [Li][2], J. Gubeladze [G][6], R.G. Swan [Sw][4], S.M. Bhatwadekar [Bh][16, 17, 19]).

⋆ Bass-Quillen-Suslin conjectures (M. P. Murthy [MP][11], H. Lindel [Li][1,3,4], N. Mohan Kumar [MK][6, 10], A. Suslin [Su][6, 7]. M. Roitman [Rt][2], S.M. Bhatwadekar [Bh][16-20], D. Popescu [P][2], R.G. Swan [Sw][2, 5, 6], R. A. Rao [RR][3, 4, 11]).

⋆ Defining new groups related to unimodular vectors, and establishing group structures on orbits of unimodular vectors (R.A. Rao & Selby Jose, in preparation).

⋆ Establishing Quillen-Suslin theory for Euler classes, other groups (M. K. Das [D][1], Selby Jose & R. A. Rao, preprint).

Chapter 5

Lemma D.0.4 *Let ξ be a soft vector bundle over X and $x \in X$.*

a) For any $a \in F_x$ there is a global section s of ξ with $s(x) = a$.

b) There are an open neighbourhood U of x and global sections $s_1, \ldots, s_n$ such that $s_1(y), \ldots, s_n(y)$ is a basis of $F_y(\xi)$ for every $y \in U$. (i.e. $s_1|_U, \ldots, s_n|_U$ is a basis over U.)

Proof. a) Let $U_1, \ldots, U_m$ be a finite open covering of X, which trivializes ξ and $e_1, \ldots, e_m$ a subordinate envelope of unity. There is an j with $e_j(x) = 1$. Over U_j there exist a section s' with $s'(x) = a$. Then define a global section $s : X \to E(\xi)$ by

$$s(y) := \begin{cases} e_j(y)s'(y) & \text{for } y \in U_j \\ 0 & \text{for } y \notin U_j \end{cases} \text{ .}$$

Since $\text{Supp}(e_j) \subset U_j$, this is continuous. Clearly $s(x) = s'(x) = a$.

b) Let U_i, U_j, e_i, e_j be as above, and let $s'_1, \ldots, s'_n$ be a base over U_j. (i.e. $s'_i(y), \ldots, s'_n(y)$ is a base of $F_y(\xi)$ for every $y \in U_j$. Remember that $\xi|_{U_j}$ is trivial.) Then define global sections s_i by $s_i(y) := e_j(y)s'_i(y)$ for $y \in U_j$ and $:= 0$ elsewhere. Then $s_1, \ldots, s_n$ is a base over $U := \{y \in U_j \mid e_j(y) \neq 0\}$.

Corollary D.0.5 *Let $f, g : \xi \to \eta$ be homomorphisms of soft vector bundles over X with $\Gamma(f) = \Gamma(g)$. Then $f = g$.*

Proof. Let $a \in E(\xi)$, say $a \in F_x(\xi)$. By Lemma D.0.4 there is a global section s with $s(x) = a$. Since by assumption $\Gamma(f)(s) = \Gamma(g)(s)$ we see $f(a) = \Gamma(f)(s)(x) = \Gamma(g)(s)(x) = g(a)$. $\square$

Lemma D.0.6 *Let ξ be a soft vector bundle over X and $x \in X$, $s \in \Gamma(\xi)$ with $s(x) = 0$. Then there are $s_0, \ldots, s_n \in \Gamma(\xi)$, $a_0, \ldots, a_n \in \mathcal{C}(X)$ with $a_i(x) = 0$ and $s = \sum_{i=0}^{n} a_i s_i$.*

Proof. Let $\sqcap = (U_j)_{1 \leq j \leq m}$ be a finite open covering of X and $(e_j)_{1 \leq j \leq n}$ be a subordinate envelope of unity such that $\xi|_{U_j}$ is trivial for all j. Choose j_0 such that $e_{j_0}(x) = 1$. Define continuous monotone maps $h, g : [0,1] \to [0,1]$ by

$$h(t) := \text{Min}\{1,\, 2t\} \quad \text{and} \quad g(t) := \text{Min}\{1, 2 - 2t\} \ (= h(1 - t) \ )$$

Let $s_1, \ldots, s_n \in \Gamma(\xi)$ be a basis of ξ over U_{j_0}. There are $b_i \in \mathcal{C}(U_{j_0})$ with $s(y) = \sum_{i=1}^{n} b_i(y)s_i(y)$ for all $y \in U_{j_0}$. Then $b_i(x) = 0$ for $i = 1, \ldots, n$. Set

$$a_i(y) := \begin{cases} (h \circ e_{j_0})(y) \cdot b_i(y) & \text{for } y \in U_{j_0} \\ 0 & \text{for } y \notin U_{j_0}. \end{cases}$$

The a_i are continuous, since $\text{Supp}(e_{j_0}) \subset U_{j_0}$. Set $s_0 := s - \sum_{i=0}^{n} a_i s_i$. Then $s_0(y) = 0$ for all $y \in X$ where $e_{j_0}(y) \geq 1/2$. Therefore $s_0 = (g \circ e_{j_0}) \cdot s_0$, since $g \circ e_{j_0}(y) = 1$ for all $y \in X$ where $e_{j_0}(y) \leq 1/2$. On the other hand $g \circ e_j(x) = 0$. So finally $s = \sum_{i=0}^{m} a_i s_i$ with $a_i(x) = 0$, if one defines $a_0 := g \circ e_{j_0}$. $\square$

Corollary D.0.7 *For $x \in X$ let $I_x := \{a \in \mathcal{C}(X) \mid a(x) = 0\}$. (It is a two-sided ideal.) Then $s \mapsto s(x)$ induces an isomorphism $\Gamma(\xi)/I_x\Gamma(\xi) \to F_x(\xi)$.* $\square$

Theorem D.0.8 *Let ξ, η be soft vector bundles over X. To every $\mathcal{C}(X)$-linear map $F : \Gamma(\xi) \to \Gamma(\eta)$ there exists a unique bundle homomorphism $f : \xi \to \eta$ with $\Gamma(f) = F$.*

Proof. The uniqueness has been proved in Corollary D.0.5. The homomorphism $F : \Gamma(\xi) \to \Gamma(\eta)$ induces for every $x \in X$ a map f_x of fibres

$$F_x(\xi) = \Gamma(\xi)/I_x\Gamma(\xi) \to \Gamma(\eta)/I_x\Gamma(\eta) = F_x(\eta).$$

These fit together to a map of the total spaces $f : E(\xi) \to E(\eta)$ over X, whose continuity will be proven later. Note first that $\Gamma(f)$ would be equal to F. Namely by the definition of f_x we have for $s \in \Gamma(\xi)$:

$$f(s(x)) = f_x(s(x)) = (F(s))(x).$$

Now to the continuity of f. Let $U \subset X$ be open such that $\xi|_U$ is trivial and $s_1, \ldots, s_n$ be a local basis over U. Any $e \in p_\xi^{-1}U$ can be written in the form $e = \sum a_i(e)s_i(p_\xi(e))$, where a_i and $s_i \circ p_\xi$ are continuously dependent of e. Now $f(s_i(p(e))) = F(s_i)(p(e))$ by the definition of f and $f(e) = \sum a_i(e)f(s_i(p(e)))$. So f is continuous. $\square$

D.0.9 Now we will show that to every finitely generated projective $\mathcal{C}(X)$-module P there is a soft vector bundle ξ over X such that $\Gamma(\xi) \cong P$.

P is a direct summand of some $\mathcal{C}(X)^n$, so it is the image of a (not necessarily unique) idempotent endomorphism α of $\mathcal{C}(X)^n$. $\square$

Lemma D.0.10 *The map $\alpha \mapsto \sqrt{\alpha}$ is continuous.*

Proof. Let (β_k) be a sequence of semi-positive hermitian matrices with $\lim_{k \to \infty} \beta_n = \alpha$. Then the sequence $(\sqrt{\beta_k}$ is bounded since $|\sqrt{\beta_k}| = \sqrt{|\beta_k|}$. Let γ be an accumulation point of the sequence $(\sqrt{\beta_?})$ and $(\sqrt{\beta_{k_j}}$ be a subsequence converging to γ. Then γ is also semi-positive hermitian and $\gamma^2 = \alpha$. Therefore γ is the only accumulation point of the bounded sequence $(\sqrt{\beta_k})$, whence $\lim_{k \to \infty} \sqrt{\beta_k} = \sqrt{\alpha}$. $\square$

Therefore, if an endomorphism α of $C(X)^n$ is semi-positive hermitian, i.e. $\alpha(x)$ is so for every $x \in X$, then there exists a semi-positive hermitian endomorphism γ of $C(X)^n$ with $\gamma^2 = \alpha$.

Proposition D.0.11 *Every finitely generated projective module P over $C(X)$ is isomorphic to the image of a hermitian idempotent endomorphism ε of $C(X)^n$ for some n. (That ε is hermitian idempotent means $\varepsilon = \varepsilon^2 = \varepsilon^*$.)*

From $\alpha^2 = \alpha$ we derive $(\alpha^*)^2 = \alpha^*$ and

$$(I + (I - 2\alpha)(I - 2\alpha^*))\alpha^* = 2\alpha\alpha^*\alpha(I + (I - 2\alpha)(I - 2\alpha^*).$$

Since $(1 - 2\alpha)(1 - 2\alpha^*)$ is positive hermitian, there is an γ with $\gamma^2 = I + (I - 2\alpha)(I - 2\alpha)^*)$. As γ^2 is invertible, also γ is so.

We set $\varepsilon := \gamma^{-1}\alpha\gamma$. Then $\varepsilon^2 = \gamma^{-1}\alpha^2\gamma = \gamma^{-1}\alpha\gamma$ And from $\gamma^2\alpha* = \alpha\gamma^2$ we infer $\varepsilon^* = \gamma\alpha^*\gamma^{-1} = \gamma^{-1}\alpha\gamma = \varepsilon$.

If $s = 1$, the A-module A/I has the composition series

$$0 \subset (X_1^{k_1-1}, X_2, \ldots, X_r)/I \subset (X^{k_1-2}, X_2, \ldots, X_r) \subset \cdots \subset \mathfrak{p}/I \subset A/I$$

The factors of this composition series all are isomorphic to $R/\mathfrak{p}$. By Proposition 8.4.7 b) the statement is true in this case.

Now let the statement be true for some $s < r$ and prove it for the case s. The A-module A/I has the compositon series $0 \subset (X_1^{k_1}, \ldots, X_s^{k_s}, X_{s+1}^{k_{s+1}-1}, X_{s+2}, \ldots, X_r)/I \subset \cdots \subset (X_1^{k_1}, \ldots, X_s^{k_s}, X_{s+1}, \ldots, X_r)/I \subset A/I$. All factors are isomorphic to $E := A/(x_1^{k_1}, \ldots, X_s^{k_s}, X_{s+1}, \ldots, X_r)$. By induction assumption $\mathrm{Ass}(E) = \mathfrak{p}$. Again use Proposition 8.4.7. $\qquad\square$

D.0.12 Construction of Vector Bundles

We shall assume that the reader has some familiarity with the notion of a smooth manifold. We show how vector bundles arise in practice. The material here is standard, but we liked its presentation in [65], which we follow in this appendix.

Smooth vector bundles formalize the notion of "smooth family of vector spaces". For example given a smooth manifold M and a vector space F we can think of the Cartesian product $F_M := F \times M$ as a smooth family $(F_x)_{x \in M}$ of vector spaces. This trivial example is not surprisingly called the trivial vector bundle with fibre F and base M.

We can obtain more interesting examples by gluing these simple ones using *gluing data*. These consist of
(i) an open cover (U_α) of a smooth manifold M.

B. a *gluing cocycle* i.e. a collection of smooth maps $g_{\beta\alpha} : U_{\alpha\beta} \to \text{Aut } (F)$, (where $U_{\alpha\beta} = U_\alpha \cap U_\beta$), such that

$$g_{\alpha\alpha}(x) \equiv 1_F, \ g_{\gamma\alpha} \equiv g_{\gamma\beta}(x) \cdot g_{\beta\alpha}(x) \ \forall \ x \in U_{\alpha\beta\gamma} := U_\alpha \cap U_\beta \cap Y_\gamma \neq \emptyset.$$

The open cover U_α is also know as a *trivializing cover*. We will also say it is the *support* of the $g_{\beta\alpha}$.

The map $g_{\beta\alpha}$ describes the "transition from $F_\alpha := F_{U_\alpha}$ to $F_{U_\beta} = F_\beta$" in the sense that for every $x \in U_{\alpha\beta}$ the element $(v, x) \in F_\alpha$ is identified with the element $(g_{\beta\alpha}(x)v, x) \in F_\beta$. Pasting together the trivial bundles F_α following the instructions given by the gluing cocycle we obtain a smooth manifold E (called the *total space*), a smooth map $\pi : E \to M$ (called the *canonical projection*) and diffeomorphisms $\psi_\alpha : \pi^{-1}(U_\alpha) \to F_\alpha$, (called local *trivializations*) such that for all $x \in U_{\alpha\beta}, v \in V$

$$\psi_\beta \circ \psi_\alpha^{-1}(v, x) = (g_{\beta,\alpha}(x)v, x).$$

$E \xrightarrow{\pi} M$ as above is called a *vector bundle* over M. The rank of E is by definition the dimension of the standard fibre F (over its field of scalars). Rank one bundles are of course known as *line bundles*.

Example 1. Consider the projective space $\mathbb{CP}^n$ defined as the set of one-dimensional complex subspaces of $\mathbb{C}^{n+1}$. There is natural projection

$$\pi : \mathbb{C}^{n+1} \backslash \{0\} \to \mathbb{CP}^n$$

where $\pi(x) :=$ *one-dimensional subspaces spanned by* x. The fibres

$$\pi^{-1}(p) \cdot p \in \mathbb{CP}^n,$$

are vector subspaces of $\mathbb{C}^{n+1}$. The family $\pi^{-1}(p)$ is indeed a smooth family of vector spaces in the sense described above. It is called the *tautological (or universal) line bundle* over the projective space and is denoted by $\mathbb{U}_n$.

Suppose that $X \xrightarrow{f} Y$ is a smooth map and $E \to Y$ is a smooth vector bundle given by a gluing cocyle $g_{\beta\alpha}$ supported by an open cover (U_α) of Y. Then f induces a vector bundle on X called the **pull-back** of E by f and denoted by f^*E. It is given by the open cover $(V_\alpha = f^{-1}(U_\alpha))$ and gluing cocycle $h_{\beta\alpha} = g_{\beta\alpha} 0 f$.

The following (exercise) describes a very general procedure of constructing smooth vector bundles.

Consider a smooth map P from a compact connected smooth manifold X to the space End (V) of endomorphisms of a vector space V such that $P^2(x) = P(x) \ \forall \ x \in X$, i.e. $P(x)$ is a smooth family of projectors of V.

(a) Show that $\dim \ker P(x)$ is independent of $x \in X$. Denote by k this common dimension.

(b) Show that the assignment $x \mapsto \ker P(x)$ defines a rank k smooth vector bundle over X.

(c) Provide a projective description of the tautological line bundle over $\mathbb{CP}^n$.

(d) Show that any map $X \to V^* \backslash \{0\}$ defines in a canonical way a vector bundle over X of rank $\dim V - 1$.

Denote by $G_k(V)$ the Grassmannian of k-dimensional subspaces of an n-dimensional vector space V. Assume V is equipped with an inner product. For each k-dimensional subspace $U \subset V$ denote by P_U the orthogonal projection onto $U^\perp$. The smooth family $G_k(V) \ni U \mapsto P_U$ defines according to the previous construction a rank k vector bundle over $G_k(V)$ called the *universal vector bundle* and denoted by $\mathbb{U}_{k,n}$. When $k = 1$ this is precisely the tautological line bundle over $\mathbb{RP}^{n-1}$ or $\mathbb{CP}^{n-1}$.

A smooth map s from a smooth manifold X to a vector space F is a smooth selection of an element $s(x)$ in each fibre $F \times x$ of F_X. In other words, it is smooth map $s : X \to F_X$ such that $\pi \circ s = \mathbb{1}_X$ w here $\pi : F_X \to X$ is the natural projection. Replacing F_X with any smooth vector bundle $E \xrightarrow{\pi} X$ we get the notion of smooth section of E. The space of smooth sections of E will be denoted by $\Gamma(E)$ or $C^\infty(E)$. In terms of gluing cocycles we can describe a section as a collection of smooth maps

$$s_\alpha : U_\alpha \to F$$

such that

$$s_\beta(x) = g_{\beta\alpha}(x) s_\alpha(x), \quad \forall \, x \in U_\alpha \cap U_\beta$$

The functorial operations in linear algebra have a vector bundle counterpart. Suppose $E_i \xrightarrow{\pi_i} X (i = 1, 2)$ are two vector bundles over X with standard fibres F_i, $i = 1, 2$, given by gluing cocycles $G_{\beta\alpha}(i)$ along the same support. For example the direct sum $F_1 \oplus F_2$ corresponds to the *direct (Whitney) sum* $E_1 \oplus E_2$ given by the gluing cocycle $G_{\beta\alpha}(1) \oplus G_{\beta\alpha}(2)$.

The dual vector bundle E_1^* is defined by the gluing cocycle $(G_{\beta\alpha}^*(1))^{-1}$ where "$*$" denotes the conjugate transpose.

We can form tensor products, symmetrix, exterior products of vector bundles, etc. In particular, the bundle $E_1^* \otimes E_2$ will be denoted by Hom (E_1, E_2). Its sections are *bundle morphisms*, i.e. smooth maps $T : E_1 \to E_2$ mapping the fibre $E_1(x)$ of E_1 linearly to the fibre $E_2(x)$ of E_2. When $E_1 = E_2 = E$ we use the notation End(E). If the induced morphisms $T(x)$ are all isomorphisms then T is called a bundle isomorphisms. A bundle automorphism of a vector bundle E is also called a **gauge transformation**. The group of bundle automorphisms of E is denoted by $\mathrm{G}(E)$ and is known as the **gauge group of** E.

The line bundle $\wedge^{\mathrm{rank}(E_1)} E_1$ is called the **determinant line bundle of** E_1 and is denoted by $\det E_1$.

In the exercises we will see how to use sections to prove that any complex line bundle over a compact manifold is the pullback of the universal line bundle over a complex projective space.

For every smooth manifold M denote by $\mathrm{Pic}^\infty(M)$ the space of isomorphism classes of smooth complex line bundles over M and by $[M, \mathbb{CP}^n]_\infty$ the set of (smooth) homotopy classes of smooth maps $M \to \mathbb{C}^n$. This is an inductive family, $[M, \mathbb{CP}^1]_\infty \hookrightarrow [M, \mathbb{CP}^2]_\infty \hookrightarrow \ldots$ and we denote by $[M, \mathbb{C}^\infty]_\infty$ its inductive limit. If M is compact we have a bijection. $\mathrm{Pic}^\infty(M) \cong [M, \mathbb{CP}^\infty]_\infty$. The tensor product of line bundles induces a structure of Abelian group on $\mathrm{Pic}^\infty(M)$ This group has a cohomological interpretation (for the more advanced reader) viz.

Since the inductive limit $\mathbb{CP}^\infty$ of the $\mathbb{CP}^n$'s is a $K(\mathbb{Z}, 2)$-space we can conclude that we have isomorphism of groups

$$c_1^{\mathrm{top}} : \mathrm{Pic}^\infty(M) \to H^2(M, \mathbb{Z}).$$

For any $L \in \mathrm{Pic}^\infty(M)$ the element $c_1^{\mathrm{top}}(L)$ is called the **topological first Chern class of** L.

One is often led to study families of vector spaces satisfying additional properties such as vector spaces in which vectors have lengths and pairs of vectors have definite angles (as in Euclidean geometry). According to Felix Klein's philosophy, this is the same as looking at the *symmetry group*, i.e. the subgroup of linear maps which preserve these additional features. In the above case this is precisely the orthogonal group. If we want to deal with families of such spaces then we must impose restrictions on the gluing maps; they must be valued in the given symmetry group. Here is one way to formalize this discussion. Suppose we are given the following data.

- A Lie group G and a representation

$$\rho : G \to \mathrm{End}\ (F).$$

- A smooth manifold X and open cover U_α.
- A *G-valued gluing cocycle*, i.e. a collection of smooth maps $g_{\beta\alpha} : U_{\alpha\beta} \to G$, such that $g_{\alpha\alpha}(x) = 1 \in G \ \forall \ x \in U_\alpha$ and

$$g_{\gamma\alpha}(x) = g_{\gamma\beta}(x) \cdot g_{\beta\alpha}(x) \ \forall \ x \in U_{\alpha\beta\gamma}.$$

Then the collection $\rho(g_{\beta\alpha}) : U_{\alpha\beta} \to \mathrm{End}\ (F)$ defines a gluing cocycle for a vector bundle E with standard fibre F and symmetry group G. The vector bundle E is said to have a G-structure.

Differential geometers usually phrase the above construction in terms of **principal G-bundles**. Given a gluing G-cocycle as above we can obtain a smooth manifold P as follows. Glue the product $G \times U_\alpha$ to $G \times U_\beta$ along $U_{\alpha\beta}$ using the following prescription: for each $x \in U_{\alpha\beta}$ the element (g, x) in $G \times U_\alpha$ is identified with the element $(g_{\alpha\beta}(x) \cdot g \cdot x)$ in $G \times U_\beta$. We obtain a smooth manifold P and a smooth map $\pi : P \to X$ whose fibres $\pi^{-1}(x)$ are diffeomorphic to the Lie group G. This is called the principal G-bundle determined by the gluing G-cocycle $g_{\beta\alpha}$. The above vector bundle E is said to be *induced from* P via the representation ρ and we write this as $P \times_\rho F$. □

Chapter 6

Theorem D.0.13 (Finiteness of minimal prime over-ideals *(cf. Corollary 6.2.6)*

Let I be an ideal of a Noetherian ring R, then $\sqrt{I} = \bigcap_{i=1}^n \mathfrak{p}_i$, for some prime ideals $\mathfrak{p}_i$ of R, $1 \leq i \leq n$. Consequently, I has only a finite number of minimal prime over-ideals $\mathfrak{p}_1, \dots, \mathfrak{p}_n$.

Proof. Let S be the set of all radical ideals which are **not** a finite intersection of prime ideals. Suppose $S \neq \emptyset$. If $I_0 \in S$ is a maximal element, then I_0 is not a prime ideal and so there are $a, b \in R$, with $a \notin I_0$, $b \notin I_0$, $ab \in I_0$. It is easy to verify that $I_0 = \sqrt{(I_0, a)} \cap \sqrt{(I_0, b)}$. But then, by the maximality of I_0 in S we get $\sqrt{(I_0, a)} \notin S$ and $\sqrt{(I_0, b)} \notin S$, and so $\sqrt{(I_0, a)}$, $\sqrt{(I_0, b)}$ are finite intersections of prime ideals. Hence I_0 too is such an intersection, which contradicts that $I_0 \in S$. □

Corollary D.0.14 *Let I be an ideal of a Noetherian ring R and $\mathfrak{p}$ be a prime ideal containing I then $\mathfrak{p}$ contains a minimal prime over-ideal of I.*

Proof. Since $\mathfrak{p}$ is prime $\mathfrak{p} \supset \sqrt{I} = \bigcap_{i=1}^r \mathfrak{p}_i$. If no $\mathfrak{p}_i \not\subset \mathfrak{p}$ and $x_i \in \mathfrak{p}_i \setminus \mathfrak{p}$, $1 \leq i \leq r$, then $x = x_1 x_2 \cdots x_r \in \bigcap_{i=1}^r \mathfrak{p}_i \setminus \mathfrak{p}$, a contradiction. Hence $\mathfrak{p}_{i_0} \subset \mathfrak{p}$ for some i_0. □

Chapter 7

Proposition D.0.15 *Let P be a finitely generated projective A-module of rank $\geq r$ and $s \in A$ such that P_s is free. For any $p \in P$ there is a $q \in P$ such that $\operatorname{ht}(A_s \mathcal{O}_P(p + sq)) \geq r$. In particular, if the residue class $\bar{p}$ of p is unimodular in the A/s-module P/sP, then $\operatorname{ht}(\mathcal{O}_P(p + sq) \geq r$.*

Proof. Let B be a ring and $b = (b_1, \ldots, b_r) \in B^r$. Then the order ideal of b is generated by $b_1, \ldots, b_r$.

Let $x_1, \ldots, x_r \in P$ such that $(x_1)_s, \ldots, (x_r)_s$ form a base of P_s. Then there is an $n \in \mathbb{N}$ with $s^n P \in Ax_1 + \cdots + Ax_r$. Let $s^n P = a_1 x_1 + \cdots + a_r x_r$, $a_i \in A$. By the usual prime avoiding procedure we find $c_1, \ldots, c_r \in A$, such that the ideal $I := A_s(a_1 + s^{n+1}c_1) + \cdots + A_s(a_r + s^{n+1}c_r)$ of A_s has height $\geq r$. Set $q := \sum c_i x_i$. Then $A_s \mathcal{O}_P(p + sq) = A_s \mathcal{O}_P(s^n p + s^{n+1}q) = I$.

To show the 2nd statement, we use that P is projective. Since $\bar{p}$ and hence $\overline{p + sq}$ is unimodular, there is a homomorphism $P/sP \to A/sA$ with $\overline{p + sq} \mapsto \bar{1}$. By projectivity this lifts to homomorphism $P \to A$ with $p + sq \mapsto 1 + as$ for some $a \in A$. So no prime ideal containing s contains $\mathcal{O}_P(p + sq)$, whence ht $A_s \mathcal{O}_P(p + sq) = $ ht $\mathcal{O}_P(p + sq)$. □

A corollary of this is a weak form of Serre's Splitting Theorem:

If A, P are as above, rank $P > \dim A$, then P has a unimodular element.

Proof. Induction on $\dim A$, the case $\dim A = 0$ being clear. We may assume that A is reduced with connected spectrum. Then $S^{-1}P$ is free, if S is the set of all non-zero-divisors of A. This implies, P_s is free for some non-zero-divisor s. By induction hypothesis there is a unimodular $\bar{p}$ in P/sP. Then apply Proposition D.0.15 to a representant p of $\bar{p}$.

$$\frac{I}{(a)} \oplus \frac{(aI, b_2, \ldots, b_r)}{(a)} = \frac{I}{aI}.$$

To see this we need to show that $(a) \cap (aI, b_2, \ldots, b_r) = aI$: This is easily checked by using the fact that I/I^2 has $\{\bar{a}, \bar{b_2}, \ldots, \bar{b_r}\}$ as a basis. By Proposition 8.3.7 we have $\mathrm{hd}_{R/(a)}I/(a) < \infty$.

Recall the definition of the norm

$$N := N_{K/Q(R)} : K \longrightarrow Q(R).$$

Let $[K : Q(R)] = n$ and Ω be an algebraic closed field containing K. There are exactly n different field embeddings $\sigma_i : K \hookrightarrow \Omega$ which leaves $Q(R)$ fixed. And the norm of $\alpha \in K$ is defined by $N(\alpha) := \prod_{i=1}^n \sigma_i(\alpha)$. One easily sees by Galois Theory that indeed $N(\alpha) \in Q(R)$, and trivially one has $N(\alpha\beta) = N(\alpha)N(\beta)$. Further $N(\alpha) = 0 \iff \alpha = 0$ and $\alpha \in A^\times \iff N(\alpha) \in R^\times$.

Then $a := N(\alpha) \in I \cap R$ and $\#(A/aA) = \#(R^n/aR^n) < \infty$. Therefore also $\#(A/I) < \infty$. To show Theorem 8.9.1 we prove some lemmas. □

Lemma D.0.16 *Let I be a non-zero ideal of a domain A of dimension 1 and P a (finitely generated) projective A-module of rank n. Then $P/IP \cong (A/I)^n$ as A-modules. In particular, if $J \neq (0)$ is an invertible ideal of A, then $J/IJ \cong A/I$ as A-modules.*

Consequently in this case $\#(A/IJ) = \#(A/I)\#(A/J)$, i.e. $\|IJ\| = \|I\| \cdot \|J\|$.

Proof. Since $I \neq (0)$ and A is a domain of dimension 1, there are only finitely many maximal ideals, say $\mathfrak{n}_1, \ldots, \mathfrak{n}_m$ containing I. Set $S := A \setminus \bigcup_{i=1}^{m} \mathfrak{n}_i$. Then

$$S^{-1}(P/IP) \cong P/IP,$$

since the elements of S act bijectively on P/IP. Thus we may assume $A = S^{-1}A$, in which case A is semilocal, hence P free. Then the claim is trivial. $\square$

Chapter 8

Let A be a Dedekind ring, F a finitely generated free module and $U \subset F$ a submodule. Assume that there is a $c \in A \setminus (0)$ with $cF \subset U$. Show that there are a basis $x_1, \ldots, x_n$ of F and a decreasing series $I_1 \supset \cdots \supset I_n$ of ideals, such that $U = I_1 x_1 + \cdots + I_n x_n$. (Consider all epimorphisms $\varphi : F \to A$ and let φ_1 be one such that $\varphi_1(U) =: I_1$ is maximal under all $\varphi(U)$. First show that $U \subset I_1 F$. One can identify F with A^n, such that $\varphi = \mathrm{pr}_1$. Namely every stably free module over a Dedekind ring is free. If $U \not\subset I_1 F$ then $\mathrm{pr}_i(U) \not\subset I_1$ for some $i > 1$, say $i = 2$. If not, let $\sigma : I_1 \to U$ be a section (right inverse) of the epimorphism $\varphi | U : U \to I_1$. The existence of a section follows from the projectivity of I_1. We will extend this section to a section $\sigma' : A \to F$ of φ. Let $S := A \setminus (0)$ and consider F as a submodule of $S^{-1}F$ in the canonical way. Choose some $a \in I_1 - (0)$ and define $\sigma' : A \to S^{-1}F$ by $\sigma'(1) := a^{-1}\sigma(a)$. Once we have shown $\sigma'(1) \in F$, we see immediately that $\sigma' : A \to F$ is a section of φ. $\square$

Proposition D.0.17 *(cf. Proposition 8.8.7) Let P, Q be rank 1 projective modules over a 1-dimensional noetherian ring R. Then*

$$P \oplus Q \cong (P \otimes Q) \oplus R.$$

Proof. By Serre's Splitting Theorem $P \oplus Q \cong L \oplus R$ for some projective module L of rank 1. Further $P \otimes Q \cong \bigwedge^2(P \oplus Q) \cong \bigwedge^2(L \oplus R) \cong L$. $\square$

Proof. P and Q are isomorphic to invertible ideals I, resp. J of R, contained in R.

CLAIM. There is an ideal $I' \cong I$ with $I' + J = R$.

Namely since $\dim(R) = 1$ and J contains a non-zero-divisor, there are only finitely many prime ideals containing J. Let $\mathfrak{p}_1, \ldots, \mathfrak{p}_n$ be the prime ideals which either contain J or belong to $\mathrm{Ass}(R)$ and $S := R \setminus \bigcup_{i=1}^{n} \mathfrak{p}_i$. Then $S^{-1}(I^{-1})$ is a principal fractional ideal of $S^{-1}R$. Let $z \in I^{-1}$ be a generator of $S^{-1}(I^{-1})$. Then $I' := zI \subset R$ and $S^{-1}I' = S^{-1}R$. Therefore I' is not contained in any of the prime ideals containing J, whence $I' + J = R$. Now write I instead of I'.

We have an exact sequence

$$0 \longrightarrow I \cap J \longrightarrow I \oplus J \longrightarrow R \longrightarrow 0,$$

(Justify!) which clearly splits. By part of the Chinese Remainder Theorem $I \cap J = IJ \cong P \otimes Q$. And so $(P \otimes Q) \oplus R \cong P \oplus Q$. $\square$

We now continue where we left off in Chapter 8, §8.3, by proving a famous formula of Auslander and Buchsbaum relating the depth and homological dimension as codimensions.

We begin with the notion of a **minimal free resolution** of a module M over a local ring $(R, \mathfrak{m})$.

A free resolution

$$L. := \cdots \to L_i \overset{d_i}{\to} L_{i-1} \overset{d_{i-1}}{\to} \cdots \overset{d_1}{\to} L_0 \to M \to 0$$

is called a **minimal free resolution** of M if $d_i : L_i \to \ker(d_{i-1})$ is **minimal**, for each $i \geq 0$, i.e. $d_i \otimes R/\mathfrak{m}$ is an isomorphism, or equivalently, d_i is onto and $\ker(d_i) \subseteq \mathfrak{m} L_i$, for all i.

Lemma 1. *Let M be a finitely generated module over a local ring $(R, \mathfrak{m})$. Let $0 \to K \to F \to M \to 0$ be an exact sequence with F a finitely generated free R-module.*

If $\operatorname{depth}(R) > 0$, $\operatorname{depth}(M) = 0$, then $\operatorname{depth}(K) = 0$.

Proof. Let $x \in \mathfrak{m}$ be a non-zero-divisor of R. Tensoring with the right exact functor $\otimes_R R/(x)$ we get an exact sequence

$$0 \to M' \to K/xK \to F/xF \to M/xM \to 0.$$

Since $\operatorname{depth}(M) = 0$, $\mathfrak{m} \in \operatorname{Ass}(M)$, equivalently there is a $m \in M \setminus \{0\}$, such that $\mathfrak{m} \cdot m = 0$.

Clearly, $m \in M'$, whence $\mathfrak{m} \in \operatorname{Ass}(M')$. But then $\mathfrak{m} \in \operatorname{Ass}(K/xK)$, whence $\operatorname{depth}(K/xK) = 0$. Since x is not a zero-divisor of K (Why?) $\operatorname{depth}(K) = 1$. $\square$

Theorem D.0.18 *(Auslander-Buchsbaum [2])*
Let M be a finitely generated module over a Noetherian local ring $(R, \mathfrak{m})$. If $\operatorname{hd}(M) < \infty$ then

$$\operatorname{hd}(M) + \operatorname{depth}(M) = \operatorname{depth}(A)$$

Proof. Let $n = \mathrm{hd}(M)$, and

$$L. : L_n \overset{d_n}{\to} L_{n-1} \overset{d_{n-1}}{\to} \cdots \overset{d_1}{\to} L_0 \to M \to 0$$

be a minimal free resolution. Let $K_i = Im(d_{i-1})$, for $i = 1, \ldots, n$.

We argue by induction on $\mathrm{depth}(R)$. If $\mathrm{depth}(R) = 0$ there is a non-zero $x \in R$ with $\mathfrak{m} \cdot x = 0$. If $n > 0$, then $L_n \subseteq \mathfrak{m}L_{n-1}$ implies that $xL_n \subseteq x\mathfrak{m}L_{n-1} = (0)$, so $L_{n-1} = (0)$, a contradiction. Therefore, $n = 0$, and M is free, so $\mathrm{depth}(M) = \mathrm{depth}(R) = 0$.

Now let $\mathrm{depth}(R) > 0$. If $\mathrm{depth}(M) > 0$ there is $x \in \mathfrak{m}$ that is not a zero-divisor of R or M by Prime Avoidance as $\mathfrak{m} \subsetneq \bigcup \mathfrak{p}$, for $\mathfrak{p}$ in the finite set $\mathrm{Ass}(M) \bigcup \mathrm{Ass}(R)$. Therefore, $\mathrm{depth}(R/(x)) = \mathrm{depth}(R) - 1$, and $\mathrm{depth}(M/xM) = \mathrm{depth}(M) - 1$. By Proposition 8.3.7 $\mathrm{hd}_{R/(x)}(M/xM) = \mathrm{hd}_R(M)$. The Auslander Buchsbaum formula is established by induction in this case.

Let $\mathrm{depth}(R) > 0$. Then by Lemma 1 $\mathrm{depth}(K_1) = 0$. But $\mathrm{hd}(K_1) = \mathrm{hd}(M) - 1$. Since $\mathrm{hd}(K_1) + \mathrm{depth}(K_1) = \mathrm{depth}(R)$, by above, it follows that $\mathrm{hd}(M) = \mathrm{depth}(R)$, as required. Induction establishes the Auslander Buchsbaum formula. $\qquad\square$

Chapter 9

Let us give an interesting example, which shows that one can expect a certain class of ideals to have a bounded number of generators – namely the class of **"local complete intersection ideals"**, i.e. those proper ideals I for which $I_{\mathfrak{p}}$ is generated by a regular sequence for all prime ideals $\mathfrak{p} \supset I$.

Proposition D.0.19 *Let R be a noetherian ring of dimension d and I a local complete intersection ideal. Then $\mu(I/I^2) \leq d$. In particular $\mu(I) \leq d + 1$.*

Proof. Let $\mathfrak{p} \in V(I)$. Clearly $\mu_{\mathfrak{p}}(I/I^2) \leq \mu_{\mathfrak{p}}(I)$. Since I is a local complete intersection ideal, we have

$$\mu_{\mathfrak{p}}(I) = \mathrm{ht}(I_{\mathfrak{p}}) \leq \mathrm{ht}(\mathfrak{p}R_{\mathfrak{p}}) = \mathrm{ht}(\mathfrak{p}).$$

Therefore, for the Forster-Swan bound

$$b(I/I^2) \leq \mathrm{Max}_{\mathfrak{p}}\{\mu_{\mathfrak{p}}(I/I^2) + \dim(R/\mathfrak{p})\} \leq \mathrm{ht}(\mathfrak{p}) + \dim(R/\mathfrak{p}) \leq \dim(R).$$

(Alternatively one can show that $X_k(I/I^2) = \emptyset$, if $k > \mathrm{ht}(I)$ and derive this.) By the Forster-Swan Theorem we get $\mu(I/I^2) \leq \dim(R)$, hence by (10.2.1)

$$\mu(I) \leq \mu(I/I^2) + 1 \leq b(I/I^2) + 1 \leq \dim(R) + 1.$$

$\qquad\square$

Chapter 10

Examples: Set-Theoretical Complete Intersection Curves

D.0.20 Let k be a field, $A := k[x_1, x_2, x_3]$ and $f_1, f_2, f_3 \in A$ defined as

$$f_1 := x_1^{\alpha_1} - x_2^{\alpha_{12}} x_3^{\alpha_{13}}, \quad f_2 := x_2^{\alpha_2} - x_1^{\alpha_{21}} \alpha_3^{\alpha_{23}}, \quad f_3 := x_3^{\alpha_3} - x_1^{\alpha_{31}} x_2^{\alpha_{32}} \quad \text{(D.1)}$$

where α_i, α_{ij} are positive integers with

$$\alpha_1 = \alpha_{21} + \alpha_{31}, \quad \alpha_2 = \alpha_{12} + \alpha_{32}, \quad \alpha_3 = \alpha_{13} + \alpha_{23} \quad \text{(D.2)}$$

$\square$

Lemma D.0.21 *We have the following equality of ideals in A:*

$$(f_1, f_2, f_3) \cap (x_1^{\alpha_{21}}, x_2^{\alpha_{12}}) = (f_1, f_2, x_1^{\alpha_{21}} f_3, x_2^{\alpha_{12}} f_3) = (f_1, f_2) \quad \text{(D.3)}$$

Proof. First equality: Since $\alpha_1 \geq \alpha_{21}$, $\alpha_2 \geq \alpha_{12}$ we have $f_1, f_2 \in (x_1^{\alpha_{21}}, x_2^{\alpha_{12}})$ So $(f_1, f_2, f_3) \cap (x_1^{\alpha_{21}}, x_2^{\alpha_{12}}) \supset (f_1, f_2, x_1^{\alpha_{21}} f_3, x_2^{\alpha_{12}} f_3)$

Let $f = \sum_{i=1}^3 f_i g_i \in (f_1, f_2, f_3) \cap (x_1^{\alpha_{21}}, x_2^{\alpha_{12}})$. Since $f_1, f_2 \in (x_1^{\alpha_{21}}, x_2^{\alpha_{12}})$, we have $f_3 g_3 \in (x_1^{\alpha_{21}}, x_2^{\alpha_{12}})$. It is an exercise to show that $\text{Ass}(A/(x_1^{\alpha_{21}}, x_2^{\alpha_{12}}) = \{(x_1, x_2)\}$. So, since $f_3 \notin (x_1, x_2)$ we see $g_3 \in (x_1^{\alpha_{21}}, x_2^{\alpha_{12}})$. So $g_3 f_3 \in (x_1^{\alpha_{21}} f_3, x_2^{\alpha_{12}} f_3)$ whence $f \in (f_1, f_2, x_1^{\alpha_{21}} f_3, x_2^{\alpha_{12}} f_3)$.

Second equation: '$\supset$' is trivial. Using Equations (D.2) we see:

$$x_2^{\alpha_{32}} f_1 + x_3^{\alpha_{13}} f_2 + x_1^{\alpha_{21}} f_3 =$$

$$(x_1^{\alpha_1} x_2^{\alpha_{32}} - x_2^{\alpha_2} x_3^{\alpha_{13}}) + (x_2^{\alpha_2} x_3^{\alpha_{13}} - x_1^{\alpha_{21}} x_3^{\alpha_3}) + (x_1^{\alpha_{21}} x_3^{\alpha_3} - x_1^{\alpha_1} x_2^{\alpha_{32}}) = 0$$

Analogously we compute

$$x_3^{\alpha_{23}} f_1 + x_1^{\alpha_{31}} f_2 + x_2^{\alpha_{12}} f_3 = 0$$

By these equations $x_1^{\alpha_{21}} f_3$ and $x_2^{\alpha_{12}} f_3$ both belong to (f_1, f_2), which establishes the second equation. $\square$

Appendix B

Lemma 2. *The square*

is a pull-back diagram, then the sequence

$$0 \longrightarrow Y \xrightarrow{\{f',g'\}} M \oplus N \xrightarrow{\langle f,g \rangle} X$$

is exact.

Proof. Note that $\langle f, -g \rangle \circ \{f', g'\} = ff' - gg' = 0$. Hence only to show that $\ker\{f, g\} = 0$. Since $Y = \{(m, n) \in M \times N : f(m) = g(n)\}$, we have $(0, 0) = \{f', g'\}(m, n) = (f'(m, n), g'(m, n)) = (m, n)$. □

(i) Note that $\ker f' = \{(0, n) \mid n \in N \text{ and } g(b) = 0\}$. So we can define a map $\varphi : \ker f' \to \ker g$ by $\varphi(0, n) = n$. Clearly φ is an isomorphism.

(ii) We have the exact sequence

$$0 \longrightarrow Y \xrightarrow{\{f',g'\}} M \oplus N \xrightarrow{\langle f,g \rangle} X$$

Suppose $m \in M$. Since g is epimorphic there exists $n \in N$ such that $f(m) = g(n)$. Hence $(m, n) \in \ker(\langle f, -g \rangle) = \operatorname{im}(\{f', g'\}$ by exactness. Thus there exists $y \in Y$ with $m = f'(y)$ (and $n = g(y)$). Hence f' is epimorphic.

(i) Note that $\ker f' = \{(0, n) \mid n \in N \text{ and } g(b) = 0\}$. So we can define a map $\varphi : \ker f' \to \ker g$ by $\varphi(0, n) = n$. Clearly φ is an isomorphism.

(ii) We have the exact sequence

$$0 \longrightarrow Y \xrightarrow{\{f',g'\}} M \oplus N \xrightarrow{\langle f,g \rangle} X$$

Suppose $m \in M$. Since g is epimorphic there exists $n \in N$ such that $f(m) = g(n)$. Hence $(m, n) \in \ker(\langle f, -g \rangle) = \operatorname{im}(\{f', g'\}$ by exactness. Thus there exists $y \in Y$ with $m = f'(y)$ (and $n = g(y)$). Hence f' is epimorphic.

□

Definition D.0.22 *Let I be a finitely generated ideal of a commutative ring A with $\bigcap_{n \in \mathbb{N}} I^n = (0)$. The I-Baer submodule of $A^{\mathbb{N}}$ consists of those elements which, considered as sequences, converge to 0 with respect to the I-adic topology.*

Lemma D.0.23 *Let A, I be as in the Definition and H be the I-Baer sub-module of $A^{\mathbb{N}}$. Then $A^{(\mathbb{N})} \subset H$ and $A^{(\mathbb{N})} + IH = H$.*

Proof. The first claim being trivial, let $a = (a_n)_{n \in \mathbb{N}} \in H$ and let N be the maximum of the n with $a_n \notin I$. Then $a' \in A^{(\mathbb{N})}$ if $a' = (a'_n)_{n \in \mathbb{N}}$ and $a'_n = a_n$ for $n \leq N$, $a'_n = 0$ for $n > N$. We have to show that $a - a' \in IH$ and know $a_n - a'_n \in I$.

Let I be generated by $x_1, \ldots, x_r$. Then every element of I^{k+1} can be written as a sum $x_1 b_1 + \cdots + x_r b_r$ with $b_j \in I^k$. Then write $a_n - a'_n = x_1 b_{1n} + \cdots + x_r b_{rn}$ in this manner. Then the sequences $b_1 := (b_{1n})_{n \in \mathbb{N}}, \ldots, b_r = (b_{rn})_{n \in \mathbb{N})}$ belong to H, whence $a - a' = x_1 b_1 + \cdots + x_r b_r \in IH$. □

Theorem D.0.24 *Let A be a ring, possessing two finitely generated ideals I_1, I_2 with $I_1 + I_2 = A$ and $\bigcap_{n \in \mathbb{N}} I_j^n = (0)$ for both j. Then $A^{\mathbb{N}}$ and $A^{(\mathbb{N})}$ are reflexive A-modules.*

Proof. For $j = 1, 2$ let H_j be the I_j-Baer submodule of $A^{\mathbb{N}}$.

We have $H_1 + H_2 = A^{\mathbb{N}}$. Namely let $(a_n)_{n \in \mathbb{N}} \in A^{\mathbb{N}}$. Then, since $I_1^n + I_2^n = A$, for every n there are $b'_n \in I_1^n$, $b''_n \in I_2^n$ with $a_n = b'_n + b''_n$. Clearly $(b'_n)_n \in H_1$, $(b''_n)_n \in H_2$.

Now set $H'_j := H_j / A^{(\mathbb{N})}$. Then $H'_1 + H'_2 = A^{\mathbb{N}} / A^{(\mathbb{N})}$.

Claim: $\mathrm{Hom}_A(A^{\mathbb{N}} / A^{(\mathbb{N})}, A) = 0$

Proof of this: Let $f : A^{\mathbb{N}} / A^{(\mathbb{N})} \to A$ be a homomorphism. Then $f(H'_j) = f(I_j H'_j) = I_j f(H'_j)$ by Lemma D.0.23. So also $f(H'_j) = I_j^n f(H'_j) \subset I_j^n$ for every n. Therefore $f(H'_j) = 0$. But then $\mathrm{im}(f) = f(H'_1 + H'_2) = f(H'_1) + f(H'_2) = 0$. –

This means that every homomorphism $f : A^{\mathbb{N}} \to A$ is already given by its restriction to $A^{(\mathbb{N})}$. i.e. for f there is a sequence $(b_n)_n \in A^{\mathbb{N}}$ with $f((a_n)_n) = \sum_n a_n b_n$. But this defines a hononorphism, i.e. the homomorphism $A^{(\mathbb{N})} \to A$, given by $(b_n)_n$ extends to $A^{\mathbb{N}}$, if and only if $b_n = 0$ for almost all n.

If A is a complete discrete valuation ring (see later), which is principal but not countable, then $A^{\mathbb{N}}$ is a free A-module. □

E

Exercises

Chapter 1

1. Let k be the field of 2 elements and $A := k[X,Y]/(X^2, XY, Y^2)$. Call x, y the residue classes of X, resp. Y and set $I := (y)$. Then $x+I = \{x, x+y\} \subset (x) \cup (x+y)$, but neither $I \subset (x)$ nor $I \subset (x+y)$. Hence the assumption that the $\mathfrak{p}_i$ are prime ideals in Lemma 1.1.8 a) cannot be weakened to the extent that at most two of them need not be prime. We do not know, whether part a) of this lemma holds true under the assumption that at most *one* of the $\mathfrak{p}_i$ is not prime.

2. Prove the theorem in Remark 1.1.9 d). First reduce to the case $U = V$ by intersecting the $x_i + U_i$ with U. (If $U \cap (x_i + U_i) \neq \emptyset$ one may assume $x_i \in U$.) Then assume, there is a counterexample. Take one with minimal r. Then $x_1 + U_1 \not\subset \bigcup_{i=2}^r (x_i + U_i)$. If now $a \in (x_1 + U_1) \setminus \bigcup_{i=2}^r (x_i + U_i)$ there is a line $a + Kv$ with $v \in V \setminus \{0\}$ through a not contained in $x_1 + U_1$. It intersects every $x_i + U_i$ in at most one point and so is not contained in the union of the $x_i + U_i$.

3. (M. Henriksen) A commutative ring R has no maximal ideal iff
   (i) it has no ideal I with R/I a field,
   (ii) $R^2 + pR = R$ for every prime number p.

4. (Nagata) Let R be an integral domain. Let $b \neq 0$, and a be a non-zero-divisor modulo (b). If $(b, a) \neq R$ then show that $(a + bX)$ is a prime ideal of $R[X]$.

5. Let A be a commutative ring with 1. Let I_1, I_2 be two ideals of A with $I_1 + I_2 = A$. Then, as A-modules,

$$\frac{I_1 \cap I_2}{(I_1 \cap I_2)^2} \xrightarrow{\sim} \frac{I_1}{I_1^2} \oplus \frac{I_2}{I_2^2}$$

6. Let A be a reduced ring with only finitely many minimal prime ideals $\mathfrak{p}_1, \dots, \mathfrak{p}_n$ and set $S := A \setminus \bigcup_{i=1}^{n} \mathfrak{p}_i$. Show $S^{-1}A \cong \prod_{i=1}^{n} Q(A/\mathfrak{p}_i)$.

7. If $\mathfrak{p}, \mathfrak{q}$ are different prime ideals disjoint to a multiplicative set S then $S^{-1}\mathfrak{p} \neq S^{-1}\mathfrak{q}$. This does not hold for general ideals: If $S := \mathbb{Z} \setminus (2)$, then $S^{-1}(2) = S^{-1}(6)$ in $S^{-1}\mathbb{Z}$.

8. Solve the equation

$$\text{CHINA} \cdot \text{CHINA} = *****\text{CHINA},$$

where the letters represent digits in the decimal system and the stars may be replaced by arbitrary digits. (Find all solutions!)

9. Let $a, b, c, d \in \mathbb{Q}$. Show that the following system of linear equations always is solvable:

$$x + 5y + az = b$$
$$(a-1)x - 4y = c$$
$$(a+7)y + z = d$$

(Use a determinant and the fact that $\mathbb{Z}$ is integrally closed.)

10. Let $f(X) \in R[X]$ have leading coefficient $a \notin \mathfrak{p}$, for some $\mathfrak{p} \in \text{Spec}(R)$. Prove that there are only finitely many prime ideals $P \in \text{Spec}(R[X])$ such that $f(X) \in P$ and $P \cap R = \mathfrak{p}$.

11. Let I be an ideal of a polynomial ring containing a monic polynomial. Let J be an ideal of A such that $I + J[X] = A[X]$. Then show that $(I \cap A) + J = A$.

12. Let M be a R-module, N a submodule of M and I an ideal of R. If $x, y \in I$ and $N :_N (x) = N = N :_N (y)$ then show that

$$\frac{(xM + N :_M I)}{xM + N} \simeq \frac{(yM + N :_M I)}{yM + N},$$

13. Let C be the ring of continuous real valued functions on the unit interval $[0, 1]$.

a) Show that for every $x \in [0, 1]$ the set of $f \in C$ with $f(x) = 0$ is a maximal ideal of C. (Here $[0, 1]$ may be replaced by any topological space.) (See e).)

b) Show that the set of the $f \in C$ with $f([0, \varepsilon]) = \{0\}$ for some $\varepsilon > 0$ is an ideal but not a prime ideal of C.

c) Show that there are non-maximal prime ideals in C. (Use Lemma 1.3.3. We do not know whether one can describe any non-maximal prime ideal explicitly in any sense.)

d) One may also show the existence of an infinite strictly descending chain (sequence) of prime ideals

$$\mathfrak{p}_0 \supsetneq \mathfrak{p}_1 \supsetneq \cdots$$

in C. Namely consider the following sequence of functions (f_k), defined by

$$f_0(x) := x, \qquad f_k(x) := \begin{cases} 0 & \text{for } x = 0 \\ \exp(-x^{-k}) & \text{for } x > 0 \end{cases}$$

Now let $\mathfrak{p}_0$ be the maximal ideal attached to the point $0 \in [0, 1]$ as in a). Now assume you have already found $\mathfrak{p}_k$. Then define S to be the multiplicative set generated (multiplicatively) by $C \setminus \mathfrak{p}_k$ and f_k and let $I := \sum_{i>k} C f_i$. Show $I \cap S = \emptyset$ and apply Lemma 1.3.3 to find $\mathfrak{p}_{k+1}$.

e) Show that for every maximal ideal $\mathfrak{m}$ of C there exists an $x \in [0, 1]$ with $\mathfrak{m} = \{f \in C \mid f(x) = 0\}$. (Otherwise for every $x \in [0, 1]$ there where an $f \in \mathfrak{m}$ with $f(x) \neq 0$. Since the set $O(f) := \{y \in [0, 1] \mid f(y) \neq 0\}$ is open in $[0, 1]$ and $[0, 1]$ is compact, there would exist finitely many $f_1, \dots, f_n$ in $\mathfrak{m}$ with $O(f_1) \cap \cdots \cap O(f_n) = \emptyset$. But then $f_1^2 + \cdots + f_n^2$ would belong to $\mathfrak{m}$ but also be a unit in C. Here $[0, 1]$ may be replaced by any compact space.)

14. Let M be a left A-module.

a) If $x \in A$ is a zero divisor of M, the ax is so for every $a \in A$.

b) If $u \in A$ operates bijectively (resp. injectively) on M and $b \in A$ nilpotently (i.e. there is an $n \in \mathbb{N}$ with $b^n x = 0$ for all $x \in M$), then $u + b$ operates bijectively (resp. injectively) on M.
(Assume: For every $x \in M$ there is an n with $b^n x = 0$. Does the above assertion hold also under this weaker condition?)

c) Let additionally A be commutative. Show that the set of zero divisors of M is the union of (suitable) prime ideals. (The non-zero-divisors make up a multiplicative set. If a is a zero divisor of M, say $ax = 0$, then the whole ideal $\mathrm{Ann}(x)$ consists of zero divisors of M. Use Lemma 1.3.3.)

d) Give an example that not every multiplicative set is the complement of a union of prime ideals.

e) Give examples which show that the set of zero divisors is not always an ideal.

15. a) Show that the ideals $I := \{0\} \times \mathbb{Z}$ and $J := \mathbb{Z} \times \{0\}$ of the ring $R :=$ $\mathbb{Z} \times \mathbb{Z}$ are not isomorphic as R-modules. Show that also $R/I \not\cong R/J$ as R-modules. (In spite of this, the latter are isomorphic as rings.) (Clearly $\mathbb{Z}$ can be replaced by any non-zero ring.)

   b) Let I, J be different ideals of a commutative ring R (or, more generally, two-sided ideals of any ring). Show that R/I and R/J are not isomorphic as R-modules.

   c) In the noncommutative ring $R = M_n(K)$ where $n > 1$, there are different left ideals I, J such that R/I and R/J are isomorphic R-modules. Namely let for e.g. I (resp. J) consist of the matrices all of whose columns but the first (resp. the second) vanish.

16. Infinite direct products may have surprising properties.

   a) Would you expect that a direct product of finite groups could have a factor group which is a non-zero $\mathbb{Q}$-vector space? Indeed this happens. Let P be the set of all prime numbers and let $G := \prod_{p \in P} \mathbb{Z}/(p)$. Show that the torsion subgroup of G is $H := \bigoplus_{p \in P} \mathbb{Z}/(p)$, i.e. consists of those families $(a_p)_{p \in P}$, $a_p \in \mathbb{Z}/(p)$ with $a_p = 0$ for almost all $p \in P$. Then show that every $n \in \mathbb{Z} \setminus \{0\}$ operates bijectively on G/H. This means that G/H is a $\mathbb{Q}$-vector space.

   b) A consequence of a) is that $G \otimes_{\mathbb{Z}} \mathbb{Q} \neq 0$, whereas $(\mathbb{Z}/(p)) \otimes_{\mathbb{Z}} \mathbb{Q} = 0$ for every p.

   c) Show that a family $(f_i)_{i \in I}$ of homomorphisms $f_i : M \to N_i$ defines canonically a homomorphism $f : M \to \prod_{i \in I} N_i$. This gives us a canonical map $\varphi : (\prod_{i \in I} E_i) \otimes_A F \to \prod_{i \in I}(E_i \otimes_A F)$ for A-modules E_i and F, which by b) is not always injective.

   d) The map φ in c) also need not be surjective. Indeed, let A be a non-zero ring and show that the canonical map $(A^{\mathbb{N}})^{(\mathbb{N})} \to (A^{(\mathbb{N})})^{\mathbb{N}}$ is not surjective. Then identify $(A^{\mathbb{N}})^{(\mathbb{N})}$ with $A^{\mathbb{N}} \otimes A^{(\mathbb{N})}$ and $(A^{(\mathbb{N})})^{\mathbb{N}}$ with $(A \otimes A^{(\mathbb{N})})^{\mathbb{N}}$.

17. Note that the above constructed G/H also is a ring A. Show that A has the following properties:

   a) An element of A is a unit if and only if an arbitrary representing element $\in G$ has only finitely many zero-components. Otherwise it is a zero-divisor.

   b) A has uncountably many zero-divisors, but no non-zero nilpotent element.

   c) $A^{\times}$ has elements of every (finite or infinite) order.

d) For every prime number p there is an injective group homomorphism of the additive group of the p-adic numbers into $A^\times$.

For c) and d) one needs Dirichlet's theorem on prime numbers in arithmetic progressions or rather the weak form of it: For any integer $m > 1$ there are infinitely many prime numbers $p \equiv 1(\bmod\, m)$. (This weak form can be proven much easier – without analytic tools – than the full Dirichlet's theorem.) Also one uses that any group $(\mathbb{Z}/p)^\times$ is cyclic, i.e. that for an arbitrary divisor m of $p-1$ there is an element of order m in $(\mathbb{Z}/p)^\times$.

18. VALUATIONS.

a) An **ordered abelian group** is an abelian group G (here written additively) together with a structure as follows: there is given a subset $P \subset G$, a so called **positivity domain**, such that
1. $P \cup -P = G$,   2. $P \cap -P = \{0\}$,   3. $P + P \subset P$ ($-P := \{-x \mid x \in P\}$, $P + P := \{x + y \mid x, y \in P\}$.)

This defines a total ordering '$\leq$' on G by $x \leq y : \Longleftrightarrow y - x \in P$, with the property 4. $x \leq y \Rightarrow x + z \leq y + z$.

Conversely, given a total ordering on G with property 4., one may define a positivity domain P by $P := \{x \in G \mid 0 \leq x\}$, which obeys 1., 2., 3.

Examples of ordered abelian groups are the subgroups of the additive group of $\mathbb{R}$. An ordered abelian group is isomorphic as an ordered group to a subgroup of $\mathbb{R}$ if and only if it is Archimedean. This means that for every $x, y \in G$ with $x > 0$ there is an $n \in \mathbb{N}$ with $nx > y$.

b) If G is an ordered abelian group, then by $G \cup \{\infty\}$ we denote G together with an extra element ∞ with $\infty \geq x$ and $\infty + x = \infty$ for every $x \in g \cup \{\infty\}$.

Let K be a field. By a valuation of K one means a map $v : K \to G \cup \{\infty\}$ obeying:
1. $v(a) = \infty \Longleftrightarrow a = 0$,   2. $v(ab) = v(a) + v(b)$,   3. $v(a + b) \geq \text{Min}(v(a), v(b))$.

Show that the set $\mathcal{O}_v := \{a \in K \mid v(a) \geq 0\}$ is a local subring of K with maximal ideal $\mathfrak{m}_v = \{a \in K \mid v(a) > 0\}$.

c) Show that $\mathcal{O}_v$ is a principal domain if and only if $v(K^\times) \cong \mathbb{Z}$ as (ordered) groups. If you know already the definition of a Noetherian ring, show that this holds if and only if $\mathcal{O}_v$ is Noetherian. In this case v is called a **discrete valuation**.

d) Let A be a domain and $p \in A$ a prime element, i.e. Ap is a non-zero prime ideal. Then every non-zero $x \in Q(A)$ can be written in the form $x = p^n \frac{a}{b}$ with $a, b \in A$, $n \in \mathbb{Z}$ and $p \nmid ab$. Here n is uniquely determined by x. The map $Q(A) \to \mathbb{Z} \cup \{\infty\}$ defined by $v(0) = \infty$, $v(x) = n$ for $x \neq 0$ is a discrete valuation with $\mathcal{O}_v = A_{Ap}$.

e) A ring A is of the form $\mathcal{O}_v$ with a discrete valuation v, if and only if it is a local principal domain and not a field. These rings are called **discrete valuation rings**.

19. We will give an example of a discrete valuation ring A and a finite field extension $Q(A) \subset L$ such that the integral closure B of A in L is not finite over A, i.e. not a finitely generated A-module. (E. Artin, and independently O. Zariski.)

   a) Let k be a field. Note that the quotient field of the formal power series ring $k[[X]]$ can be understood as the ring of Laurent series with finite principal part, i.e. as Laurent series of the form $\sum_{i=n}^{\infty} a_i X^i$ where $n \in \mathbb{Z}$ is allowed to be negative. This field will be denoted by $k((X))$.

   b) Let $K \subset L$ be a radicial (i.e. purely inseparable) field extension of degree $p = \mathrm{char}(K) \, (> 0)$. Show that for every $\alpha \in K - L$ the minimal polynomial of α over K equals $X^p - \alpha^p$. Consequently, if $A \subset K$ is integrally closed and $Q(A) = L$, then $B := \{\alpha \in L \mid \alpha^p \in A$ is the integral closure of A in L.

   c) Now we construct the announced example. Let k be a field of characteristic $p > 0$. The construction will take place within $k((X))$. Let $\eta := \sum_{i > 0} c_i X^i \in k[[X]]$ be transcendental over $k(X)$. Such an η exists by a cardinality argument. (For simplicity you may assume that k is finite (or countable), hence $k(X)$ and even its algebraic closure are countable. But $k[[X]]$ is not countable.) Now, set $Y := \eta^p = \sum_{i > 0} c_i^p X^{pi}$. Then define $K := k(X, Y)$, $L := k(X, \eta) = K(Y^{1/p})$ and $A := k[[X]] \cap K$. Show that $Q(A) = K$ and A is a discrete valuation ring with prime element X. (Restrict the 'canonical' valuation of $k((X))$ to K.) Let C denote the integral closure of A in L. For every $n \in \mathbb{N}$ one can write Y in the form

   $$Y = f_n^p + X^{pn} Y_n$$

   with a polynomial $f_n \in k[X]$ of degree $< n$ and $Y_n \in k[[X]]$. Then $Y_n \in K$, hence $\in A$ and further $Y^{1/p} = X^{-n}(\eta - f_n) \in L$, hence in C.

   Finally let $B := A + A\eta + \cdots + A\eta^{p-1}$, which is a subring of C with $Q(B) = Q(C) = L$. (Note that $1, \eta, \dots, \eta^{p-1}$ is a basis of L over K as well as one of B over A.) If C were finite over A there would be a $d \in A$ with $dC \subset B$. Especially $dY_n^{1/p} \in B$. But $Y_n^{1/p} = X^{-n}(\eta - f_n)$ would imply $X^n | d$ for every n.

Chapter 2

1. Let A be a Noetherian ring. Let I be an ideal of A such that $I = I^2$. Then I is generated by an idempotent element e (i.e. $e^2 = e$).

2. Let I be a finitely generated ideal of R. Prove that if I/I^2 is generated by r elements as an R/I-module then I is generated by $(r+1)$ elements.

3. Let (R, m) be a local ring. Let $x \in m$. Suppose that m is generated by k elements. When can you say that the least number of generators of the maximal ideal $m/(x)$ of $R/(x)$ is $k - 1$?

4. (N. Mohan Kumar) Let A be a Noetherian ring. Let I be an ideal of A such that $I = (a_1, \ldots, a_n) + I^2$, for some $a_i \in A$. Then $I = (a_n, \ldots, a_n, e)$, with $e(1 - e) \in (a_1, \ldots, a_n)$.

5. Let M be a finitely presented module over a reduced ring R. Show that M is a projective R-module if $\mu_{\mathfrak{p}}(M)$ is constant on $\mathrm{Spec}(R)$. (Here $\mu_{\mathfrak{p}}(M)$ denotes the minimal number of generators of the module $M_{\mathfrak{p}}$ over the ring $R_{\mathfrak{p}}$. This equals the dimension of the vector space $M_{\mathfrak{p}}/\mathfrak{p}M_{\mathfrak{p}}$ over the field $R_{\mathfrak{p}}/\mathfrak{p}R_{\mathfrak{p}}$.)

6. Let R be a principal (ideal) domain. Show that every submodule of any free R-module is also free. Consequently, every projective module over a principal domain is free. You may use the following hints.

    a) Let $f : M \to R$ be R-linear. Since the image is free, hence projective, one obtains $M \cong \ker(f) + \mathrm{im}(f)$.

    b) Let $M \subset R^n$ and $p : R^n \to R$ be the projection to the first direct summand. Then $(M \cap \ker(p)) \oplus p(M) \cong M$ by a). So M is free by induction on n.

    c) Now let F be a – not necessarily finitely generated – free R-module with a basis B. Usind a well-ordering of B one may prove that any submodule of F is free analogously to the proof in the case where $F \cong R^n$.

    d) But one may also use Zorn's Lemma. Consider pairs (C, D) where $C \subset B$ and D is a basis of $M \cap \sum_{c \in C} Rc$.

7. Let M be a finitely generated torsion free module over a principal domain A. Show that M is free. 'Torsion free' means that $am = 0$ for $a \in A, m \in M$ implies $a = 0$ or $m = 0$. Let $S = A \setminus \{0\}$. By torsion freeness we get an imbedding $M \to S^{-1}M$. Since the latter is a (finite dimensional) $S^{-1}A$-vector space, it has a basis, which we may assume to lie in M, i.e. of the form $\frac{m_1}{1}, \ldots, \frac{m_n}{1}$. Since M is finitely generated, there is an $s \in S$ with $sM \subset Am_1 \oplus \cdots \oplus Am_n$. Then use Exercise 6 and that $M \cong sM$.

8. Is $\mathbb{Z}^{\mathbb{N}}$ (i.e. the group of infinite sequences of integers) a free $\mathbb{Z}$-module? The answer is: 'No'.

   Show for every countable principal domain A which is not a field that $A^{\mathbb{N}}$ is not a free A-module.

   Use that submodules of free modules over a principal domain are again free. (Exercise 1) Choose any prime element p of A and consider the submodule H of sequences $(a_0, a_1, a_2, \dots)$ having the following property:

   For every m almost all a_n are divisible by p^m.

   Show that H is uncountable. If H were free it would have an uncountable basis and so would have the A/pA-vector space H/pH. Then show that H/pH has a countable basis, consisting of the residue classes of the sequences $(1,0,0,0,\dots)$, $(0,1,0,0,\dots)$, $(0,0,1,0,\dots)$, $\dots$.

9. (S.M. Bhatwadekar, Raja Sridharan) Let A be a Noetherian ring. Let I be an ideal of A. Let $I_1 \subset I$, $I_2 \subseteq I^2$ be two ideal of A such that $I_1 + I_2 = A$. Then $I = I_1 + (e)$, for some $e \in I_2$, and $I_1 = I \cap K$, with $I_2 + K = A$.

Chapter 3

1. Let $v = (a_1, \dots, a_n)$ be a unimodular row over a commutative ring A. Suppose that there exists a unimodular row $(c_1, \dots, c_n)$ with $\sum a_i c_i = 0$. Then show that the projective module P_v corresponding to $v = (a_1, \dots, a_n)$ has a unimodular element, i.e. there is a $p \in P$ such that $Ap \oplus Q \simeq P$, for some submodule Q of p.

2. Let $v = (a_1, \dots, a_n)$ be a unimodular row over a commutative ring A.
   (i) (Bass) If n is even, prove that P_v has a unimodular element,
   (ii) (R.A. Rao-Raja Sridharan). Suppose that n is odd, and some a_i is a square. Prove that P_v has a unimodular element.
   (iii) Give an example to show that the condition that some a_i is a square is necessary.

3. There is a good reason why, in the definition of stably free, we require that the free summand is finitely generated Namely for every projective module P there is a – generally not finitely generated – free module F such that $P \oplus F$ is free, as we see by the following trick of Eilenberg:

   Let P, Q be projective such that $P \oplus Q = E$ is free and let $F = E \oplus E \oplus E \oplus E\dots$ which is also free. Then $P \oplus F \cong P \oplus E \oplus E \oplus \dots$
   $\cong P \oplus (Q \oplus P) \oplus (Q \oplus P) \oplus \dots \cong (P \oplus Q) \oplus (P \oplus Q) \oplus \dots \cong E \oplus E \oplus E \oplus \dots \cong F$

4. We know the following theorem of Gabel:

   If P is stably free, but not finitely generated, then P is actually free.

   Say $P \oplus R^m \cong F$, where F is free with basis $\{e_i\}_{i \in I}$. Since P is not finitely generated, I must be an infinite set. View P as $\ker(F \xrightarrow{f} R^m)$ for some epimorphism f. Let $\{x_1, ..., x_n\}$ be a basis of R^m. Then there are $y_i \in F$ such that $f(y_i) = x_i$. But every y_i is a linear combination of $\{e_i\}_{i \in I_i}$ where I_i is a finite subset of I. Take $I_0 = \cup_{i=1}^m I_i$. Then I_0 is a finite subset of I.

   Define $F_0 = \sum_{i \in I_0} (Re_i)$ Then $f|_{F_0} : F_0 \to R^m$ is surjective. Thus $F = P + F_0$. Put $Q = P \cap F_0$, we have two short exact sequences

   1. $0 \to Q \to P \to P/Q \to 0$.

   2. $0 \to Q \xrightarrow{g} F_0 \xrightarrow{h} R^m \to 0$.

   (2) is exact because $\ker h = \{x \in F_0 : h(x) = 0\} = \{x \in F_0 : x \in P\} = P \cap F_0 = Q = \operatorname{img}$.

   By the second isomorphism theorem $F_0/P \cap F_0 \cong P + F_0/P$, i.e. $F_0/Q \cong F/P$. So $F/F_0 = P/Q$. But $F/F_0 = \sum_{i \in I - I_0} (Re_i)$. Since $I \setminus I_0$ is infinite we can write $P/Q \cong R^m \oplus F_1$ for some free module F_1.

   Now the first exact sequence splits naturally. Also second one splits since R^m is free. So we have $P \cong Q \oplus P/Q$ and $F_0 \cong Q \oplus R^m$. Now $P \cong Q \oplus P/Q \cong Q \oplus (R^m \oplus F_1) \cong (Q \oplus R^m) \oplus F_1 = F_0 \oplus F_1$ which is free.

5. (T.Y. Lam) Let R be a ring. let P be a stably free prjective R-module with $P \oplus R^m \simeq R^n$ for $n > m$. Let $P \simeq P^1 \oplus R^k$ for some $k > 0$. Let $r \geq \{m/k\} + \{m/(n-m)\}$, where $\{a\}$ denotes the least integer $\geq a$. Show that $_rP = P \oplus \cdots \oplus P$ (r times) is a free R-module.(Hint: By induction mi, $iP \oplus R^m \simeq R^{i(n-m)} \oplus R^m$. If $r = s + t$, $s \geq m/(n-m)$, and $t \geq m/k$, then use the free summands coming from tp to 'liberate' sP then use the resulting free summands to 'liberate' tP.)

6. (T.Y.Lam) Let R be right noetherian. Let $\{P_i, i \geq 1\}$ be a sequence of R-modules such that $P_i \oplus R^{Mi} \simeq R^{ni}$. Assume that $ni > m_{i+1}$ for all i. Then, for r sufficiently large, the partial sums $P_1 \oplus \ldots \oplus P_r$ are all free (Hint: Imitate the proof of Theorem 4.4)

7. Let P, Q be R-modules. Assume that $P \oplus R^m \simeq R^n$, and $Q \oplus R^s \simeq R^t$. If $n \geq s$ then show that $Q \oplus P \oplus R^m$ is free.

   Deduce Whitehead's lemma for reactangular matrices from the proof of the above.

8. The following is a generalized version of Whitehead's lemma due to L.N. Vaserstein: Let $m, n \geq 1$ be integers, y an $m \times n$ matrix over a ring A, x an $n \times m$ matrix over A. Assume that $I_m + xy \in \operatorname{GL}_m A$. Then

$$I_n + yx \in \mathrm{GL}_n A, \begin{pmatrix} I_m + xy & 0 \\ 0 & (I_n + yx)^{-1} \end{pmatrix}, \begin{pmatrix} I_m + xy & 0 \\ 0 & I_n \end{pmatrix}, \begin{pmatrix} (I_n + yx)^{-1} & 0 \\ 0 & I_m \end{pmatrix}$$

$\in E_{m+n}(A)$.

9. For any $\alpha \in \mathrm{GL}_n A$ and $\beta \in \mathrm{GL}_n A$, we have $\begin{pmatrix} [\alpha, \beta] & 0 \\ 0 & I_n \end{pmatrix} \in E_{2n}(A)$. (Hint: Take $x = \alpha(\beta - 1_n) \in M_n A$, $y = \alpha^{-1} \in M_n A$ in the previous exercise.)

Chapter 4

1. In Lemma 4.4.2 one may drop the hypothesis that c is a non-zero-divisor. We can still use the procedure of the proof to get explicit formulas for the entries of M, thus avoiding the use of c in the denominator.

   We introduce new indeterminates y, z and write

   $$f_i(t + yz) = f_i(t) + y\varphi_i(t, y, z)$$
   $$g_i(t + yz) = g_i(t) + y\psi_i(t, y, z)$$
   $$b' = b + cd \quad (d \in A).$$

   We shall try to define $M = (m_{ij})_{2 \times 2} \in \mathrm{SL}_2 A$. The $(1,1)$ entry in the product of the two matrices considered above is

   $$g_1(b) f_1(b') + f_2(b) g_2(b')$$
   $$= g_1(b)[f_1(b) + c\varphi_1(b, c, d)] + f_2(b)[g_2(b) + c\psi_2(b, c, d)]$$
   $$= c + cg_1(b)\varphi_1(b, c, d) + ef_2(b)\psi_2(b, c, d)$$
   $$= c[1 + g_1(b)\varphi_1(b, c, d) + f_2(b)\psi_2(b, c, d)].$$

   Thus we can define in the expression in the bracket to be m_{11}. Similarly we can define m_{12}, m_{21}, m_{22}. We can conclude on formal grounds, that $\det(m_{ij}) = 1$ and that $f(b)(m_{ij}) = f(b')$.

2. a) Let M be an A-module, I an ideal of A and $z \in \mathrm{Um}(M)$. Then $\bar{z} \in \mathrm{Um}(M/IM)$, where $\bar{z}$ denotes the residue class of z in M/IM and the latter is regarded as an (A/I)-module.

   b) Conversely, if, with the above notations, I consists of nilpotent elements and M is projective, then $\bar{z} \in \mathrm{Um}(M/IM)$ implies $z \in \mathrm{Um}(M)$.

3. (D. Quillen) Let R be an algebra (not necessarily comutative) over A, $f \in A$, and let $\theta \in (1 + TR_f[T])^*$ (=the group of invertible elements in $R_f[T]$ which are congruent to 1 modulo T). Then there exists an integer $k \geq 0$ such that for any $g_1, g_2 \in A$ with $g_1 - g_2 \in f^k A$, there exists $\psi \in (1 + TR[T])^*$ such that $\psi_f(T) = \theta(g_1 T)\theta(g_2 T)^{-1}$.

4. (D. Quillen) Let M be a finitely presented $A[T]$-module suppose that M_m is an extended $A_m[T]$-module for each maximal ideal m of A, then M is extended from A.

5. (Roitman) Let $(x_0, \dots, x_n) \in Um_{n+1}(A), n \geq 2$. Let $y_{n-1}, y_n \in A$ such that $x_{n-1}y_{n-1} + x_n y_n$ is invertible modulo $(x_0, \dots, x_{n-2})$. Show that $(x_0, \dots, x_n)$ and $(x_0, \dots, x_{n-2}, y_{n-1}, y_n)$ are in the same elementary orbit.

6. (Roitman) Let $(x_0, \dots, x_n) \in Um_{n+1}(A), n \geq 2$. Define J(I) to be the intersection of all the maximal ideals containing I. Let $I_j = (x_0, \dots, x_{j-1}, x_{j+1}, \dots, x_{n-1})$ for $0 \leq j \leq n-1$. Show that if $x \equiv x_n$ modulo $(I_0 + \dots + I_{n-1})$ then $(x_0, \dots, x_n)$, $(x_0, \dots, x_{n-1}, x)$ are in the same elementary orbit.

7. (Roitman) Let $(x_0, \dots, x_n) \in Um_{n+1}(A)$, $n \geq 2$. Let t be an elemnet of A which is invertible modulo $(x_0, \dots, x_{n-2})$. Show that $(x_0, \dots, x_n)$ and $(x_0, \dots, x_{n-1}, t^2 x_n)$ are in the same elementary orbit.

8. Open Questions: (Roitman)
   i) Can you replace $I_0 + \dots + I_{n-1}$ by $J(I_0 + \dots + I_{n-1})$ above ?.
   ii) Can you replace $I_0 + \dots + I_{n-1}$ by $\sqrt{(x_0, \dots, x_{n-1})}$ above?.

9. (Vaserstein) Let $(x_0, \dots, x_n), (y_0, \dots, y_n) \in Um_{n+1}(A), n \geq 2$ be in the same elementary orbit.
   (1) For any $m \geq 1$, $(x_0^m, x_1, \dots, x_n), (y_0^m, y_1, \dots, y_n)$ are in the same elementary orbit.
   (2) If $x_0 x_0 = 1$ modulo $(x_1, \dots, x_n)$, $y_0 y_0 = 1$ modulo $(y_1, \dots, y_n)$, then $(x_0, x_1, \dots, x_n), (y_0, y_1, \dots, y_n)$ are in the same elementary orbit.

10. Let $\begin{pmatrix} x_1, \dots, x_n \\ y_1, \dots, y_n \end{pmatrix} \in M_{2,n}(A)$ have right inverse. Let $\delta \in GL_2(A)$. Show that $\delta \begin{pmatrix} x_1, \dots, x_n \\ y_1, \dots y_n \end{pmatrix} = \begin{pmatrix} x_1, \dots, x_n \\ y_1, \dots, y_n \end{pmatrix} \varepsilon$, for some $\varepsilon \in E_n(A)$.

11. (Roitman) Let $(x_0, \dots, x_n) \in Um_n(A)$, $n \geq 2$. Let k be an integer between 0 and $n-1$. Choose $y_k, \dots, y_n \in A$, such that
$$I_2 \begin{pmatrix} x_k \dots x_n \\ y_k \dots y_n \end{pmatrix} + \sum_i^{k-1} A x_i = A.$$
Prove that $(x_0, \dots, x_{k-1}, x_k, \dots, x_n)$ and $(x_0, \dots, x_{k-1}, y_k, \dots, y_n)$ are in the same elementary orbit.

12. Let $\alpha = \begin{pmatrix} x_1, \dots, x_n \\ y_1, \dots, y_n \end{pmatrix} \in M_{2,n}(A)$. Let $I_2(A)$ be the ideal of A generated by 2×2 minors of α, i.e. $I_2(A)$ is the ideal generated of A which is by $\{x_i y_i - x_j y_j \mid 1 \leq i \neq j \leq n\}$. Show that for any $\alpha \in I_2(D)$ there exists β such that $\alpha\beta = dI_2$.

13. **(Karoubi Squares or Patching diagrams)**
   Let $\varphi : B \to A$ be a homomorphism of rings. Let $s \in B$ such that

(i) s is a non-zero-divisor in B.

(ii)$\varphi(s)$ is a non-zero-divisor in A.

(iii) φ indcues an isomorphism $B/sB \to A/\varphi(s)A$.

Then the commutative diagram.

resulting from a situation as above will be called a patching diagram. One also says $B \to A$ is an **analytic isomorphic along** s.

Show that

(a) (ii) and (iii) imply $B/s^nB \to A/\varphi(s)^nA$ is an isomorphism for all n.

(b) B is the fibre product of B_s and A over A_s, i.e. the square is cartesian.

(c) If B is noetherian then (i) follows from (ii) and (iii). This is not true in general; give an example.

Examples. Show that the following are patching diagrams.

(a) Let $Rs + Rt = R$. Then $R \to R_t$ is an analytic isomorphism along s. (Such diagrams are called covering diagrams.)

(b) Let $B = k[[t_1, \cdots, t_{n-1}]][t_n]$, $A = k[[t_1, \cdots, t_n]]$, where k is a field. Let $f \in B$ which is a distinguished monic in t_n, i.e. it is a monic polynomial in t_n with its lower degree coefficients belonging to the maximal ideal of $k[[t_1, \ldots, t_{n-1}]]$. Show that $B \to A$ is an analytic isomorphism along f.

(c) Let $(R, \mathfrak{m})$ be a local ring. Let $f \in R[t]$ be a Weierstrass polynomial, i.e. f is a monic polynomial in t with its lower degree coefficents in $\mathfrak{m}$. Then $R[t] \to R[t]_{(\mathfrak{m},t)}$ is an analytic isomorphism along f.

(d) Let B be a noetherian ring, and s a non-zero divisor in B. Let $A = \widehat{B}^s$ denote the (s)-adic completion of B. (See remarks on completion of a ring, and I-adic completion of rings for the definition). Then $B \to A$ is an analytic isomorphism along s.

(e) Let Λ be a flat $\mathbb{Z}$-algebra. Show that applying $\otimes_{\mathbb{Z}}\Lambda$ to a patching diagram gives a new patching diagram.

Splitting property: Let $B \to A$ be an analytic isomorphism along s. Let $\alpha \in E_n(A_s)$,

(1) If $n \geq 3$, then there exists $\alpha_1 \in E_n(A)$, $\alpha_2 \in E_n(B_s)$ such that $\alpha = (\alpha_1)_s(\alpha_2 \otimes 1)$.

(2) If $n = 2$, there exist $\alpha_1 \in Sl_2(A)$, $\alpha_2 \in E_n(B_s)$ such that $\alpha = (\alpha_1)(\alpha_2 \otimes 1)$.

Patching property Let $\mathbb{P}(R)$ denote the category of all finitely generated projective R-modules show that if B to A is an analytic isomorphism along s then the corresponding square

$$\begin{array}{ccc} \mathbb{P}(B) & \longrightarrow & \mathbb{P}(A) \\ \downarrow & & \downarrow \\ \mathbb{P}(B_s) & \longrightarrow & \mathbb{P}(A_s) \end{array}$$

Chapter 5

1. **An example of a nonsoft vector bundle.**

   Define the subsets D_+, D_- of $S^1 := \{(x,y) \in \mathbb{R}^2 \mid x^2 + y^2 = 1\}$ by $D_+ := \{(x,y) \in S^1 \mid y > 0\}$ and $D_- := \{(x,y) \in S^2 \mid y < 0\}$. On S^1 consider the topology given by the following definition of "open":

   A subset of S^1 is called open, iff it is of the form $\emptyset$ or $S^1 - F$ or $D_+ - F$ or $D_- - F$, where F is a finite set.

   It is easy to check the axioms of a general topological space. Now consider the two open sets

   $$U_1 := D_+ \cup D_- \cup \{(1,0)\} = S^1 - \{(-1,0)\},$$
   $$U_{-1} := D_+ \cup D_- \cup \{(-1,0)\} = S^1 - \{(1,0)\}$$

   We can glue the trivial line bundles over U_1 and U_{-1} by the continuous function

   $$\alpha : U_1 \cap U_{-1} \to \mathbb{R}, \quad \text{with } \alpha(u) = 1 \text{ for } u \in U_1, \ \alpha(U) = -1 \text{ for } u \in U_{-1}$$

   to get some kind of Möbius bundle. But there is no partition of unity subordinate to the open covering (U_1, U_{-1}). Namely every continuous function $U_1 \to \mathbb{R}$, resp. $U_{-1} \to \mathbb{R}$ is constant.

2. Let $R = \mathbb{Z}[2i]$, $S = \mathbb{Z}[i]$. Computer the conductor $\mathfrak{C}(S/R)$.

3. Let $n > 0$ be an integer. Then $\mathbb{Z}[X]/(X^n - 1)$ is isomorphic to the so called group ring of the cyclic group of n elements over $\mathbb{Z}$.

   The exercises which follow are based on the material found in Odds and Ends, Chapter 5.

4. Describe a gluing cocycle for $\mathbb{U}_n$.

5. Suppose that $x \mapsto P(x)$ is a smooth family of projectors of a vector space V parameterized by a connected smooth manifold X. Set $k = \dim \ker P(x)$ and $n = \dim V$ and denote by f the map

   $$f : X \to G_k(V), \ x \mapsto \ker P(x) \in G_k(V).$$

   Show that f is smooth and that the pull-back of $\mathbb{U}_{k,n}$ by f coincides with the vector bundle defined by the family of projections $P(x)$.

6. Suppose $L \to X$ is a smooth complex line bundle over X. Show that $G(L) \cong C^{\infty}(M.\mathbb{C}^*)$.

7. Suppose M is a smooth compact manifold and $E \to M$ is a complex line bundle. A subspace $V \subset C^{\infty}(E)$ is said to be *ample* if for any $x \in M$ there exists $u \in V$ such that $u(x) \neq 0$.

   (a) show that there exist finite-dimensional ample subspaces $V \subset C^{\infty}(E)$.

   (b) Let V be a finite-dimensional ample subspace of $C^{\infty}(E)$. For each $x \in M$ set $V_x = \{v \in V : v(x) = 0\}$. Equip V with a Hermitian metric and denote by $P(x) : V \to V$ the orthogonal projection onto V_x. Show that $\dim \ker P_x = 1$ and the family of projections $\{p(x) : x \in M\}$ is smooth. we thus obtain a complex line bundle $E_v \to V$.

   (c) Show that the line bundle E is isomorphic to E_V. In particular, this shows that E is the pull-back of a universal line bundle over a projective space.

   (d) Suppose that $f.g : M \to \mathbb{CP}^n$ are two (smoothly) homotopic maps. Denote by E_f (resp. E_g) the pull-backs of the universal line bundle $\mathbb{U}_n$ via f (resp. g). Show that $E_f \cong E_g$.

8. Show that the manifold P described above comes with a natural free right G-action and the space of orbits can be naturally identified with X.

9. Regard S^{2n+1} as a real hypersurface in $\mathbb{C}^{n+1}$ given by the equation $|z_0|^2 + |z_1|^2 + \cdots + |z_n|^2 = 1$. The group $S^1 = \{e^{it}; t \in \mathbb{R}\} \subset \mathbb{C}^*$ acts on S^{2n+1} by scalar multiplication. The quotient of this action is obvious $\mathbb{CP}^n$.

   (a) Show that $S^{2n+1} \to \mathbb{CP}^n$ is a principal S^1-bundle. (It is known as the **Hopf bundle**).

   (b) Show that the line bundle associated to it via the tautologival representation $S^1 \to \mathrm{Aut}(\mathbb{C}^1)$ is precisely the universal line bundle $\mathbb{U}_n$ over $\mathbb{CP}^n$.

10. If $E \to X$ is $\mathbb{R}$-vector bundle then as *metric* on E is a section h of $\mathrm{Symm}^2(E^*)$ such that $h(x)$ is positive definite for every $x \in X$. If E is complex one defines similary Hermitian metrics on E. A *Hermitian bundle* is a vector bundle equipped with a Hermitian metric. Show that any metric on a rank n real vector bundle naturally defines an $O(n)$-structure.

Chapter 6

1. Let $A = k[X_1, X_2, \ldots, X_d]$, $B = k[X_1, X_2]/[X_2^2 - X_2 X_1)$, and let $C = k[X_1, X_2, \ldots, X_{d+1}]/(f)$, where f is a homogeneous polynomial in

$X_1, \ldots, X_{d+1}$, monic in X_{d+1} and of degree n.

(1) Show that A, B, C are graded rings with zero-th component $= k$.

(2) Consider the following product of formal power series

(a) $(1 + X_1 + X_1^2 + \ldots)(1 + X_2 + X_2^2 + \ldots) \ldots (1 + X_d + X_d^2 + \ldots)$

(b) $(1 + X_1 + X_1^2 + \ldots)(1 + X_2)$

(c) $(1 + X_1 + X_1^2 + \ldots) \ldots (1 + X_d + X_d^2 + \ldots)(1 + X_{d+1} + \ldots + X_{d+1}^{n-1})$

(i) Show that each monomial of A (resp. B,C) occurs precisely once in the power series (a) (resp. (b), (c)).

(ii) By specializing each $X_j = T$, and comparing the coefficient of T^i in the resulting formal power series in T, find the dimension of each graded component of the rings A,B, C (Hint: The monomials (which survive) form a k-basis.)

(iii) The Hilbert series of a graded ring $A = A_0 \oplus A_1 \oplus \ldots$, with $A_0 = k$, a field, is the power series $\sum_{i=0}^{\infty} \dim_k A_i \, t^i \in \mathbb{Z}[[t]]$. Show that the Hilbert series of the rings A, B, C is a rational polynomial of type $p(t)/(1-t)^d$, where d is the Krull dimension of the ring, and $p(t) \in \mathbb{Q}[t]$.

2. (Rees) Let A be a Noetherian ring, $I = (a_1, \cdots, a_r)$ be an ideal of A. The subring $A[a_1 t, \cdots, a_r t, t^{-1}]$ of $A[t, t^{-1}]$ is called the Rees ring of A with respect to I, and denoted by $R(A, I)$. Show that

(i) $R(A, I) = \{ \sum_{r=-p}^{q} c_r t^r \mid c_r \in I^r, \text{ if } r \geq 0 \}$ is a graded Noetherian ring.

(ii) $(t^{-n}) R(A, I) \cap A = I^n$, for all $n > 0$.

3. (Krull's Intersection theorem) Let A be a Noetherian ring, I an ideal of A. Then x in $\cap_{n \geq 1}^{\infty} I^n$ if and only if $x = ax$ for some $a \in I$.

(Hint: Prove this when I is principal. Deduce the general case by considering $IR(A, I)$.)

4. (Artin-Rees Lemma) Let A be a Noetherian ring, and I, J be two ideals of A. Show that there is an integer k such that if $n > k$,

$$I^n \cap J = (I^k \cap J) I^{n-k}.$$

(Hint: Show that $JR[t, t^{-1}] \cap R(A, I)$ has a finite set of generators of the form $j_r t^r$. Take k to be the greatest exponent of t occuring amongst these elements).

5. (The Principal Ideal Theorem) Let A be a local domain with maximal ideal $\mathfrak{m}$ such that there exists an $\mathfrak{m}$-primary ideal generated by a single element x. Then $\mathfrak{m}$ is the only non-zero prime ideal of A.

(Hint: Show that if $y(\neq 0) \in A$ then (y) is not primary i.e., $x^k \in (y)$ for k chosen satisfying Artin-Rees lemma with $I = (x), J = (y)$. It suffices to show $(x^k, y) = (x^{k+1}, y)$, which follows if $\ell((x^k, y)/(x^{k+1})) = \ell((y)/(x^{k+1}) \cap (y))$. Prove this.)

6. Any injective (resp. surjective) endormorphism φ of an Artinian (resply. Noetherian) R-module M is surjective (resply. injective).

7. Let A be an affine domain over a field k. If A is not a field then its quotient field $Q(A)$ is never algebraically closed.

8. (Rings of Invariants) Let G be a finite subgroup of $GL_n(\mathbb{C})$. Let G act linearly on $\mathbb{C}[X_1,\dots,X_n]=$ Any matrix $\alpha \in GL_n(\mathbb{C})$ acts on the variables by $(X_1,\dots,X_n) \mapsto \alpha(X_1,\dots,X_n)^T$. Extend this action to $\mathbb{C}[X_1,\dots,X_n]$ so that $f \mapsto \alpha(f)$ is an automorphism of $\mathbb{C}[X_1,\dots,X_n]$.

Let $\mathbb{C}[X_1,\dots,X_n]^G = \{f \in \mathbb{C}[X_1,\dots,X_n] \mid \alpha(f) = f, \text{ for all } f \in \mathbb{C}[X_1,\dots,X_n]\}$ $\mathbb{C}[X_1,\dots,X_n]^G$ is a subring of $\mathbb{C}[X_1,\dots,X_n]$, called the **ring of invariants** of G.
(a) Show that dimension $\mathbb{C}[X_1,\dots,X_n]^G = n$.
(b) (Reynolds) Define the map ρ is called the Reynolds operator. Show that ρ is $k[X_1,\dots,X_n]^G$ linear, and $\rho \mid k[X_1,\dots,X_n]^G =$Identity.
(c) If I is an ideal of $\mathbb{C}[X_1,\dots,X_n]^G$,
show that $I\mathbb{C}[X_1,\dots,X_n] \cap \mathbb{C}[X_1,\dots,X_n]^G = I$. Deduce that $\mathbb{C}[X_1,\dots,X_n]^G$ is a (graded) noetherian ring.
(d) Let $f_1,\dots,f_s$ be homogeneous elements of positive degree which generate the maximal graded ideal of positive degree. Show that $R = \mathbb{C}[f_1,\dots,f_s]$. Deduce that $\mathbb{C}[X_1,\dots,X_n]^G$ is a f.g. $\mathbb{C}$-algebra.
(e) (Molien) Show that the Hilbert series of $\mathbb{C}[X_1,\dots,X_n]^G$ is

$$\frac{1}{|G|} \sum_{\alpha \in G} \frac{1}{\det(I - t\alpha)}.$$ (Hint: The Reynolds operator induces a linear map

ρ_i from $k[X_1,\dots,X_n]_i \to k[X_1,\dots,X_n]_i^G$, for all i show that and rank $(\rho_i) = $ trace $(\rho_i) = \dim(R^G)i$.)

9. Let R be a Noetherian ring. Let $\mathfrak{m}$ be a maximal ideal of $R[X_1,\dots,X_n]$. Show that $\mathfrak{m} \cap R$ is a prime ideal of height equal to (height $m - n$).

10. (Zariski) Let k be a field. Assume that E is a f.g. k-algebra which is a field. Then E is an algebraic extension of k, and is a finite dimensional vector space over k. Deduce Hiblert's Nullstellensatz from this.

11. (Artin) Let R be a Noetherian ring, and $\mathfrak{m}$ a maximal ideal of $R[X_1,\dots,X_n]$, $n \geq 1$. Then $R/\mathfrak{m} \cap R$ is a semi-local ring of dimension ≤ 1.

12. (Davis-Geramita) Let R be a noetherian ring. Let $\mathfrak{m}$ be a maximal ideal in $R[X,Y]$. Show that $\mathfrak{m} \cap R[X - Y^s]$ is a maximal ideal for large s.

13. (R.C. Cowsik) Let k be a field. Let $A = k[T^5, T^6, T^7, T^8] \subseteq k[T]$. Let φ be the obvious homomorphism from $k[X,Y,Z,W]$ onto A. Prove that ker φ is a prime ideal generated by $XZ - Y^2$, $YW - Z^2$, $X^3 - ZW$, $W^2 - X^2Y$ and $XW - YZ$. Show that $(\ker \varphi)^{3+n} \neq (\ker \varphi)^{(3+n)}$ for $n \geq 0$. (Hint: It suffices to find $f \in \ker \varphi$, $f \notin (X,Y,Z,W)^{6+2n}$, with $Yf \in (\ker \varphi)^{3+n}$. If

$$I_n = \{XZ - Y^2)^2(W^2 - XY) + (XW - YZ)^2(YW - Z^2)\}(XW - YZ)^n$$

then $g_n \in (\ker \varphi)^{3+n}$, and $Y \mid g_n$. Let $f = g_n Y^{-1}$. check that $f \notin (X, Y, Z, W)^{6+2n}$).

14. Let R be a ring such that every ideal of R is countably generated.

   a) Show that this also holds for the polynomial ring $R[X]$. (Let J be an ideal of $R[X]$ and I_k for $k \in \mathbb{N}$ denote the ideal of R, consisting of 0 and the leading coefficients of the polynomials of degree k in J. Choose a countable generating set E_k of every I_k and to every $a \in E_k \setminus \{0\}$ a polynomial of degree k whose leading coefficient is a. The chosen polynomials make up a generating set of J.)

   b) Show that this also holds for the polynomial ring $A := R[X_i \mid i \in \mathbb{N}]$ in countably infinitely many indeterminates. (Consider the rings $A_n := R[X_i \mid i \leq n]$. Since every polynomial envolves only finitely many indeterminates, we see $A = \bigcup_{n \in \mathbb{N}} A_n$. Hence $I = \bigcup (I \cap A_n)$. So the union of countable generating sets of the $I \cap A_n$ generates I.)

   c) On the other hand there are uncountable chains of ideals in the ring A of b). (One can take the rational numbers ρ as indices of the indeterminates instead of the natural ones. Having done this, for every real number r let I_r be generated by the X_ρ with $\rho \in \mathbb{Q}$, $\rho \leq r$.)

15. Let A be a discrete valuation ring (i.e. a principal domain with only one maximal ideal $\neq (0)$). Or more generally let A be a commutative Noetherian ring of dimension 1 with only finitely many maximal ideals. Show that in the polynomial ring $A[X]$ there are maximal ideals as well of height 2 as of height 1.

16. Let $A \subset B$ be an extension of finite type of domains. Show that there are an $s \in A - (0)$ and 'variables' $x_1, \ldots, x_n \in B$ such that B_s is integral over $A_s[x_1, \ldots, x_n]$. (The meaning of $x_1, \ldots, x_n$ being 'variables' is that the ring homomorphism from the polynomial ring $A[X_1, \ldots, X_n]$ to $A[x_1, \ldots, x_n]$, given by $X_j \mapsto x_j$, is an isomorphism.) (Use Noether normalization.)

17. Let $A \subset K$ be a ring extension of finite type, where K is a field. Show:

   a) there is an $s \in A \setminus (0)$ with $Q(A) = A_s$;

   b) $[K : Q(A)] < \infty$;

   c) (0) is in A not the intersection of the non-zero prime ideals of A.

18. Let A be a Noetherian domain. Show that there is an $x \in Q(A)$ with $Q(A) = A[x]$ if and only if A is semilocal of dimension 1.

19. A ring is called a **Hilbert ring** (or a **Jacobson ring**) if everyone of its prime ideals is an intersection of maximal ideals. (If K is a field, every K-algebra of finite type is a Hilbert ring.)

   Let A be a Hilbert ring and $\varphi : A \to B$ be a homomorphism of finite type. Show:

   a) B is a Hilbert ring.

   b) If $\mathfrak{n}$ is a maximal ideal of B, then $\mathfrak{m} := \varphi^{-1}(\mathfrak{n})$ is a maximal ideal of A and the induced field extension $A/\mathfrak{m} \subset B/\mathfrak{n}$ is finite.

   (Prove b) first. Then let $\mathfrak{p} \in \mathrm{Spec}(B)$, $s \in A \setminus \mathfrak{p}$ and $\mathfrak{q} \supset \mathfrak{p}$ be a prime ideal of B such that $\mathfrak{q}_s$ is maximal in B_s. Note that $(B/\mathfrak{p})_s \neq 0$. Since B_s is of finite type over A, the counterimage $\varphi^{-1}(\mathfrak{q})$ is maximal in A. By Hilbert's Nullstellensatz $B/\mathfrak{q}$ is a Hilbert ring. Hence $\mathfrak{q}$ is an intersection of maximal ideals. Since $s \notin \mathfrak{q}$ we have shown that no $s \in B \setminus \mathfrak{p}$ belongs to all maximal ideals that contain $\mathfrak{p}$.)

20. Let A be a principal domain with infinitely many maximal ideals. Show that every maximal ideal of $A[X_1, \ldots, X_n]$ is of height $n + 1$ and can be generated by $n + 1$ elements.

21. Let R be a discrete valuation ring (i.e. a local principal domain which is not a field) and let p be a generator of its maximal ideal. We consider the ring $A := R[X]$. It is factorial. So its prime ideals of height 1 are principal.

   Some of them are maximal ideals, for e.g. $(pX - 1)$. More general every irreducible polynomial of the form $\sum_{j=0}^{n} a_j X^j$ with $a_j \in R$, $p \nmid a_0$, $p | a_j$ for $j \geq 1$ generates a maximal ideal. Special such polynomials are $p^m X^n - 1$ with $\gcd(m, n) = 1$. (See [45] VI.9.) So there are infinitely many maximal ideals of height 1.

   If $\mathfrak{m}$ is a maximal ideal of height 2 of A. Then $\mathfrak{m} \cap R = pR$. Otherwise there were a chain of three prime ideals lying over the ideal (0) of R. So $\mathfrak{m}$ is of the form (p, f) where $f \in A$ is irreducible modulo p.

   We see: All prime ideals $\mathfrak{p}$ of A are so called complete intersections, i.e. they are generated by the 'right' number, namely $\mathrm{ht}(\mathfrak{p})$, of generators. (See Section 9.1.)

22. Let k be a field and V be a k-vector space of infinite dimension. Define a multiplication on the additive group $k \oplus V$ by $(a, v)(b, w) := (ab, aw + bv)$. Show that $A := k \oplus V$ together with this multiplication is a ring. Show further that A is not Noetherian, but has a Noetherian spectrum. Namely this consists of the single prime ideal $0 \oplus V$.

Chapter 7

1. Let A be a Noetherian ring. Let $I = (a_1, \ldots, a_n, s)$ be an ideal of A, with $s \in I^2$. Show that there exist $a_1, \ldots, a_n \in A$ such that $(a_1 + sc_1, \ldots, a_n + sc_n) = I \cap I'$, where $I' + As = A$, and with height $I' \geq n$.

Chapter 8

1. Make precise the following intuitive definitions of dimension of an affine algebra $\Gamma(X)$ of functions on a subset $X \subset \mathbb{C}^n$.

   - Define $\dim X$ by induction on n; it being clear for $n = 1$. If $X = \mathbb{C}^n$, then its dimension is n. If not, after a change of variables, show that $p = (0, \ldots, 1) \notin X$, and $q = (0, \ldots, a) \in X$, for some $a \in \mathbb{C}$. Show further that the closure $\bar{Y}$ of the image Y of X in $\mathbb{C}^{n-1}$ under the projection map, is an affine algebra. Define $\dim X = \dim \bar{Y}$.

   - Show that there is a point $p \in X$ where $\Gamma(X)_{\mathfrak{m}_p}$ is a regular local ring. (Where $\mathfrak{m}_p$ denotes the maximal ideal corresponding to the point p). In this case X will be a manifold at p. Show that there exist local coordinates $x_1, \ldots, x_n$ of $p \in \mathbb{C}^n$, such that X is locally defined by the vanishing of $x_1 = x_2 = \cdots = x_k = 0$. Define $\dim X = n - k$.

2. (Serre, Murthy) Let M be a finitely generated R-module with $\mathrm{hd}(M) = 1$. Then the following natrual numbers are equal.
   (i) $\min \{t \mid \exists$ an exact sequence $0 \to R^t \to P \to M \to 0$, with P finitely generated projective $\}$.
   (ii) $\min \{\mu(P_1) \mid \exists$ an exact sequence $0 \to P_1 \to P_0 \to M \to 0$, with P_0, P_1 f.g. projective $\}$.
   (iii) $\mu(\mathrm{Ext}^1_R(M, R))$.

3. Let I be an ideal of a commutative ring R. Suppose that I/I^2 is a free R/I module and that $\mathrm{hd}_R I$ is finite. If $\mu(I) = \mu(I/I^2)$ then I is generated by a regular sequence.

4. Let R be a Noetherian ring. Assume that all f.g. projective R-module are free. Show that the following conditions are equivalent.
   (a) I is generated by a regular sequence.
   (b) (i) $\mathrm{hd}_R(I) < \infty$.
       (ii) for any set $y_1, \ldots, y_n \in I$ whose images $\bar{y}_1, \ldots, \bar{y}_n \in I/I^2$ is a basis of I/I^2 the associated unimodular row $(y_1, \ldots, y_n)$ over $R_{ss'}$ can be "lifted" to a unimodular row over $R_{s'}$, for some $s, s' \in R$, $(s, s') = R$. (i.e. the projective $R_{ss'}$-module corresponding to $(u_1, \ldots, u_n)$ is extended

from a projective $R_{s'}$ -module corresponding to a unimodular row over $R_{s'}$.

(c) (i) $\mathrm{hd}_R(I) < \infty$

(ii) I/I is a free R/I-module.

(iii) there is a set $y_1, \ldots, y_n \in I$ with $\bar{y}_1, \ldots, \bar{y}_n \in I/I^2$. a basis of I/I^2 and such that the associated unimodular row $(y_1, \ldots, y_n)$ over $R_{ss'}$ (for some s and s') can be lifted to a unimodular row over $R_{s'}$.

Give an example to show that the conditions (b)(iii) and (c)(iii) are essential.

5. (Mount) Let R_m denote the polynomial ring $\bar{k}[x_1, \ldots, x_m]$ over an algebraically closed field $\bar{k}$. Let $f = (f_0, \ldots, f_n) \in R_m^{n+1}$. f is said to be a **determinantal sequence** if there is a $n \times (n+1)$ matrix M with maximal minors f_i, for each i. Show that if the ideal generated by $f_0, \ldots, f_n$ has homological dimension less than 2, then f is a determinantal sequence.

6. (Ischebeck)

a) We give a new example of a principal domain A with $\mathrm{SK}_1 A \neq 0$. Consequently A is far from being a Euclidean ring. This example seems to be simpler than those in [Ba, 9.2] and [Is]. ($\mathrm{SK}_1 A :=$ $\lim_{\to} \mathrm{SL}_n(A)/E_n(A)$.)

Let $A := S^{-1}\mathbb{R}[x, y]$, where S is generated by $x^2 + y^2$ and all polynomials f with $f(0, 0) \neq 0$. (S is not of the form S_X for some set $X \subset \mathbb{R}^2$.) Then A is a principal domain with $\mathrm{SK}_1 A \neq 0$.

Idea of proof: Being a localization of a factorial ring, A is factorial. Further its dimension is 1, since it is the localization of the local ring of the origin in the real affine plane by $x^2 + y^2$. So A is a principal ideal domain.

Now we consider the matrix

$$\mu := \begin{pmatrix} (x^2 + y^2)^{-1} & 0 \\ 0 & 1 \end{pmatrix} \begin{pmatrix} x & y \\ -y & x \end{pmatrix} \in \mathrm{SL}_2(\mathbb{R}[x, y]_{x^2+y^2}).$$

We argue as follows: If μ became zero in $\mathrm{SK}_1 A$, it would already become zero in $\mathrm{SK}_1(\mathbb{R}[x, y]_{(x^2+y^2) \cdot f})$ for some f with $f(0, 0) \neq 0$. Then there would be a 1–sphere around the origin which would not meet the real zero set of f. But if one restricts μ to this 1–sphere, it would give a nontrivial element of SK_1 of this sphere.

Remark. There is even a principal ideal domain A such that $\mathrm{SK}_1 A_f \neq 0$ for every $f \in A - \{0\}$. Namely let $A := S^{-1}\mathbb{R}[x, y]$, where S is generated by all $(x - a)^2 + (y - b)^2$, $a, b \in \mathbb{R}$ and all polynomials without real zeros.

b) If B is a discrete valuation ring, then it is Euclidean. Therefore $SL_n(B) = E_n(B)$ for every n. So the above matrix μ is locally elementary but not globally.

7. In the Theorem on the Finiteness of Class Number, it is essential that the ring is a domain (or at least reduced). (Swan)

   Let $A := \mathbb{Z}[X]/(X^2) \cong \mathbb{Z} \oplus \mathbb{Z}$ where in the latter additive group multiplication is defined by $(a, b)(a', b') = (aa', ab' + ba')$. For $n \in \mathbb{N}$ define ideals $I_n := n\mathbb{Z} \oplus \mathbb{Z} \subset \mathbb{Z} \oplus \mathbb{Z}$. Then $(0, 1)I_n = 0 \oplus n\mathbb{Z}$, hence $I_n/(0, 1)I_n \cong \mathbb{Z} \oplus (\mathbb{Z}/(n))$. So for $n \neq m$ the ideals I_n and I_m cannot be isomorphic.

8. Show that that every regular, not necessarily local ring is reduced.

9. Let A be a commutative ring such that there are elements $a, b \in A$ with $(a) = \text{Ann}(b)$ and $(b) = \text{Ann}(a)$. Show that $A/(a)$ as an A-module has infinite projective dimension if neither $A/(a)$ nor $A/(b)$ is projective (i.e. a direct factor of A). One can construct an explicit infinite free resolution.

   This is enough to classify those residue class rings of $\mathbb{Z}$ (or of any other principal domain) whose modules have finite projective dimension. One does not need Theorem 8.5.8.

10. a) Show that a left A-module F is flat if and only if for any (finitely generated) right ideal I of A the group homomorphism $I \otimes_A F \to F$, induced by the inclusion $I \hookrightarrow A$ is injective.

    b) Conclude that a module over a Dedekind ring is flat if and only if it is torsion free. (Show first that flatness localizes.)

11. a) Let k be a field. Show: If F is flat over $k[X_1, \dots, X_n]$, then X_i is a non-zero-divisor of $F/(X_1, \dots, X_{i-1})F$ for every $i = 1, \dots, n$. (If $i = 1$, then $(X_1, \dots, X_{i-1})F = 0$.)

    b) Show that the ring C_n of continuous functions $f : \mathbb{R}^n \to \mathbb{R}$ is not flat over the polynomial ring $\mathbb{R}[X_1, \dots, X_n]$ for $n > 1$. Clearly it is flat for $n = 1$.

12. Let $\mathbb{P}$ be the set of prime numbers. The the torsion group of the abelian group $G := \prod_{p \in \mathbb{P}} \mathbb{Z}/p$ is $T := \bigoplus_{p \in \mathbb{P}} \mathbb{Z}/p$. We know already by Exercise I.9 that the residue class group G/T is a $\mathbb{Q}$-vector space. Derive that the canonical projection $\kappa : G \to G/T$ admits no cross section, i.e. there is no homomorphism $\alpha : G/T \to G$ with $\kappa \circ \alpha = \text{id}_{G/T}$. So G is *not* a direct sum of its torsion and its torsion free part.

13. Consider the ring $A := \mathbb{Z}[\sqrt{-5}] = \mathbb{Z} + \mathbb{Z}\sqrt{-5}$. It is the integral closure of $\mathbb{Z}$ in $\mathbb{Q}(\sqrt{-5})$, hence a Dedekind ring.

a) Show that $1, -1$ are the only units of this ring. (Use the norm $N : \mathbb{Q}(\sqrt{-5}) \to \mathbb{Q}$, $a + b\sqrt{-5} \mapsto (a + b\sqrt{-5})(a - \sqrt{-5}) = a^2 + 5b^2$ for $a, b \in \mathbb{Q}$. It is multilicative, since the map $\sigma : A \to A$, $a + b\sqrt{-5} \mapsto a - b\sqrt{-5}$ is an automorphism.

b) Show (using again the norm) that the elements $2, 3, 1 \pm \sqrt{-}, 2 + \sqrt{-5}$ are irreducible, i.e. no nontrivial products.

c) Show that A is not factorial by computing $2 \cdot 3$, 3^2, $(1 + \sqrt{-5})(1 - \sqrt{-5})$, $(2 + \sqrt{-5})(2 - \sqrt{-5})$.

d) Prove the following equalities of ideals of A:

$$(3, 2 + \sqrt{-5}) = (3, 1 - \sqrt{-5}) , \quad (2, 1 + \sqrt{-5}) = (2, 1 - \sqrt{-5})$$
$$(3, 2 + \sqrt{-5})^2 = (2 + \sqrt{-5}) , \quad (3, 2 + \sqrt{-5})(3, 2 - \sqrt{-5}) = (3),$$
$$(2, 1 + \sqrt{-5})^2 = (2), \quad (2, 1 + \sqrt{-5})(3, 1 + \sqrt{-5}) = (1 + \sqrt{-5})$$

e) Show that $(3, 2 + \sqrt{-5})$, $(2, 1 + \sqrt{-5})$ are maximal ideals of A. What are the decompositions in prime ideals of (6) and (9)?

14. FINITE EXTENSIONS OF DEDEKIND RINGS Let $A \subset B$ be a finite extension of Dedekind rings. If $S := A \setminus (0)$, then $S^{-1}B$ is a finite extension of $S^{-1}A = Q(A)$, hence a field, hence $S^{-1}B = Q(B)$. Since $B \supset A$ is finite, $n := [Q(B) : Q(A)] < \infty$. As a finitely generated torsion-free A-module, B is projective and finitely generated. Further its rank is constant; for A is a domain and so $\mathrm{Spec}(A)$ is connected. Let $\mathrm{rk}_A B = n$.

Now let $\mathfrak{m}$ be any maximal ideal of A. Then $B_\mathfrak{m}(:= T^{-1}B$ with $T = A\setminus\mathfrak{m})$ is a free $A_\mathfrak{m}$-module of rank n. Therefore

$$\dim_{A/\mathfrak{m}} B/\mathfrak{m}B = \dim_{A_\mathfrak{m}/\mathfrak{m}A_\mathfrak{m}} B_\mathfrak{m}/\mathfrak{m}B_\mathfrak{m} = n.$$

Let $\mathfrak{p}_1, \ldots, \mathfrak{p}_r \in \mathrm{Spec}(B)$ be the prime ideals over $\mathfrak{m}$, i.e. those which contain $\mathfrak{m}B$. Then $A/\mathfrak{m} \hookrightarrow B/\mathfrak{p}_i$ are field extensions. Let f_i denote their degrees: $[B/\mathfrak{p}_i : A/\mathfrak{m}] = f_i$.

In the above situation let $\mathfrak{m}B = \mathfrak{p}_1^{e_1} \cdots \mathfrak{p}_r^{e_r}$ be the factorization of $\mathfrak{m}B$ in B. Then $\sum_{i=1}^r e_i f_i = n$.

Idea of the proof. In B we have the following finite sequence of ideals:

$$B \supset \mathfrak{p}_1 \supset \mathfrak{p}_1^2 \supset \cdots \supset \mathfrak{p}_1^{e_1} \supset \mathfrak{p}_1^{e_1}\mathfrak{p}_2 \supset \cdots \supset \mathfrak{p}_1^{e_1}\mathfrak{p}_2^{e_2} \supset \cdots \supset \mathfrak{p}_1^{e_1} \cdots \mathfrak{p}_r^{e_r-1} \supset \mathfrak{m}B.$$

Every factor module of subsequent members of this sequence is of the form $I/\mathfrak{p}_i I$ with some ideal I of B and some $i \leq r$. But $I/\mathfrak{p}_i I \cong B/\mathfrak{p}_i$. So for every $i = 1, \ldots, r$ we have e_i factors isomorphic to $B/\mathfrak{p}_i$. And $B/\mathfrak{p}_i$ is a vector space of dimension f_i over $A/\mathfrak{m}$. This finishes the proof.

15. Let R be a principal domain and M a finitely generated R-module. Let $M \cong R/(f_1) \oplus \cdots \oplus R/(f_r)$ with $f_1|\cdots|f_r$. Prove $\mu(M) = r$. (The divisibility condition is necessary, since for e.g. $\mathbb{Z}/(6) \cong \mathbb{Z}/(2) \oplus \mathbb{Z}/(3)$.)

Show further that $I_r(M) = R$, if and only if $\mu(M) \leq r$.

16. Let R be a domain. In the polynomial ring $A := R[X_1,\ldots,X_n]$ consider the ideal $I := (X_1^{k_1},\ldots,X_r^{k_r})$ with positive integers $r \leq n, k_1,\ldots,k_r$. Show that $\mathrm{Ass}(A/I)$ consists of exactly one element, namely $\mathfrak{p} := (X_1,\ldots,X_r)$.

(To see the general scheme you could first assume $r = 2$. Then consider first the case $k_2 = 1$. You have the following composition series of A/I:

$$0 \subset (X_1^{k_1-1}, X_2)/I \subset (X_1^{k_1-2}, X_2) \subset \cdots \subset \mathfrak{p}/I \subset A/I$$

The factors of this composition series all are isomorphic to $R/\mathfrak{p}$. By Proposition 8.4.7 b) the statement is true in this case.

For general $k_2 \geq 1$ you have the composition series

$$0 \subset (X_1^{k_1}, X_2^{k_2-1})/I \subset (X^{k_1}, X_2^{k_2-2}) \subset \cdots \subset \mathfrak{p}/I \subset A/I$$

Every factor of this is isomorphic to $A/(X_1^{k_1}, X_2)$, and you know already $\mathrm{Ass}(A/(X_1^{k_1}, X_2)) = \{\mathfrak{p}\}$. Again use (8.4.7).

In the general case use induction on s for the claim:
The statement is true in the case $k_{s+1} = \cdots = k_r = 1$.
The case $s = 0$ is obvious. The case $s = 1$ is proved as the first case above, the step $s \to s+1$ as the second step above.)

17. a) Let A be a regular ring of dimension ≤ 2 and N a finitely generated A-module. Show that $N^* := \mathrm{Hom}_A(N, A)$ is projective. Especially every finitely generated reflexive A-module is projective. (An A-module M is called reflexive if the canonical map $M \to M^{**}$ is an isomorphism.) Note that from a projective resolution $Q \to P \to N \to 0$ one derives an exact sequence $0 \to N^* \to P^* \to Q^* \to E \to 0$ with a suitable E. Since $\mathrm{hd}(E) \leq 2$ we get that N^* is projective.

b) Let A be as above and $S \subset A$ be multiplicative. Show that then $\mathbb{P}(A) \to \mathbb{P}(S^{-1}A)$ is surjective. (If P is finitely generated projective over $S^{-1}A$ there is a finitely generated A-submodule $M \in P$ with $S^{-1}M = P$. Then also $S^{-1}M^{**} = P$.)

18. Let k be a field and S a multiplicative subset of the polynomial ring $A := k[X_1,\ldots,X_n]$. Show that every finitely generated projective module P over $S^{-1}A$ is stably free.

(There is a finitely generated A-module M with $P \cong S^{-1}M$. There is a projective resolution

$$0 \longrightarrow Q_n \longrightarrow \cdots \longrightarrow Q_0 \longrightarrow M \longrightarrow 0$$

with finitely generated Q_j, whic is actually free. This induces a free resolution

$$0 \longrightarrow S^{-1}Q_n \longrightarrow \cdots \longrightarrow S^{-1}Q_0 \longrightarrow P \longrightarrow 0.$$

over $S^{-1}A$. This 'splits' everywhere; so show that P is stably free.)

19. A ring R is called hereditary, if all of his ideals are projective R-modules. (So Dedekind rings are hereditary.)

   a) Show that R is hereditary if and only if every submodule of any projective (or free) R-module is projective.

   b) Show that every hereditary domain is a Dedekind ring.

   c) Show that every hereditary Noetherian commutative ring is a finite direct product of Dedekind rings.

   d) Let K be a field and A the subring of the infinite product $K^{\mathbb{N}}$, consisting of all 'nearly constant' sequences. Show that A is hereditary. (An infinite sequence (a_n) is called nearly constant, if there is an $N \in \mathbb{N}$ such that $a_n = a_m$ for all $n, m \geq N$.)

Chapter 9

1. Let M be a finitely generated R-module. Let S be a multiplicatively closed subset in R, and I be an ideal of R. Assume S is disjoint from I and $\mathrm{Ann}(M)$. Show that the grade of M in I (i.e. the length of the maximal M-sequence in I) is $\leq$ grade of $S^{-1}M$ in $S^{-1}I$.

2. (McRae) Let R be a Noetherian ring. Let I be an ideal of R such that each associated prime ideal of I has grade one. Let $I^{-1} = \mathrm{Hom}_R(I, R)$. Assume that I has a finite free resolution.

   (i) Show that $\sqrt{II^{-1}}$ is contained in a associated prime ideal $\mathfrak{p}$ of I then $I_{\mathfrak{p}}$ is both projective and non-invertible!

   (ii) If $b \in \sqrt{II^{-1}}$ but not in any associated prime ideal of R then show that there $a \in I$ such that $I = aR : b^n R$, for some n. (Hint: I_b is principal).

3. (McRae) Let R be a Noetherian domain. Let I be an ideal of R such that each associated prime ideal of I has grade one. If $\mathrm{hd}_R I$ is finite then $\mathrm{hd}_R I = 0$. (Hint: Show thatif II^{-1} is a proper ideal then it has grade

atleast 2. Let $\mathfrak{p}$ be a prime idealof R containing II^{-1}. Show that $I_{\mathfrak{p}}[X]$ is not an invertible ideal of $R[X]$. Deduce that $I_{\mathfrak{p}}[X]I_{\mathfrak{p}}[X]^{-1}$ has an element $p(x)$, not belonging to any associated prime ideal of $I_{\mathfrak{p}}[X]$, and which is a prime. Now use McRae's characterisation to show that $I_{\mathfrak{p}}[X]$ is principal!)

4. Let R be a Noetherian domain. Let I be an ideal of finite homological dimension generated by two elements. Show that its homological dimension is atmost one.

5. (Auslander-Buchsbaum) Let $(R, \mathfrak{m})$ be a local ring, and M a R-module of finite homological dimension. Show that $hd_R(M)$ can atmost be grade of $\mathfrak{m}$, and that equality holds if and only if $\mathfrak{m}$ is an associated prime of M.

6. (Rees) Let M be a finitely generated R-module. If $\mathfrak{p}$ is an associated prime ideal of M then grade of $\mathfrak{p}$ is atmost $hd_R M$.

7. (McRae) Let R be a Noetherian ring, and M be a finitely generated R-module. Let $\mathfrak{p}$ be an associated prime ideal of M. Let S be a multiplicatively closed subset of R which is disjoint from $\mathfrak{p}$ and $Ann(M)$. Show that if $hd(M)$ is finite then grade of $\mathfrak{p}$ equals grade of $S^{-1}\mathfrak{p}$.

8. Here is an example of a curve in affine 3-space which is not an ideal theoretic complete intersection, but is a set theoretic one (even not locally). Let k be an algebraically closed field, and let $C := \{(t^3, t^4, t^5) \mid t \in k\}$. Set $I := I(C) \subset k[X, Y, Z]$ and $\mathfrak{m} := (X, Y, Z) \subset k[X, Y, Z]$. Show that:

   a) If $f_1 := XZ - Y^2$, $f_2 := X^3 - YZ$, $f_3 := X^2Y - Z^2$, then $f_1, f_2, f_3 \in I$.

   b) If $f(X, Y.Z) \in I$, then the linear part of f is zero. So $I \subset \mathfrak{m}^2$.

   c) f_1, f_2, f_3 are linearly independent modulo $\mathfrak{m}^3$, hence also modulo $\mathfrak{m}I$. So I is at the point, defined by $\mathfrak{m}$ not a complete intersection, i.e. $\mu_{\mathfrak{m}}(I) \geq 3$.

   d) C is set-theoretically defined by $XZ - Y^2$, $X^5 + Z^3 - 2X^2YZ$.

9. Let I be generated by a R-sequence a_1, a_2. Consider the Koszul resolution $K(a_1, a_2) : 0 \to R \xrightarrow{\phi} R^2 \xrightarrow{\psi} I \to I \to 0$, where $\phi(1) = (a_2, -a_1) \in R^2$, $\psi(e_1) = a_i$, $i = 1, 2$. Analyse the induced $\mathrm{Hom}(-, I)$ sequence

$$\mathrm{Hom}_R(R^2, R) \xrightarrow{\alpha} \mathrm{Hom}(R, R) \xrightarrow{\beta} \mathrm{Ext}^1(I, R) \longrightarrow 0$$
$$\| \qquad\qquad\quad \Big\| e \qquad\qquad\quad \Big\downarrow \gamma$$
$$\mathrm{Hom}(R^2, R) \longrightarrow R \xrightarrow{\eta} R/I \longrightarrow 0$$

(a) Show that the top row it is exact.
(b) $\alpha(1) = [K(a_1, a_2)]$ generates $\mathrm{Ext}^1(I, R)$.
(c) Show that the bottom row is exact.

(d) Define a map $\gamma : \mathrm{Ext}^1(I, R) \to R/I$ so that the above diagram is commutative (Hint: $\gamma([K(a_1, a_2)]) = \bar{1} \in R/I$).

(e) Deduce that $\mathrm{Ext}(I, R) \simeq R/I$.

10. Let R be a Noetherian ring, and $I \neq R$ an ideal of R. Let $\mathfrak{p}$ be a prime ideal of R containing I. Show that if $I_\mathfrak{p}$ is generated by a regular sequence of length n, then there exists $s \notin \mathfrak{p}$ such that I_s is generated by a regular sequence of length n.

11. Show that R is Cohen-Macaulay if and only if every complete intersection ideal of R is generated by an R-sequence if and only if every complete intersection ideal of R is unmixed.

12. Show that I is a complete intersection ideal of R of $\mathrm{hd}_R I = 1$ if and only if I is generated by a R-sequence of length 2.

13. (Sally-Vasconcelos) Let R be an affine algebra over a field k of dimension ≤ 2. Then $\{\mu(\mathfrak{p}) \mid \mathfrak{p} \in \mathrm{Spec}(R)\}$ is bounded.

14. (**Fitting ideals**) Let $\varphi = R^n \to R^m$ be a linear map, and let $M(\varphi) \in M_{m,n}(R)$ denote the matrix of φ. $F_r(\varphi)$ denotes the ideal generated by $t \times t$ minors of $M(\varphi)$.

   a) Show that $F_t(\varphi)$ only depends on φ, and independent of choice of bases. They are called the Fitting ideals of M.

   b) Let $R^m \overset{\varphi}{\to} R^n \to M \to 0$, $R^n \overset{\psi}{\to} R^\mathfrak{p} \to M \to 0$ be finite free presentations of the R-module M. Show that $F_{n-k}(\varphi) = F_{\mathfrak{p}-k}(\psi)$, for all $k \geq 0$. (Hence $I_{n-k}(\varphi)$ is also called the k-th Fitting invariant of M).

   c) If S is an R-algebra then show that $F_t(\varphi \otimes_R S) = F_t(\varphi)S$. Let M have a finite free presentation $R^n \overset{\varphi}{\to} R^m \to M \to 0$.
   Let $\mathfrak{p}$ be a prime ideal of R.

   d) Show that the following are equivalent:

   (i) $F_t(\varphi) \not\subset \mathfrak{p}$.

   (ii) $(\mathrm{im}(\varphi))_\mathfrak{p}$ contains a unimodular element.

   (iii) $\mu(M_\mathfrak{p}) \leq m - t$.

   e) Show that the following are equivalent :

   (i) $F_t(\varphi) \not\subset \mathfrak{p}$ and $F_{t+1}(\mathfrak{p})_\mathfrak{p} = 0$.

   (ii) $(\mathrm{Im}\varphi)_\mathfrak{p}$ is a free direct summand of $R^m_\mathfrak{p}$ of rank 1. (iii) $M_\mathfrak{p}$ is free of rank $(m - t)$.

   Consequently, M is a R-projective of rank $m - r$ if and only if $F_r(\varphi) = R$ and $F_{r+1}(\varphi) = 0$.

f) Prove that $\varphi = R^n \to R^m$ is an homomorphism then rank (φ) (=rank $M(\varphi)$) $= r$ if and only if $F_r(\varphi)$ has a R-regular element, and $F_{r+1}(\varphi) = 0$.

15. Let M be a finitely generated R-module. Let $F_i(M)$ be its Fitting ideals.

a) If R is a local ring with maximal ideal $\mathfrak{m}$ then show that $\mu(M) = r$ if and only if $F_{r-1}(M) \subset \mathfrak{m}$, and $F_r(M) = R$.

b) M is projective of rank r if and only if $F_0(M) =?$, $F_{r-1}(M) = 0$, $F_r(M) = R$.

16. (Geyer) Consider the subring $S = k[a^2, b, c, ab, ac]$ of $k[a, b, c,]$. Show that $S \approx k[X_1, X_2, X_3, X_4, X_5]/\mathfrak{p}$, where $\mathfrak{p}$ is the prime ideal $(X_4^2 - X_1 X_2^2,\ X_5^2 - X_1 X_3^2,\ X_5 X_2 - X_3 X_4,\ X_5 X_4 - X_1 X_2 X_3)$.

Show that $\bar{S} = S/S X_2$ is not a Cohen-Macaulay ring. (Hint: $\mathrm{Ann}(\bar{X}_4) = (\bar{X}_3, \bar{X}_4, \bar{X}_5) \subset \mathrm{Ass}(\bar{S})$, and contains the minimal prime ideal $(\bar{X}_4)$. Deduce that S is not a Cohen-Macaulay ring.

17. (Geyer) Let char $k = 2$ in above example. Let $f_1 = X_5^2 - X_1 X_3^2$, and $f_2 = X_4^2 - X_1 X_2^2$. Show that
(i) f_1, f_2 are relatively prime.
(ii) $V(f_1, f_2) \subseteq A_k^5$ is a complete intersection.
(iii) $\bar{R} = k[X_1, \ldots, X_5]/(f_1, f_2)$ is a Cohen-Macaulay ring.
(iv) $\bar{R}/\sqrt{0} \simeq S$.

18. Let A be a Noetherian ring. Let I be an ideal of A. Then $a_1, \ldots, a_n \in I$ generate I if (and only if) $a_1, \ldots, a_n$ generate I modulo I^2 and $\sqrt{I} = \sqrt{(f_1, \ldots, f_n)}$, i.e. $V(I) = V(f_1, \ldots, f_n)$.

19. Let A be a Noetherian ring of dimension d. Let I be an ideal of A such that I/I^2 is generated by $n \geq d+1$ elements. Then I can be generated by n elements.

20. Let A be a Noetherian ring. Let I be an ideal of A. If $f_1, \ldots, f_n \in I$ generate I modulo I^2 and if every maximal ideal which contains $(f_1, \ldots, f_n)$ also contains I, then show that $f_1, \ldots, f_n$ generate I.

21. (Northcott) Let M be a Noetherian R-module, I an ideal of R, and $r_0 \in R$. Show that there is an integer $q \geq 0$ such that $(I^n M :_M r_0) = I^{n-q}(I^q M :_M r_0) + (0 :_M r_0)$, for all $n \geq q$. Consequently, for $n \geq q$,

$$(I^n M :_M r_0) \subset I^{n-q} M + (0 :_M r_0),$$

22. (Northcott) Let $M_1, \ldots, M_s$ be submodules of a Noetherian R-module M, and let I be an ideal of R. Show that there exists an integer $q \geq 0$ such that $\cap_{i=1}^s I^n M_i = I^{n-q}(\cap_{i=1}^s I^q M_i)$.

23. Let $a_1, \ldots, a_s$ be a R-sequence on M. Let $s \geq 2$. Show that $(a_1, \ldots, a_{s-2}, a_s) M :_M (a_{s-1}) = (a_1, \ldots, a_{s-2}, a_s) M$. Deduce that $a_1, \ldots, a_{s-2}, a_s, a_{s-1}$ is a R-sequence on M if and only if $(a_1, \ldots, a_{s-2}) M :_M (a_s) = (a_1, \ldots, a_{s-2}) M$.

24. Let M be a R-module, and let I be an ideal of R. Let $x, y \in I$ which are not zero-divisors on M. Show that $(xM :_M I)/xM \approx (yM :_M I)/yM$ as R-modules.

25. Let M be an Noetherian R-module, I be an ideal of R such that $IM \neq M$. Let $a_1, \ldots, a_m$, and $b_1 \ldots, b_n$ be R-sequence on M in I. If $m < n$ then show that there exist $a_{m+1} \ldots, a_n$ so that $a_1, \ldots, a_m, a_{m+1}, \ldots, a_n$ is a R-sequence on M in I.

26. Let M be a Noetherian R-module and I be an ideal of R with $IM \neq M$. If $J \subset \sqrt{I}$ is an ideal with $\sqrt{J} = \sqrt{I}$ then show that $JM \neq M$ and the grade of J w.r.t. M is equal to the grade of I w.r.t. M.

27. (Cohen) Let k be a field. Show that $A = k[X^2, X^3, Y, XY] \subset k[X, Y]$ is not Cohen -Macaulay.

    Let $\varphi = k[x, y, u, v] \to A$ be the onto homomorphism which takes x, y, u, v, to X^2, Y, X^3, XY respectively. Show that $\ker \varphi$ is generated by 2×2 minors of

    $$\begin{pmatrix} x & y & u & v \\ u & v & x^2 & xy \end{pmatrix}$$

28. (Upper semicontinuity of μ) Let $I, \mathfrak{p}$ be ideals of a ring R, with $\mathfrak{p}$ a prime ideal, and I f.g. Show that there is a $s \in R - \mathfrak{p}$ such that $\mu(IR_s) = \mu(IR_\mathfrak{p})$.

29. (Davis-Geramita) Let R be a commutative (not necessarily noetherian) ring. Let $\mathfrak{m}$ be a maximal ideal of $R[X_1, \ldots, X_n]$ and let $\mathfrak{p} = \mathfrak{m} \cap R$. Assume that $\mathfrak{p}$ is f.g. Then

    (1) If $\mathfrak{p}$ is maximal $\mu(\mathfrak{m}) \leq \mu(\mathfrak{p}) + n$.

    (2) In general, $\mu(\mathfrak{m}) \leq \mu(\mathfrak{p}R_\mathfrak{p}) + n + 1$

    (3) If $n > 0$, and $\mathfrak{p}$ maximal, $\mu(\mathfrak{m}) \leq \mu(\mathfrak{p}R_\mathfrak{p}) + n$

30. (Hochster) Let $t < r < s$ be integers, and let K be a field of characteristic zero. Let $A = K[X_{ij}]$ be the ring of polynomials in rs variables, and let $I_t(X)$ be the ideal generated by the $t \times t$ minors of the $r \times s$ matrix (X_{ij}). Then show that $I_t(X)$ is not set theoretically a complete intersection.

31. Prove the following version of the Eisenbud-Evans theorem: Let R be a commutative ring, M be a R-module, $\mathcal{P}$ be a subset of $\mathrm{Spec}(R)$, and $\delta : \mathcal{P} \to \mathbb{N} \cup \{0\}$ be a generalized dimension function on $\mathcal{P}$. Let $M' \subset M$ be $(\delta(\mathfrak{p}) + 1)$-fold basic in M at all $\mathfrak{p} \in \mathcal{P}$. Then M' contains an element m' which is basic in $M_\mathfrak{p}$ for every $\mathfrak{p} \in \mathcal{P}$.

In particular, if $(r,m) \in R \oplus M$ is basic at all $\mathfrak{p} \in \mathcal{P}$, then there exists an element $m' \in M$ such that $m + rm'$ is basic at all primes of $\mathcal{P}$.

32. Prove the Eisenbud-Evans estimates for a ring R having a g.d.f. δ.
    EE I: A projective R-module P of rank $>$ g.d. R has a unimodular element.
    EE II: A projective R-module P of rank $>$ g.d. R is cancellative.
    EE III: A R-module M is generated by $e_\delta(M) = \sup_\mathfrak{p}\{\mu_\mathfrak{p}(M) + \delta(\mathfrak{p})\}$ elements.

33. (Plumstead) Prove the Eisenbud-Evans conjectures over a polynomial ring: Let R be a commutative noetherian ring of dimension d, $A = R[X]$, P be a finitely generated projective A-module of rank $\geq d+1$, and M be a finitely generated A-module. The show that P has a unimodular element, and is cancellative, and deduce from these two facts that

$$\mu(M) \leq \sup{}_\mathfrak{p}\{\mu_\mathfrak{p}(M) + \dim A/\mathfrak{p} \mid \mathfrak{p} \in \mathrm{Spec}(A), \dim(A/\mathfrak{p}) < \dim(A)\}.$$

Chapter 10

1. Let C be the affine space curve with parametric equations $X = t^3$, $Y = t^4$, $Z = t^5$. Show that the ideal $I(C)$ of polynomial functions vanishing on $C = \{(t^3, t^4, t^5), t \in \bar{k}\}$ is generated by the maximal minors of the matrix
$$\begin{pmatrix} X & Y & Z \\ Y & Z & X^2 \end{pmatrix}$$

   Let $\mathfrak{m} = (X, Y, Z)$ and observe that $I \subset \mathfrak{m}^2$, but $I \not\subset \mathfrak{m}^3$. Show that $I/\mathfrak{m}^3$ is a k-vector space of dimension three. Deduce $\mu(I) = 3$, and that C is not a complete intersection (i.e. ht $I \neq \mu(I)$).

   Show that $C = V(XZ - Y^2) \cap V(X^5 + Z^3 - 2X^2YZ)$ is a set-theoretic complete intersection.

2. Show that the ideal $I = (Y^2 - X^3)$ in $k[X, Y]$ is locally a complete intersection ideal at primes $\mathfrak{p} \supset I$, but that $k[X, Y)/I$ is not a regular ring.

3. (Scheja-Storch) Let $C \subseteq A^n_k$ be a non-singular curve. Show that after applying a suitable k-automorphism of $\bar{k}[X_1, \ldots, X_n]$ the canonical injection $\bar{k}[X_1, X_2, X_3]/I(C) \cap \bar{k}[X_1, X_2, X_3] \hookrightarrow \bar{k}[X_1, \ldots, X_n]/I(C)$ is an isomorphism.

   Deduce C is an complete intersection if $C' = V(I(C) \cap \bar{k}[X_1, X_2, X_3])$ is.

4. Let I be generated by a R-sequence a_1, a_2. Show that $\wedge^2(I/I^2)$ is a free R/I-module of rank 1 generated by $\bar{a}_1 \wedge \bar{a}_2$. Deduce that the evaluation map $e : \mathrm{Hom}_{R/I}(\wedge^2(I/I^2), R/I) \to R/I$ defined by $e(h) = h(\bar{a}_1 \wedge \bar{a}_2)$ is an R/I-module isomorphism.

5. Let R be a Noetherian ring. Let $I \neq R$ be an ideal of which is locally generated by a regular sequence of length 2 at every prime ideal containing I. Prove that $\operatorname{Ext}^1(I, R) \simeq \operatorname{Hom}_{R/I}(\Lambda^2(I/I^2), R/I)$.

6. Let $X \subset \mathbb{C}^3$ be a complete intersection of two equations, say f and g. Let $J = \begin{pmatrix} f_x & f_y & f_z \\ g_x & g_y & g_z \end{pmatrix}$ be the Jacobian matrix. Using the Implicit function theorem, or otherwise, show that there is a point $p \in X$ at which atleast one of the 2×2 minors of J does not vanish. Deduce that there exist $a, b, c \in \mathbb{C}[x, y, z]$ such that $aJ_{12} + bJ_{23} + cJ_{31}$ does not vanish anywhere on X (i.e. the 1-form $\omega = a\,dx + b\,dy + c\,dz$ does not vanish on X). Show that the existence of a nowhere vanishing 1 form on X implies that X is a complete intersection. Construct examples of non-complete intersection curves.

7. Two surfaces (f) and (g) through a point $\underline{a} \in A_k^3$ are said to meet transversally at $\underline{a}$ if
   (1) $\underline{a}$ is a simple point on each,
   (i.e. $f\left(\frac{\partial f}{\partial x}, \frac{\partial f}{\partial y}, \frac{\partial f}{\partial z}\right)|_{\underline{a}} \neq 0$, $\left(\frac{\partial g}{\partial x}, \frac{\partial g}{\partial y}, \frac{\partial g}{\partial z}\right)|_{\underline{a}} \neq 0$.
   (2) The surfaces have distinct tangent planes at $\underline{a}_0$.
   Show that a curve $C \subseteq A_k^3$ is a set-theoretic complete intersection of two surfaces $f = 0$, $g = 0$ which are transversal at every point of C if and only if C is non-singular and a complete intersection of f, g.

8. Let $C \subseteq A_k^3$ be the affine space curve with parametric equation $x = t$, $y = t^2$, $z = t^3$. Show that the ideal $I(C)$ of polynomial functions vanishing on $C = \{(t, t^2, t^3) \mid t \in \bar{k}\}$ is generated by the maximal minors of $\begin{pmatrix} 1 & X & Y \\ X & Y & Z \end{pmatrix}$
   (Such ideals are called determinental ideals).

9. Show that the ideal of non-singular curve in A_k^3 has a generating set consisting of 2×2 subdeterminants of a 2×3 matrix over $k[X, Y, Z]$.

10. Let I be a height 2 unmixed ideal of $k[X_1, \ldots, X_n]$ which is a local complete intersection ideal at every prime containing I. Then I is a complete intersection ideal if and only if $\operatorname{Ext}^1(I, R)$ is cyclic.

11. Given an example of a complete intersection ideal I which is not generated by a R-sequence.

12. (J. Herzog) The curve $C = \{(t^{n_1}, t^{n_2}, t^{n_3})) \ t \in \bar{k}\}$, with g.c.d. of $(n_1, n_2, n_3) = 1$ is a complete intersection if and only if the subsemigroup of $(\mathbb{N}, +)$ generated by n_1, n_2, n_3 is "symmetric" (i.e. there is an $m \in \mathbb{Z}$ such that for all $z \in \mathbb{Z}$, $z \in H$ if and only if $m - z \notin H$).

13. (Hilbert-Burch) Let R be a Noetherian ring, and I an ideal with a free resolution.
$$0 \to R^n \xrightarrow{\varphi} R^{n+1} \to I \to 0 \qquad (*)$$

Then there exists an R-regular element a such that $I = aF_n(\varphi)$. If I is projective then $I = (a)$. If hd $I = 1$, then $F_n(\varphi)$ has a regular sequence of length 2.

Conversely, if $\varphi : R^n \to R^{n+1}$ is an R-linear map, and $F_n(\varphi)$ has a R-sequence of length ≥ 2, then $I = F_n(\varphi)$ has a free resolution (*).

14. Let M be a Noetherian R-module, and let I be an ideal of R. Let $a_1, \ldots, a_m$, and $b_1, \ldots, b_m$ be two R-sequences on M in I. Then show that there is an R-module isomorphism

$$((a_1, \ldots, a_m)M :_M I)/(a_1, \ldots, a_m)M \overset{\sim}{\to} ((b_1, \ldots, b_m)M :_M I)/(b_1, \ldots, b_m)M.$$

15. (Geyer) Give an example in characteristic 0 of a complete intersection X whose co-ordinate ring is not Cohen-Macaulay.

    Question: Are there examples of this type in dimension 2?

16. Let A be a semilocal Noetherian ring. Let I be an ideal of A. Show that $\mu(I) = \mu(I/I^2)$.

17. Let A be a local Noetherian ring. Let I be an ideal of $A[X]$. If I has a monic polynomial then show that $\mu(I) = \mu(I/I^2)$.

18. Let A be a Noetherian ring. Let I be an ideal of A. If $\mu(I/I^2) \geq \dim(A/J(A)) + 1$ then show that $\mu(I) = \mu(I/I^2)$. (Here $J(A)$ denotes the Jacobson radical of A).

19. (Forster's Conjecture) Let k be a field, and $\mathfrak{p} \subset A = k[X_1, \ldots, X_n]$ be a prime ideal such that $A/\mathfrak{p}$ is regular. Show that $\mathfrak{p}/\mathfrak{p}^2$ is generated by n elements. Show that $\mathfrak{p}$ is generated by n elements. (This was settled by N. Mohan Kumar in [60]).

20. Let A be a commutative ring with 1. Let I_1, I_2 be two ideals of A with $I_1 + I_2 = A$. If I_1/I_2, I_2/I_2^2 are generated by n elements, then so is $(I_1 \cap I_2)/(I_1 \cap I_2)^2$.

21. Show that the maximal ideal $m = (\bar{X} - \bar{1}, \bar{Y})$ of the coordinate ring of real circle $A = \mathbb{R}[X, Y]/(X^2 + Y^2 - 1)$ is not a principal ideal.

    (Hint: $A \overset{i}{\hookrightarrow} B = \mathbb{C}[X, Y]/(X^2 + Y^2 - 1) \simeq \mathbb{C}[u, v]/(uv - 1) \simeq \mathbb{C}[u, u^{-1}]$, for $u = X + iY$, $v = X - iY$. Check that $mB = (u - 1)$. Show that $\lambda u^n(u - 1) \notin \text{Im}\ (i)$, for all $\lambda \in \mathbb{C}^*$. Deduce that m is not principal.)

22. Let R be a commutative Noetherian ring. Let I be an ideal of R. Let S be the multiplicative closed subset of R consisting of all $s \in R$ such that $(I, s) = R$. Let n be the least number such that there exist $x_1, \ldots, x_n \in R$ with $\sqrt{I} = \sqrt{(x_1, \ldots, x_n)}$. If m denotes the least number such that there exists $y_1, \ldots, y_m \in S^{-1}R$ with $\sqrt{S^{-1}I} = \sqrt{(y_1, \ldots, y_m)}$, then show that $m \leq n \leq m + 1$.

    Moreover, if all finitely generated projective R-modules are free then prove that $m = n$.

23. Let I be an ideal of R. Let $s, s' \in R$ with $Rs + Rs' = R$, and with $s' \in I$. Suppose that $I_s = (x_1, \ldots, x_n)$ where $x_i \in R_s$ for $1 \le i \le n$. If there exists a unimodular row $(a_1, \ldots, a_n)$ over $R_{s'}$ and an elementary matrix $\gamma \in E_n(R_{ss'})$ such that $\gamma(x_1, \ldots, x_n) = (a_1, \ldots, a_n)$ over $R_{ss'}$ then $\mu(I) \le n$.

24. Let R be a Noetherian ring. Let $R = A[X]$ and let $\mathfrak{m}$ be a maximal ideal of R with $\operatorname{htm} = \dim R$, and with $R_\mathfrak{m}$ regular local. Show that there exists $t \in 1 + (\mathfrak{m} \cap A)$ such that $\mu(\mathfrak{m}_t) = \dim R$. Moreover, one can choose generators $f_1, \ldots, f_n$ of $\mathfrak{m}_t$, $n = \mu(\mathfrak{m}_t)$, with $f_1, \ldots, f_{n-1}$ generating $\mathfrak{m}_t \cap A_t[X]$. (Hint: Show that if $\mathfrak{p} = \mathfrak{m} \cap A$ then $A_\mathfrak{p}$ is regular local. Deduce $\mathfrak{p} = (f_1, \ldots, f_{n-1}, e)$ where $e(e - 1) \in (f_1, \ldots, f_{n-1})$, and $\mathfrak{m} = (f_1, \ldots, f_{n-1}, e, f_n)$. Take $t = 1 - e$, and complete the argument.

25. Let R be a commutative ring which contains a field k. Let $s \in R$ such that $R_s \simeq A[X]$ for some ring A. Let $\mathfrak{m}$ be a rational maximal ideal of R (i.e. the natural map $k \to R/\mathfrak{m}$ is an isomorphism). Assume that $s \notin \mathfrak{m}$. If $R_\mathfrak{m}$ is a regular local ring of dimension equal to $\dim R$, then $\mathfrak{m}$ is generated by a regular sequence (Hint: By a further localisation assume that A is a reduced ring. Now settle the case when $\dim R = 1$. Using an earlier exercise show that (after a further localisation if necessary) $\mathfrak{m}_s = (f_1, \ldots, f_d)$, where $d = \dim R$, and with $f_1, \ldots, f_{d-1}$ generating $\mathfrak{m}_s \cap A$. Find $s' \in \mathfrak{m}_s \cap A$ so that $(f_1, \ldots, f_d) \in e_1 E_d(R_{ss'})$.)

26. Let $R = \bar{k}[x_1, \ldots, x_n, y_1, \ldots, y_n]/(\sum_{i=1}^{n} x_i y_i - 1), n \ge 2$. Show that each maximal ideal of R, other than the "origin" $(\bar{x}_1, \ldots, \bar{x}_n, \bar{y}_1, \ldots, \bar{y}_n)$ is generated by a regular sequence.

Appendix A

1. (R. A. Rao & R. Khanna) For an almost commutative ring R, i.e. one which is finitely generated over its center, and an integer$n \ge 3$ the following properties are equivalent:

   a) *Normality:* $E_n(R)$ is a normal subgroup of $\mathrm{SL}_n(R)$;

   b) $I_n + v^t w \in E_n(R)$ if $v \in \mathrm{Um}_n(R)$ and $\langle v, w \rangle = 0$;

   c) *Local-Global Principle:* If $\alpha(X) \in \mathrm{SL}_n(R)$, $\alpha(0) = I_n$ and $\alpha_\mathfrak{m}(X) \in E_n(R_\mathfrak{m})[X]$ for all $\mathfrak{m} \in \mathrm{Spmax}(R)$, then $\alpha(X) \in E(R[X])$;

   d) *Dilation Principle:* If $\alpha(X) \in Sl_n(R[X])$, $\alpha(0) = II_n$ and $\alpha_s(X) \in E_n(R_s[X])$ for some non-zero $s \in R$, then $\alpha(bX) \in E_n(R[X])$ for $b \in (s^l)$ where L is large. Actually we mean there is some $\beta(X) \in E_n(R[X], (X))$ (i.e. β in the normalizer of the the group generated by the elementary generators which are congruent to I_n) with $\beta_s(X) = \alpha(bX)$. But since there is no ambiguity, for simplicity we are using the notation $\alpha(bX)$ instead of $\beta_s(X)$.)

e) If $\alpha(X) = I_n + X^d v^t w$, where $v \in E_n(R)e_1$ and $\langle v, w \rangle = 0$, then $\alpha(X) \in E_n(R[X])$ and is a product of the form $\prod E_{ij}^{Xh(X)}$ for big d.

f) $I_n + v^t w \in E_n(R)$ if $v \in E_n(R)e_1$ and $\langle v, w \rangle = 0$.

g) $I_n + v^t w \in E_n(R)$ if $v \in \mathrm{SL}_n(R)e_1$ and $\langle v, w \rangle = 0$.

2. Give an example of a locally elementary polynomial matrix, but which is not an elementary matrix. (Hint: As P.M. Cohn has shown,

$$C := \begin{pmatrix} 1 + xy & x^2 \\ -y^2 & 1 - xy \end{pmatrix}$$

is not elementary over $\mathbb{Z}[x, y]$. Consider the matrix $C E_{12}^T C^{-1}$.)

3. Let R be a ring, $s \in R$ a non zero divisor and $a \in R$. Using Quillen's Local Global Principle show for $n \geq 3$ and $m \geq 2$

$$\left[E_{12}\left(\frac{a}{s}X\right) , \mathrm{SL}_n(R, s^m R) \right] \subset E_n(R[X])$$

4. Using the previous exercise, prove A. Bak's theorem that the group $\mathrm{SL}_n(R)/E_n(R)$ is nilpotent of class d, where d is the dimension of R, for $n \geq 3$.

5. Suppose $\alpha(X) \in \mathrm{SL}_n R[X]$ such that $\alpha(0) = I_n$ and $\alpha(X)_{\mathrm{m}} \in [\mathrm{SL}_n(R_{\mathrm{m}}[X]), \mathrm{SL}(R_{\mathrm{m}}[X])]$ for every maximal ideal m of R. Then $\alpha(X) \in [\mathrm{SL}_n(R[X]), \mathrm{SL}_n(R[X])]$.

For the following two exercises we need some definitions: Let M be a finitely generated module over a commutative ring R.

a) *Transvection:* An automorphism of M of the form $I + \varphi_p$, for some $\varphi \in M^* = \mathrm{Hom}(M, R)$, $p \in M$, with $\varphi(p) = 0$, and with either p or φ unimodular.

b) By $\mathrm{Flip}(M \oplus R)$ we denote the subgroup of $\mathrm{Trans}(M \oplus R)$ generated by

$$\left\{ \begin{pmatrix} I_r & \varphi \\ 0 & 1 \end{pmatrix}, \begin{pmatrix} I_r & 0 \\ p & 1 \end{pmatrix} \,\middle|\, \varphi \in M^*, p \in M \right\}$$

6. (R.A. Rao) *Dilation Principle for Transvections*
Let P be a projective R-module of rank $r \geq 2$. If $\sigma \in \mathrm{Aut}(P[X] \oplus R[X])$ with $\sigma(0) = \mathrm{id}$ and $\sigma_s \in E_{r+1}(R_s[X])$ for a non zero divisor s, then there is an l such that $\sigma(bX) \in \mathrm{Flip}(P[X] \oplus R[X])$ for all $b \in Rs^l$.

7. (R.A. Rao) *Local Global Principle for Transvections*

   Let P be a projective R-module of rank $r \geq 2$. If $\sigma \in \text{Aut}(P[X] \oplus R[X])$ with $\sigma(0) = \text{id}$ and $\sigma_\mathfrak{p} \in E_{r+1}(R_\mathfrak{p}[X])$ for all $\mathfrak{p} \in \text{Spec}(R)$, then $\sigma \in \text{Flip}(P[X] \oplus R[X])$.

8. (R.A. Rao) Let P be a projective R-module of rank ≥ 2. Then $\text{Trans}(P[X] \oplus R[X]) = \text{Flip}(P[X] \oplus R[X])$.

9. (Suslin) Let $v = (v_1, v_2, \cdots, v_n) \in M_{1,n}(R)$, $n \geq 3$, with $v_1, v_2, \cdots, v_n$ a regular sequence. Let $w \in M_{1,n}(R)$ with $wv^t = 0$. Show that $I_n + v^t w \in E_n(R)$.

10. (Vaserstein) Let A be an associative ring with unity, $n \geq 2$, $v = (v_i) \in A^n$, $u = (u_i) \in (A^n)^T$, and $uv = 0$. Assume that $1 + u_k v_k \in \text{GL}_1 A$ for some index k. Then $1_n + vu \in E_n(A)$.

11. (P.M. Cohn) The 2×2 elementary subgroup need not be a normal subgroup. Prove the following steps to get an example:

    (1) Let $E(a) = \begin{pmatrix} a & 1 \\ -1 & 0 \end{pmatrix}$. Show that $E_2(A)$ is generated by $\{E(a) \mid a \in A\}$.

    (2) Let $\alpha \in D_2(A)E_2(A)$, where $D_2(A)$ is the subgroup of $\text{GL}_2(A)$ of diagonal matrices. Show that $\alpha = \text{diag}\{a, b\}E(a_1)\ldots E(a_r)$ such that $a_i \notin A^* \cup \{0\}$, for $1 \leq i < r$.

    (3) Let $A = k[X, Y]$ be a polynomial ring in 2 variables over a field k. Let $\alpha \in D_2(A)E_2(A)$ have a representation as above. Assume further that $a_r \notin k$. If $e_1 \alpha = (f, g)$, then $\deg > \deg g$.

    (4) Let $\alpha \in \text{GL}_2(k[X, Y])$, with $e_1 \alpha = (f, g)$, with $\deg f = \deg g$. Assume further that the leading forms of f and and g are linearly independent over k. Prove that $\alpha \notin D_2(k[X, Y])E_2(k[X, Y])$.

    (5) Show that $E_2(k[X, Y])$ is not a normal subgroup of $Sl_2(k[X, Y])$.

12. (Mennicke symbols). Let A be a commutative ring and $(a, b) \in \text{Um}_2(A)$. Let $c, d \in A$ such that $ad - bc = 1$. The Mennicke symbol ms (a, b) is defined to be the class of $\begin{pmatrix} a & b \\ c & d \end{pmatrix}$ in $Sl_3(A)/E_3(A)$. This is independent of choice of c, d. Prove that the Mennicke symbol has the following properties.
    (MS1) ms $(u, b) = 1$ for u unit.   (MS2) ms $(aa', b) = \text{ms}(a, b)\text{ms}(a', b)$
    (MS3) ms $(a, b) = \text{ms}(b, a)$.   (MS4) ms $(a + \lambda b, b) = \text{ms}(a, b)$, $\forall \lambda \in A$.

13. Let A be a local ring. Let $f(X) \in A[X]$ be a monic polynomial. Suppose that $(f(X), g(X)) \in \text{Um}_2(A[X])$. Show that ms $(f(X), g(X) = 1$.

14. (Murthy) (Local Global Principle for Mennicke Symbols). Let $(f, g) \in \text{Um}_2(A[X])$. Suppose that ms $(f, g) = \text{ms}(f(0), g(0))$ over $A_\mathfrak{m}[X]$, for every maximal ideal $\mathfrak{m}$ of $A[X]$. Then ms$(f, g) = \text{ms}(f(0), g(0)$ over $A[X]$.

15. (Suslin) Let $A = k[X_1, \ldots, X_n]$ be a polynomial ring over a field k. Prove that $Sl_r([X_1, \ldots, X_n]) = E_r(k[X_1, \ldots, X_n])$, for all $r \geq 3$.

16. (A. Suslin) Let $\alpha(x) \in E_n(R_s[x])$, $n \geq 3$, with $\alpha(0) = I_n$. Show that there is a number of k such that for $r_1, r_2 \in R[x]$, with $r_1 - r_2 \in s^k R[x]$, $\alpha(r_1 x)^{-1} \alpha_2(r_2 x) \in E_n(R[x])$.

17. Let $v(x) \in Um_n(R[x])$, $n \geq 3$. Suppose that for all $\mathfrak{m} \in \mathrm{Spmax}(R)$, $v(x) \equiv v(0) \pmod{E_n(R_{\mathfrak{m}}[x])}$. Then show that $v(x) \equiv v(0) \pmod{E_n(R[x])}$.

18. (R.A. Rao) Deduce Corollary 4.4.8 using the Local Global Principle for elementary action. (Hint: If $v(x) \in Um_n(R[x])$, then show $v(x) \equiv v(1) \pmod{E_n(R[x])}$. Consider $w(x^{-1}) = (x^{-d_i} v_i(x))$, where $v(x) = (v_1(x), \ldots, v_n(x))$. Show $w(x^{-1}) \in Um_n(R[x^{-1}])$, and that $w(0) \equiv e_1 \pmod{E_n(R)}$. Finish the proof.)

Appendix C

1. (Hartshorne) Let X be a connected topological space. Let y be a closed subspace of X. Let $X_y = \{x \in X \mid y \in \overline{\{x\}}\}$. Suppose that for each $y \in Y$, $X_y \setminus \{y\}$ is a non-empty and connected space. Show that $X \setminus Y$ is connected.

   The concept of a topological space being connected in codimension k was made precise by R. Hartshorne, as follows:

   Let X be a Noetherian topological space, and Y be a clsoed subset of X. Let Y' be a closed irreducible subset of Y. The codimension of Y' in Y is the supremum of $\{r \mid$ there exists a sequence of closed irreducible subspaces Z_i of X, $Y' \subseteq Z_0 \subseteq Z_1 \subseteq \ldots \subseteq Z_r \subseteq X\}$. The codimension of Y in $X = \inf\{$ codimension of $Y' \subset Y$ in X, Y' closed and irreducible $\}$.

2. (Hartshorne) Let X be a Noetherian topological space. We say that X is connected in codimension k if it satisfies any of the following equivalent conditons.

   (i) If Y is a closed subset of X with codimension of Y in $X > k$ then $X \setminus Y$ is connected.

   (ii) If X', X'' are irreducible components of X then there is a sequence $X' = X_1, X_2, \ldots, X_r = X''$ of irreducible components of X, with $X_i \cap X_{i+1}$ of codimension $\leq k$ in X, for all $1 \leq i \leq r - 1$.

   Prove that (i) and (ii) are equivalent conditions.

3. Let X be a Noetherian topological space. Show that the following conditions are equivalent.

   (i) For any $y \in X$, X_y is connected in codimension k.

   (ii) Whenever $y \in X$ is such that $\dim X_y > k$, then $X_y \setminus \{y\}$ is connected.

   (A space X is said to be **locally connected in codimension k** if it satisfies any of the above conditions).

4. Let $X = \operatorname{Spec}(A)$ be a connected, locally Noetherian space. Let $Y = V(I)$ be a closed subset of X such that A/I locally has depth ≥ 2. Show that $X \setminus \{y\}$ is connected.

5. Let $X = \operatorname{Spec}(A)$ be a locally Noetherian space. For each $\mathfrak{p} \in X$ with $\dim(A_\mathfrak{p}) > k$ assume that $\operatorname{depth}(A_\mathfrak{p}) \geq 2$. Show that X is locally connected in codimension k.

References

Local Bibliography

1. Atiyah M.F. and I.G. Macdonald: **Introduction to Commutative Algebra**. Addison-Wesley Publishing Co., Reading, Mass.-London-Don Mills, Ont. (1969).
2. Auslander M., Buchsbaum D: Homological dimension in local rings, Trans. Am. Math. Soc. **85** (1957), 390–405.
3. Bădescu L.: On the Equations defining Affine curves. (Notes given in the School on Commutative Algebra and Interactions with Algebraic Geometry and Combinatorics at I.C.T.P. in May, 2004.)
4. Basu R.: On Forster's Conjecture and Related Results. M. Phil. dissertation (2002).
5. Bhatwadekar S.M., H. Lindel, R.A. Rao: The Bass - Murthy question : Serre dimension of Laurent polynomial extension. *Invent. Math.* **81** (1985), 189–203.
6. Bhatwadekar S.M. and A. Roy: Some theorems about projective modules over polynomial rings. *J. Algebra* **86** (1984), 150–158.
7. Bhatwadekar, S. M., Rao, R. A.: On a question of Quillen. Trans. Amer. Math. Soc. **279** (1983), no. 2, 801–810.
8. Boothby W.: On two classical theorems of algebraic topology. *Amer. Math. Monthly* **78**, (1971), 237–249.
9. Boratynski M.: A note on set-theoretic complete intersection ideals. *J. Algebra*, **54** (1978), 1–5.
10. Bourbaki, N.: **Commutative algebra**. Springer, (1988).
11. Brüske R., F. Ischebeck, F. Vogel: **Kommutative Algebra**. B.I.Wissenschaftsverlag Mannheim, Wien Zürich, (1989).
12. Cartan H., Eilenberg, S.: **Homological Algebra**, Princeton (1956). With an appendix by David A. Buchsbaum. Reprint of the 1956 original. Princeton Landmarks in Mathematics. Princeton University Press, Princeton, NJ, (1999). ISBN: 0-691-04991-2.
13. Cohn P.M. : On the structure of the GL_2 of a ring. *Inst. Hautes Etudes Sci. Publ. Math. No.*30 (1966), 365–413.
14. Cohn P.M. : Some remarks on the invariant basis property. *Topology* **5** (1966), 215–228.

15. Bruns W., Herzog J.: **Cohen-Macaulay rings**, Cambridge studies in advanced mathematics **39**, Cambridge University Press, (1993, 1998).
16. Carral M.: Modules projectifs sur les anneaux de fonctions. (French) [Projective modules over rings of functions] J. Algebra **87** (1984), no. 1, 202–212.
17. Cowsik R.C and M.V. Nori: Curves in characteristic p are set theoretic complete intersections. *Invent. Math.* **45** (1978), 111–114.
18. Eisenbud D. and E.G. Evans Jr.: Generating module efficiently: Theorems from Algebraic K-Theory. *J. Algebra* **27** (1973), 278–315.
19. Eisenbud D. and E.G. Evans Jr.: Every Algebraic set in n-space is the intersection of n Hypersurfaces. *Invent. Math.* **19** (1973), 107–112.
20. Eisenbud D.: **Commutative algebra with a view toward algebraic geometry.** Graduate Texts in Mathematics **150**, Berlin Heidelberg New York, Springer (1995).
21. Ferrand D.: Suite régulière et intersection complète. *C.R. Acad. Sci. Paris* **264** (1967), 427–428.
22. Ferrand D.: Courbes gauches et fibres de rang 2, *C. R. Acad. Sc. Paris* **281**, (1975), 345–347.
23. Flenner H., O'Carroll L., Vogel, W.: **Joins and Intersections**, Springer Monographs in Mathematics, ISBN 3-540-66319-3, (1999).
24. Forster O.: Über die Anzahl der Erzeugenden eines Ideals in einem noetherschen Ring. *Math Z.* **84** (1964), 80–87.
25. Bănică C., Forster O.: Multiplicity structures on space curves. The Lefschetz centennial conference, Part I (Mexico City) (1984), 47–64, Contemp. Math., **58**, Amer. Math. Soc., Providence, RI, 1986.
26. Gabel, M. R.: Generic orthogonal stably free projectives. *J. Algebra* **29** (1974), 477–488.
27. Geyer W.D.: On the Number of Equations which are Necessary to Describe an Algebraic Set in n-Space. Atas 3^a Escola de Algebra. Brasilia (1976), 183–17.
28. Gopalakrishnan, N.: **Commutative Algebra**. Oxonian Press Pvt. Ltd. New Delhi, ISBN 81-7087-039-9, (1984, 1988).
29. Hartshorne, R.: Complete intersections and connectedness. *Amer. J. Math.* **84** (1962), 497–508.
30. Hartshorne R.: **Algebraic Geometry**. Springer New York etc. (1977).
31. Herzog J.: Generators and relations of abelian semigroups and semigroup rings. Manuscripta Math. **3** (1970) 175–193.
32. Hilton P., Stammbach U.: **A Course in Homological Algebra**, GTM 4, Springer Verlag, (1970).
33. Horrocks G.: Projective modules over an extension of a local ring. *Proc. London Math. Soc.* **14** (1964), 714–718.
34. Ischebeck F., Ojanguren M.: Another example of a projective module over a localized polynomial ring. Arch. Math. (Basel) **70** (1998), no. 1, 29–30.
35. Kaplansky I.: Projective modules. *Ann. of Math.* (2), **68** (1958), 372–377.
36. Kaplansky I.: **Fields and Rings**. Univ. of Chicago Press (1969).
37. Kaplansky I.: **Commutative Rings**. Allyn and Bacon, Boston (1970).
38. Kelley, J.L.: **General topology**. D. Van Nostrand Company, Inc., Toronto-New York-London, (1955).
39. Knus, M.A., Ojanguren, M.: **Théorie de la Descente et Algèbres d'Azumaya**. LNM **389**, Springer.
40. Kronecker L.: Grundzüge einer arithmetischen Theorie der algebraischen Größen. *J. reine angew. Math.* **92** (1882), 1–123.

41. Krusemeyer M.: Skewly completable rows and a theorem of Swan and Towber. *Communications of Algebra* **4** (7) (1975), 657–663.

42. Kunz E.: **Introduction to commutative algebra and algebraic geometry**. Birkhäuser, (1985).

43. Lam, T. Y. Series summation of stably free modules. *Quart. J. Math. Oxford Series* (2), **27** (1976), no. 105, 37–46.

44. Lam T.Y.: **Serre's Conjecture**. Springer Verlag, **635**, New York, (1978).

45. Lang S.: **Algebra** 3rd ed. Addison-Wesley Reading, Massachusetts, (1993).

46. Lindel H.: Unimodular elements in projective module. *J. Algebra* **172** (1995), 301 - 319.

47. Lindel H.: On the Bass-Quillen conjecture concerning projective modules over polynomial rings. Invent. Math. **65** (1981/82), no. 2, 319–323.

48. Lønsted K.: Vector bundles over finite CW-complexes are algebraic. Proc. Amer. Math. Soc. **38** (1973), 27–31.

49. Lyubeznik G.: The number of defining equations of affine algebraic sets. *Amer. J. Math* **114** (1992), 413–463.

50. Lyubeznik G.: The number of defining equations of affine algebraic sets. *Amer. J. Math* **114** (1992), 413–463.

51. Lyubeznik G.: A Survey of Problems and Results on the Number of Defining Equations. *Commutative algebra*, (Berkeley, CA, 1987), 375–390, Math. Sci. Res. Inst. Publ., **15**, Springer, New York, (1989).

52. Macaulay, F.S.: **Algebraic Theory of Modular Systems**. Cambridge Tracts **19** (1916).

53. Mandal S.: On efficient generation of ideals. *Invent. Math.* **75** (1984), 59–67.

54. Matsumura H.: **Commutative Algebra**. Cambridge University Press, (1986).

55. McDonald B.: **Linear Algebra over Commutative Rings**. Marcel Dekker, Inc., New York and Basel, (1984).

56. Milnor J.: Analytic proofs of the "Hairy Ball Theorem" and the Brouwer Fixed Point Theorem. *Bulletin AMS*, (1978), 521–524.

57. Milnor J.: **Topology from the differentiable viewpoint**. Based on notes by David W. Weaver. Revised reprint of the 1965 original. Princeton Landmarks in Mathematics. Princeton University Press, Princeton, NJ.

58. J. Milnor; **Introduction to Algebraic K-theory**, Ann. of Mathematics studies **72**, Princeton University Press.

59. Moh T.T.: On the unboundedness of generators of prime ideals in power series rings of three variables. *J. Math. Soc. Japan* **26** (1974), 722–734

60. Mohan Kumar N.: On two conjectures about polynomial rings. *Invent. Math.* **46** (1978), 225–236.

61. Mohan Kumar N.: Affine Geometry. appeared in Current Science.

62. Murthy M.P.: Complete intersections. In: Conf. Comm. Algebra; Kingston (1975), *Queen's papers Pure and Appl. Math.* **42** 197–211.

63. Murthy M.P.: **Suslin's work on linear groups over polynomial rings and Serre problem**. (Notes by S.K. Gupta). ISI Lecture Notes, **8**. Macmillan Co. of India, Ltd., New Delhi, 1980.

64. Nashier B.S. and W. Nichols: Ideals containing monics. *Proc. of the Amer. Math. Soc.* **99** (1987), 634–636.

65. Nicolaescu L.: **Notes on Seiberg-Witten Theory**, Graduate Studies in Mathematics **28**, Amer. Math. Soc. (2000).

66. Northcott D. G.: **A first course of homological algebra.** Cambridge University Press, Cambridge-New York, (1980), ISBN: 0-521-29976-4 18-01.

67. Northcott D. G.: **Finite free resolutions.** Cambridge Tracts in Mathematics, No. **71**, Cambridge University Press, Cambridge-New York-Melbourne, (1976).

68. Ohm J.: Space curves as Ideal-Theoretic Complete Intersection. In: *Studies in Math.* **20** Math. Assoc. Amer. (1980), Ed: A. Seidenberg, 47–115.

69. Ojanguren M. and R. Sridharan: Cancellation of Azumaya algebras. *J.Algebra* **18** (1971), 501–505.

70. Plumstead B.R. : The Conjectures of Eisenbud and Evans. Ph.D. Thesis (1979), *Amer. J. Math.* **10** (1983), 1417–1433.

71. Quillen D.: Projective modules over polynomial rings. *Invent. Math.* **36** (1976), 167–171.

72. Rao R.A.: Patching techniques in Algebra - - Stability theorems for overrings of Polynomial Rings and Extendability of Quadratic Modules with sufficient Witt Index: Doctoral thesis submitted in June, (1983) to the Bombay University, India.

73. Rao R.A.: An elementary transformation of a special unimodular vector to its top coefficient vector. *Proc. Am. Math.* **93** (1985), 21–24.

74. Rao R.A.: The Bass-Quillen conjecture in dimension three but characteristic $\neq 2,3$ via a question of A. Suslin. Invent. Math. **93** (1988), no. 3, 609–618.

75. Rees D.: Two classical theorems of ideal theory. Proc. Cambridge Philos. Soc. **52** (1956), 155–157.

76. Rees, D.: A theorem of homological algebra. Proc. Cambridge Philos. Soc. **52** (1956), 605–610.

77. Rotman M.: **Introduction to Homological Algebra**. Pure and Applied Mathematics, **85**, Academic Press, Inc. [Harcourt Brace Jovanovich, Publishers], New York-London, (1979).

78. Roy A.: Application of patching diagrams to some questions about projective modules. *J. of pure and Applied Alg* **24**, (1982), 313–319.

79. Roy A.: An approach to a question of Bass via patching diagram. *article on a talk in University of Poona, March (1981)*.

80. Roy A.: Remarks on a result of Roitman. *J. Indian Math. Soc. (N.S)* **44**, (1980), 117–120.

81. Sarges H.: Ein Beweis des Hilbertschen Basissatzes. *J. reine angew. Math.* **283/284** (1976), 436 - 437.

82. Sathaye A.: On the Forster-Eisenbud-Evans Conjecture. *Invent. Math.* **46** (1978), 211–224.

83. Serre J-P.: **Algèbre locale. Multiplicités.** (French) Cours au Collège de France, 1957–1958, rédigé par Pierre Gabriel. Seconde édition, (1965). LNM **11**, Springer-Verlag, Berlin-New York (1965). **Local algebra.** (English) Springer, 2000.

84. Serre J-P.: Faisceaux algébriques cohérents. (French) Ann. of Math. (2) **61**, (1955), 197–278.

85. Sharma P.K.: Projective modules over group rings. *J. Alg.* **19** (1971), 303–314.

86. Raja Sridharan: Commutative algebra with applications to Combinatorics. M. Sc. thesis, University of Bombay, (1989).

87. Raja Sridharan: Non-vanishing sections of Algebraic Vector Bundles. J. Algebra **176**, (1995), no. 3, 947–958.

88. Sally J. D.: **Numbers of generators of ideals in local rings**. Marcel Dekker, Inc., New York-Basel, (1978). ISBN: 0-8247-6645-8

89. Sally J. D.: Boundedness in two dimensional local rings. Amer. J. Math. **100** (1978), no. 3, 579–584.

90. Stafford J.: Projective modules over polynomial extensions of division rings. *Invent. Math.* **59** (1980), no. 2, 105–117.

91. Storch U.: Bemerkung zu einem Satz von Kneser. Arch. Math. **23** (1972), 403–404.

92. Suslin A.A.: On Stably Free modules. *Math. USSR Sb.* **31** (1977), 479–491.

93. A. Suslin; On the structure of special linear group over polynomial rings. Math. USSR. Izv **11** (1977), 221–238.

94. Suslin A.A.: On Projective modules over polynomial rings. *Math. USSR Sb.* **22** (1974), 595–602. (English translation).

95. Swan R.G.: Vector bundles and projective modules. *Trans. AMS* **105** (1962) 264–277

96. Swan R.G.: The number of generators of a module. *Math. Z.* **102** (1967), 318–322.

97. Swan R.G.: **K-Theory of Finite Groups and Orders**. LNM **149** Springer-Verlag Berlin, Heidelberg, New York (1970).

98. Swan R.G.: Cancellation theorem for the projective modules in metastable range. *Invent. Math.* **27** (1974), 23–43.

99. Swan R.G. Topological examples of projective modules. *Trans. of A.M.S.* **230** (1977), 201–234.

100. Swan R. G.: A simple proof of Gabber's theorem on projective modules over a localized local ring. Proc. Amer. Math. Soc. **103** (1988), no. 4, 1025–1030.

101. Swan R.G. and J. Towber: A class of projective modules which are nearly free. *J. of Algebra* **36** (1975), 427–434.

102. Szpiro L.: **Lectures on equations defining space curves** (Notes by N. Mohan Kumar). *Tata Inst. Lecture Notes in Math., Bombay* (1979); by Springer-Verlag, Berlin-New York, 1979, ISBN: 3-540-09544-6.

103. Vaserstein L.N.: Vector Bundles and Projective Modules. *Trans. Amer. Math. Soc.* **294** (1986), no. 2, 749–755.

104. Vaserstein L.N. and Suslin A.: Serre's problem on projective modules over polynomial rings and Algebraic K-theory. Math. U.S.S.R. Izvestija **10** (1976), 937 - 1001.

105. Vasconcelos W.V. : Ideals generated by R-sequence. *J. of Algebra* **6** (1967), 309 - 316.

106. Valla G. : On set-theoretic complete intersections. *Complete intersections* (Acireale, 1983), 85–101, LNM **1092**, Springer, Berlin, (1984).

107. Valla G.: On determinantal ideals which are set-theoretic complete intersections. Compositio Math. **42** (1980/81), no. 1, 3–11

108. Weil A. : **Foundations of Algebraic Geometry**. A.M.S Publication, Vol XXIX, (1962).

109. Zariski O.: The concept of a simple point of an abstract Algebraic Variety. *Trans. Am. Math. Soc.* **62** (1947), 1-52.

110. Zariski O. and P. Samuel : **Commutative Algebra** Vol I-II. Van Nostrand, Princeton (1958, 1960).

Global Bibliography

The citations below list authors with the Math. Reviews number of the reviews of their articles. The citation [B][2] in the text will indicate that one is refering to the paper corresponding to the second Math Review reference given in [B], namely MR0432626. Further details can then be found via Math. Sci. Net. or Math Reviews.

[A]   Asanuma, Teruo; MR0862714, MR0951192, MR0678522, MR0564418.

[B]   Bass, Hyman; MR0527279, MR0432626, MR0527279, MR0564418.

[Bk]   Bak, Anthony; MR1115826, MR1329456, MR1810843, MR1810843, MR1329456.

[Bh]   Bhatwadekar, S.M.; MR1824883, MR1626485, MR1659957, MR1775418, MR1156463, MR0517136, MR0642336, MR0765772, MR1155426, MR2017619, MR0994128, MR1199695, MR1318088, MR1421074, MR1279265, MR1992040, MR1858341, MR1156463, MR1690788, MR0893595, MR1429296.

[Bt]   Boratýnski, M.; MR0533101, MR0511453, MR0895462, MR0883967, MR0877531, MR0837812, MR0822425, MR0783093, MR0677709, MR0533101, MR0533113.

[Bo]   Bose, N.K.; MR1861117, MR1803518.

[Co]   Cowsik, R.C.; MR0472835.

[D]   Das, M. K.; MR1981423, MR2006420.

[E]   Eisenbud, D.; MR0432627, MR0327742.

[Fe]   Ferrand, D.; MR0444646

[F]   Fitchas, Noa; MR1124807.

[K]   van der Kallen, W.; MR0987316, MR0704762.

[Ko]   Kopeiko, V.I.; MR1815566, MR0497932, MR0978207.

[G]   Gubeladze, Joseph; MR1824231, MR1360174, MR1206629, MR1161570, MR1079964, MR0937805.

[Li]   Lindel, Hartmut; MR0641133, MR1322406, MR0689374, MR0796196.

[La]   Laubenbacher, Reinhard C.; MR1745576, MR1457848. MR1358266.

[Lm]   Lam, T.Y.; MR1732042, MR0485842, MR0399212, MR0302739.

[Ly]   Lyubeznik, G.; MR1156572, MR0954979, MR1015529, MR0940486, MR0857929, MR0954979.

[Ma]   Mandal, Satya; MR1417820, MR1174904, MR1428789, MR1480172, MR1966529, MR1848970, MR1626712, MR0728138, MR0691806.

[Mc]   McDonald, Bernard; MR0769104, MR0476639.

[MK]   N. Mohan Kumar; MR1444497, MR1155427, MR0977765, MR0749107, MR0828868, MR0681997, MR0499785, MR1155427, MR1044349, MR0815767.

[MP]   Murthy, Pavaman; MR0611151, MR1626712, MR1697952, MR0396591, MR0977765, MR0422276, MR0940495, MR1044346, MR1298718, MR1732046, MR0200289.

[PH]   Park, Hyungju; MR1358266.

[PR]   Parimala, R.; MR0517136, MR0712622, MR0694376, MR0644232, MR0636877, MR0732195, MR0627667, MR0610478.

[P]   Popescu, D.; MR1060701, MR0986438, MR0868439, MR0818160.

[O]   Ojanguren, M.; MR1487451, MR1048288, MR0802307, MR0776448, MR0352073.

[Ra] Raghunathan, M.S.; MR0541022, MR0813075.
[RR] Rao, R.A.; MR09555292, MR0766519, MR0952284, MR0991967, MR1317126, MR0709584, MR0784289, MR1317126, MR0784289, MR0709584, MR0919503.
[Rt] Roitman, M.; MR0801320, MR0845969.
[Ry] Roy, A.; MR0656854, MR0727374, MR0701360, MR0666638. MR0752648.
[Sa] Sathaye, A.; MR0499784, MR0533106, MR0722731.
[RS] Sridharan, Raja; MR1609901, MR1418947, MR1351372 MR1351370, MR1940666, MR1775418, MR1688449.
[Se] Serre, J.-P.; MR0068874, MR0177011, MR0244257, MR0201468.
[Su] Suslin, A.; MR0469905, MR0444647, MR0469914, MR0457525, MR0558957, MR0562623, MR0441949, MR0669571, MR0466104, MR0535484.
[St] Stafford, J.; MR0617082, MR0577357, MR0662253.
[Stm] Sturmfels, Bernd; MR1144671.
[Sw] Swan, Richard; MR0469906, MR1697953, MR1256458, MR1144038, MR0954977, MR0921488, MR0396531, MR0218347.
[Va] Vaserstein, L.N.; MR0447245, MR0840588, MR1086811, MR0982288, MR0882801, MR0619316. MR0830354, MR0830354, MR1086811, MR0982288.
[Vo] Vorst, Ton; MR0750693, MR0606650, MR0550060.
[W]   Warfield, R. B., Jr.; MR0593603, MR0550426, MR0548137.

Index

Springer Monographs in Mathematics

This series publishes advanced monographs giving well-written presentations of the "state-of-the-art" in fields of mathematical research that have acquired the maturity needed for such a treatment. They are sufficiently self-contained to be accessible to more than just the intimate specialists of the subject, and sufficiently comprehensive to remain valuable references for many years. Besides the current state of knowledge in its field, an SMM volume should also describe its relevance to and interaction with neighbouring fields of mathematics, and give pointers to future directions of research.